Guidelines for Process Hazards Analysis, Hazards Identification & Risk Analysis

by

Nigel Hyatt

DISCLAIMER

Guidelines for Process Hazards Analysis, Hazards Identification & Risk Analysis

Nigel Hyatt

1st Edition, 8th Printing - March 2004

ISBN 0849319099

Co-Published and distributed by CRC Press.

For information, write to:

Dyadem Press,
9050 Yonge Street, Suite 401
Richmond Hill, Ontario
Canada
L4C 9S6

Phone: 905-882-5055
Fax: 905-882-5057

Boca Raton London New York Washington, D.C.

Library of Congress Cataloging-in-Publication Data

Catalog record is available from the Library of Congress

Direct all inquiries to CRC Press LLC, 2000 N.W. Corporate Blvd., Boca Raton, Florida 33431.

Visit the CRC Press Web site at www.crcpress.com

International Standard Book Number 0-8493-1909-9
Printed in the Canada 1 2 3 4 5 6 7 8 9 0
Printed on acid-free paper

About the Author

Nigel Hyatt is recognized as a leading authority on Hazards Analysis, Assessment and Risk Management. Mr. Hyatt is a professional engineer with more than 35 years of industrial experience in design, operations and engineering in Petrochemical, Refinery, Oil Production, Offshore, Chemical, Environmental, Power, Biochemical and Food industries.

Over a 24-year period, Mr. Hyatt worked in a leadership role for two major engineering companies, managing and completing projects for significant multinational firms. In 1987, he was Risk Analysis Program Manager for a large tar sands expansion study. He was responsible for the creation, setup and implementation of risk assessment programs that dealt with many leading consulting companies as well as being focused towards meeting the needs of insurance companies.

His experience in the field of risk has been particularly focused on Process Hazards Analysis and facilitation, hazards identification, quantitative risk assessment and risk management. In addition, he also specializes in the field of incident investigation. Moreover, Mr. Hyatt was the originator and key designer of PHA-Pro®, one of the world's best selling hazard identification software tools.

Mr. Hyatt is used to working with, and being responsible for, multi-disciplinary teams of people. He regularly gives courses on process safety and is particularly interested in extending the boundaries and methodologies for hazards evaluation and risk assessment.

Mr. Hyatt is registered as a Professional Engineer in Ontario, is a Chartered Engineer of the U.K. and is also a Member of the Institution of Chemical Engineers. He has 3 children and resides with his wife in Richmond Hill, Ontario.

Table of Contents

DYADEM

DYADEM

DYADEM

DYADEM

DYADEM

DYADEM

DYADEM

Introduction

Guidelines for Process Hazards Analysis, Hazards Identification & Risk Analysis is a major update to Dyadem's very popular *Process Hazards Analysis Training Manual*. It comes at a time when there is ever increasing awareness of hazardous risks that need to be managed by the industrial community at large.

The guidelines are driven principally by the need to provide practical guidance to both the novice and the seasoned risk professional. The guidelines are also considered to be a useful adjunct to Dyadem's very widely used PHA-Pro® software, Internet reference www.dyadem.com.

Chapters 1 to 4 address Risk Concepts, Regulatory Developments, Risk Terminology and Process Hazards & Risk Management Alternatives. The purpose here is to familiarize the reader with the technical definition of risk, past industrial incidents and their impacts, the legislation for which these incidents have acted as catalysts, the language and terms used in the risk field, types of hazards and simple management strategies.

Chapters 5 to 10 address the different types of structured analytical techniques for conducting Process Hazards Analyses, such as HAZOP, "What if," Checklist, FMEA and so forth. The purpose here is to familiarize readers with the different methods so they understand that different techniques can be used with different applications and for different situations. The user should understand that an older facility, whose drawings are unobtainable or illegible, places different demands on a PHA team than say a new facility, where fully detailed and extensive CAD drawings are available, or a facility that is merely at a conceptual phase only without any drawings. Different situations demand different tools, and this is certainly true in the application of Process Hazards Analysis tools.

Chapters 11 and 12 deal with the subjects of revalidating PHAs and handling Management of Change (MOC) issues, where PHAs may, or may not, be required. With revalidation, it is now understood that there are many issues and concerns with the quality and validity of early PHAs. In addition, new legislation and increasingly stringent demands have to be met to bring these early efforts to an acceptable standard in very many cases. With MOC, companies are continuously updating and modifying their facilities, and the criteria demanding whether or not these changes require PHAs are proposed.

Chapter 13 provides a rapid, order-of-magnitude method of estimating the time required for PHAs. There may, of course, be considerable variance, depending on the experience of the PHA team and the level of detail considered necessary.

Chapter 14 provides guidance in relation to the Management of Hazards associated with the Location of Process Plant Buildings, as well as addressing facility siting issues. When assessing hazards and their impacts on plant personnel and equipment, the overall philosophy of plant layout has changed considerably. It was once considered to be good practice to have equipment located as close as possible, with minimum spacing to minimize pipe runs, etc. and thus minimize plant costs. Incidents, such as Flixborough, 1974, where the control center was located in the heart of the plant and where there were 100% fatalities, have largely changed this approach in favor of safer layouts.

Chapter 15 provides certain important protocols for conducting PHAs and for guidance on safeguarding, especially with respect to Administrative and Engineering Controls, as well as addressing the consequences of failures of such controls.

Chapter 16 addresses human factors. The importance here is not to believe that human error can be totally eliminated, but rather to analyze for factors that can exacerbate and increase the chances of error. Once known, these factors can be addressed in order to minimize the potential for human error.

Chapter 17 deals with Loss of Containment. The different factors to be considered are dealt with qualitatively. Examples of common hazards, e.g., the storage of anhydrous liquid ammonia, LPG, where loss of containment might occur, are presented.

Chapter 18 deals with Managing and Justifying Recommendations that result from PHAs. Since the driving force for risk mitigation and deciding which recommendations should receive priority is somewhat arbitrary, a rationale for applying financial pay-back, based on rate of return applied to the risk, is presented. Different forms of risk matrices are also presented, and their relative merits are discussed.

Chapter 19 looks at PHA Team Leadership issues. It gives direction on the role of the PHA Leader (Facilitator) as well as preparation, setting up, responsibilities, organization and documentation of PHAs. Frequently, the PHA Team-Leader-to-be is thrust into the role where he or she responds "Yes, but what am I supposed to do now?" The object of this chapter is to help such individuals cope and manage what they may regard as an intractable situation.

Chapter 20 provides an overview of the application of Safety Integrity Levels (SILs) in the process industry and the relevant standards ANSI/ISA S84.01 and IEC 61511 developed by the American National Standards Institute / Instrument Society of America, and the International Electrotechnial Commission, respectively.

Chapter 21 provides an overview of Layer of Protection Analysis (LOPA). An example is used to illustrate the concept of building scenarios in LOPA. This is associated with guidance on constructing and assigning numerical values to individual scenario components, i.e., Consequence, Initiating Event, Enabling Event and Condition, Condition Modifier and Independent Protection Layer. It also provides recommendations on the expertise required to conduct LOPA and a template for documenting LOPA.

Chapter 22 addresses some of the basics of Quantitative Risk Assessment (QRA). It is desirable to understand how hazards, once identified, can be quantified in terms of risk from the consequences, i.e., impacts, as well as determining their frequency of occurrence, as likelihood. Although QRA is considered to belong to a more complex form of risk analysis than PHAs, it is felt that an understanding of the basics of QRA are very important for the risk professional.

Appendix I presents a basic methodology for Deriving Deviations from First Principles. The corollary to this appendix is that it allows the user to apply HAZOP to various types of systems or equipment, such as Compressors, Pumps, etc., where it is currently considered to be ineffectual.

Appendix II presents information on the different forms of HAZOP technique currently being used. Although the Parametric Deviation based method is the most widely used, it is not, for example, necessarily the best method for analyzing batch processes. The alternatives, together with their relative merits and an example of Procedural HAZOP, are presented.

Acknowledgements

I would like to acknowledge the assistance of Dyadem Engineering Corporation (DEC) personnel in the preparation of these guidelines.

In addition, feedback from members of Dyadem International Ltd. (DIL) as well as DEC and DIL clientele, typically through PHA-Pro® software use, PHA Training, PHA Facilitation and QRA Projects, and from advisers that DEC has used from time-to-time, have all proven informative.

Nigel Hyatt
Richmond Hill, Ontario
May 2002

Update: To assist the users of this manual, an index has been added. In addition, to accommodate the duel needs of both SI units and English FPS units, clarification has been provided in Chapter 17 to enable both systems to be used.

March 2004

CHAPTER 1

Risk Concepts

Hazardous Event

The release of a material or energy that has the potential for causing harmful effects to:

- The plant personnel;
- The surrounding community at large;
- The environment.

What is Risk?

Risk relates two important factors:

- How *much* of *what* causes *how much damage* to *whom* (or *whatever* else) from the hazardous event, i.e., the **Consequence**.
- How *often* the hazardous event can be expected to occur, i.e., the **Frequency** or **Likelihood**.

Risk is defined as the product of Consequence and Frequency:

RISK = CONSEQUENCE × FREQUENCY

Typical Incidents that Concern Us

- Toxic gas clouds;
- Asphyxiates;
- Fires (jet fires, pool fires, fireballs);
- Explosions (VCEs, BLEVEs, mechanical/chemical explosions);
- Missile hazards;
- Hazardous liquid spills;
- Combustible dusts;
- Corrosive substances.

Industrial Incidents of Major Significance

The following industrial incidents of major significance are listed below and tabulated:

- Ludwigshafen;
- Flixborough;
- Texas City Disaster;
- Romeoville;
- Pemex;
- Bhopal;
- Ufa;
- Pasadena;
- Chernobyl (worst incident ever);
- The Great Halifax Explosion (worst Canadian incident);
- Piper Alpha;
- Visakhapatham;
- Tosco Refinery;
- Toulouse Fertilizer Complex;
- Seveso, Italy;
- Mississauga, Ontario;
- Sandoz, West Germany.

TEXAS CITY DISASTER

Location:	Texas City Harbor, French Liberty Ship S.S.Grand Camp.
Date:	April 1947
Hazardous material:	Ammonium Nitrate
Event:	2,300 tons of fertilizer in holds caught fire. Attempts to extinguish fire failed resulting in a huge explosion.
Type of incident:	Condensed phase explosion equivalent to c.700 t of TNT.
Damage:	Massive destruction causing entire ship to disintegrate. Huge damage to surrounding area, at least $1 billion by current standards. Destroyed approx. 1/3 rd of town.
Dead & Missing & Homeless:	576 dead and 178 missing, 2,000 homeless

Figure 1-1: Texas City Disaster (Ref: http://www.local1259iaff.org/disaster.html)

L U D W I G S H A F E N

Location:	Chemical facility at Ludwigshafen, Germany.
Date:	July 1948
Hazardous material released:	Dimethyl Ether
Event:	Tank car failure due to overfilling and overheating by the summer sun. The vapor cloud released was ignited 10 to 25seconds later by a welder's torch.
Type of incident:	Vapor cloud explosion equivalent to 20 to 100 t of TNT.
Damage:	Total destruction of a 230 m x 170 m area. Extensive damage over 570 m x 520 m area. $30 millions damage.
Deaths:	207 people killed and 3,818 injured.

F L I X B O R O U G H

Location:	Petrochemical plant, Nypro works, producing 70,000 t/yr. of caprolactam (raw material for nylon) at Flixborough, England.
Date:	June 1974
Hazardous material released:	Cyclohexane
Event:	Massive failure of 20-inch bypass around a cyclohexane reactor, releasing about 40 t of cyclohexane. Approximately 22 t were in the explosive range. Most likely, the ignition source would have been fired heater. Piping most likely failed at the expansion bellows from a temporary dog-leg connection joining two reactors.
Type of incident:	Vapor cloud explosion equivalent to 15 t of TNT.
Damage Onsite:	Total destruction of plant. Destruction of control room, located inside the facility. $48 millions direct damage to plant.
Damage Offsite:	Extended 13 km offsite, including 2,488 houses, shops and factories. Approximately $200 millions offsite damage.
Deaths:	28 people killed (18 in control room) and 36 injured.

R O M E O V I L L E

Location:	Union oil refinery at Romeoville, U.S.A.
Date:	July 1984
Hazardous material released:	Hydrocarbons (mainly propane)
Event:	A worker spotted a crack in a circular weld on a 55-ft monoethanolamine (MEA) tower. He attempted to isolate the feeds to the tower but a spark ignited the vapors, causing the 34 t tower to explode. The tower rocketed over 1 km and downed a 130 kV power line. Nearby towers and tanks were ruptured, including an LPG tank that BLEVEd resulting in a second explosion.
Type of incident:	Vapor cloud explosion followed by BLEVE.
Damage:	Severe blast damage within refinery. $500 millions damage.
Deaths:	14 people killed.

P E M E X

Location:	San Ixhuatepec, Mexico, LPG storage distribution center.
Date:	November 1984
Hazardous material released:	LPG
Event:	Explosion during an unloading operation, leading to two 1250 t and four 625 t spheres BLEVEing.
Type of incident:	**BLEVE** (**B**oiling **L**iquid **E**xpanding **V**apor **E**xplosion). 2nd BLEVE worst, causing a 300 to 400 m fireball. 12 explosions in 90 minutes.
Damage Onsite:	Total destruction of facility.
Damage Offsite:	200 homes destroyed and 1800 homes damaged. Homes encroached on area.
Deaths:	542 dead and 4248 injured.

B H O P A L

Location:	Union Carbide's Sevin plant, Bhopal, India.
Date:	December 1984
Hazardous material released:	Methyl isocyanate (MIC)
Event:	2,000 lb. of water entered a storage tank containing MIC. Some MIC boiled off. The vent scrubber was shut down for maintenance so that the vapor could not be neutralized and highly toxic MIC vapor escaped from a 33 m high vent line. The refrigeration system, designed to keep the stored MIC cool, was out of commission. The flare tower was not available since a corroded section of line had not been replaced. The water curtain was not designed for 33 m in height.
Type of incident:	Toxic vapor cloud.
Damage:	No damage to plant itself.
Deaths:	2,000 to 15,000 killed & 200,000 to 300,000 injured due to there being a shanty town surrounding the facility.

U F A

Location:	Ufa, U.S.S.R. NGL transmission pipeline.
Date:	June 1989
Hazardous material released:	Natural Gas Liquids (NGL)
Event:	NGL pipeline was 800 m from railroad and slightly higher. The smell of gas was reported as far as 8 km away from line rupture. Hours after the release, two trains in opposing directions headed into cloud and ignited vapor cloud. The trains derailed and collided into each other.
Type of incident:	Vapor cloud explosion.
Damage:	Trains were destroyed and trees were flattened in 4 km radius.
Deaths:	645 persons killed, many injured.

PASADENA

Location:	Petrochemical plant producing Polyethylene, Pasadena, Texas.
Date:	October 1989
Hazardous material released:	Isobutane, Ethylene and Catalyst carrier
Event:	During routine maintenance of a fluff settling leg on a high-density polyethylene reactor, the entire reactor contents were discharged to the atmosphere. The cloud ignited one minute after release.
Type of incident:	Vapor cloud explosion equivalent to 10 t of TNT.
Damage:	Two complete units were destroyed. Approximately $750 millions damage.
Deaths:	23 killed, 130 injured.

C H E R N O B Y L

Location:	Nuclear power plant, Chernobyl, Ukraine.
Date:	April 1986
Hazardous material released:	Contents of nuclear reactor
Event:	Occurred due to decision by plant management to test ability of turbine generator to power certain cooling water pumps, while generator was freewheeling to a standstill after its steam supply was cut off.
Type of incident:	Local explosion, fire and widespread release of nuclear radiation products.
Damage:	Immense financial and societal impacts, including evacuation of nearby cities.
Deaths:	31 immediate deaths and approximately 75,000 excess cancers in the northern hemisphere. Massive pollution – global impacts. Effects are ongoing.

Figure 1-2: Chernobyl Incident (Ref: http://www.ccani.com/chernob.htm)

THE GREAT HALIFAX EXPLOSION

Location:	Halifax harbor (the "narrows"), Nova Scotia.
Date:	December 1917
Event:	The Belgium ship "Imo" collided with the French freighter "Mont Blanc," which was carrying over 2,300 t of picric acid, 200 t of TNT, 35 t of benzole and 10 t of guncotton. There was a fire followed by an explosion, ***creating the world's largest explosion before the atomic bomb dropped on Hiroshima***.
Type of incident:	Massive condensed phase explosion.
Damage:	Large amount of shipping destroyed, 25,000 persons left without shelter, 6,000 lost their homes, 1,600 homes destroyed, 12,000 damaged buildings.
Total cost:	Approximately $15 billion by present-day worth.
Deaths:	1,963 killed. 9,000 injured. 199 blinded.

PIPER ALPHA

Location:	Offshore Production Platform, North Sea, U.K.
Date:	July 1988
Hazardous material released:	Natural gas condensate
Event:	Release and ignition of gas condensate from a section of piping in the gas compression module triggered a chain of fires and explosions, resulting in the almost total destruction of the Piper Alpha Offshore Production Platform. The condensate was released from the former location of a pressure relief valve, which had been removed for maintenance when over pressurizing had occurred. The severity was enhanced by the rupture of oil and gas pipelines connected to the platform, and disabling of most of the emergency systems, as a result of the initial explosion. The control was rendered useless by the explosion.
Type of incident:	Multiple fires and explosions.
Damage:	Total destruction of offshore platform. $1.2 billion.
Deaths:	165 people killed.

VISAKHAPATHAM

Location:	Refinery in India.
Date:	September 1997
Hazardous material released:	Liquefied Petroleum Gas
Event:	A leak developed in a pipeline carrying LPG from a harbor terminal to the refinery. The LPG found a source of ignition that resulted in a large vapor cloud explosion. The resulting fire engulfed 18 storage tanks, destroying 7 tanks containing LPG and crude oil.
Type of incident:	Vapor Cloud Explosion and extensive fire.
Damage:	$23.6 millions.
Deaths:	50 people killed.

TOSCO REFINERY

Location:	Tosco Refinery, Martinez, California.
Date:	February 1999
Hazardous material released:	Hydrocarbons (Naphtha)
Event:	Workers attempted to remove and replace a leaking pipe attached to a fractionating column. Over a 13-day period, repeated attempts had been made to isolate and drain the pipe, but leaking and corroded shut-off valves hampered efforts. While workers were in the process of replacing the pipe section, naphtha was released, causing a fire. At the time, five workers were positioned on scaffolding a hundred feet above the ground and were unable to escape.
Type of incident:	Fire.
Deaths:	4 people killed plus one critically injured.

TOULOUSE FERTILIZER COMPLEX

Location:	Toulouse, France.
Date:	September 2001
Hazardous material released:	Ammonium Nitrate
Event:	Blast was sparked at a site containing 300 tons of ammonium nitrate. Uncertainty as to whether the residue resulting from a leak of sulfuric acid and neutralized by whitewash and caustic soda could have contaminated the store of ammonia nitrate causing a chain reaction starting the explosion.
Type of incident:	Condensed phase explosion.
Damage:	Total destruction of fertilizer plant and significant damage to surrounding community (4,000 homes and 80 schools).
Deaths:	30 dead and 2,000 injured.

SUGGESTED READING (Note: URLs current at date of publication)
"Guidelines for Evaluating the Characteristics of Vapor Cloud Explosions, Flash Fires and BLEVE's" by AIChE, CCPS, 1994 (Chapter 2). www.aiche.org/pubcat/seadtl.asp?Act=C&Category=Sect4&Min=20
"Learning from Accidents" by T.Kletz, pub. by Butterworth-Heinemann, 2001 www.bhusa.com/gulf/us/subindex.asp?maintarget=bookscat%2Fsearch%2Fresults%2Easp&country=United+States&ref=&mscssid=GKTMNF4S2L2C8K5B017248LP4MJXFWVF
"Lessons from Disaster – How Organisations Have No Memory and Accidents Recur" by T.Kletz, pub. by IChemE, 1993 http://harsnet.iqs.url.es/library.htm#books
"What Went Wrong? – Case Histories of Process Plant Disasters" by T.Kletz, pub. by Gulf Publishing, 1998 www.processassociates.com/bookshelf/publisher/gulf_2.htm
Piper Alpha – Spiral to Disaster", AIChE, CCPS (Videotape), 2001 www.aiche.org/pubcat/seadtl.asp?Act=C&Category=Sect4&Min=60
"Loss Prevention in the Process Industries" by F.P.Lees, published by Butterworth-Heinemann, 1996. (Volume 3, Appendices 1 to 6, 16, 19, 21 & 22) www.aiche.org/pubcat/seadtl.asp?Act=C&Category=Sect4&Min=50
"Large Property Damage Losses in the Hydrocarbon-Chemical Industries – A Thirty-year Review", 18th edition, 1998, Risk Control Strategies, J&H Marsh & McLennan www.mmc.com/frameset.php?embed=risk/index.php
"Large Property Damage Losses in the Hydrocarbon-Chemical Industries – A Thirty-year Review", Trends and Analysis, 19th edition, February 2001, Marsh Risk Consulting www.mmc.com/frameset.php?embed=risk/index.php
U.S. Chemical Safety and Hazards Investigation Board – Incidents Report Center (Website) www.chemsafety.gov/circ/
"A $100-million vapor cloud fire" by R.S.Al-Ameeri et al., Hydrocarbon Processing, November 1984, pages 181 to 188 www.hydrocarbonprocessing.com/contents/publications/hp/
"HPI loss-incident case histories" by C.H. Vervalin, Hydrocarbon Processing, February 1978, pages 183 and following www.hydrocarbonprocessing.com/contents/publications/hp/
"Process Safety Analysis, An Introduction" by Bob Skelton, IChemE, 1997 www.icheme.org/framesets/aboutusframeset.htm

CHAPTER 2

Regulatory Developments

North America

Bodies and Regulatory Developments in North America

1985:

- The American Institute for Chemical Engineers (AIChE) forms the Center for Chemical Process Safety (CCPS)
- The Chemical Manufacturers Association (CMA) creates the Community Awareness Response Program (CAER) as a result of Bhopal. CAER was initiated by the Canadian Chemical Producers' Association (CCPA)

1990:

- The American Petroleum Institute (API) - Recommended Practice # 750: Management of Process Hazards
- US Environmental Protection Agency (EPA) - The Clean Air Act

1992:

- US Occupational Safety and Health Administration (OSHA) - 29 CFR 1910.119: Process Safety Management of Highly Hazardous Chemicals and Blasting Substances

1996:

- EPA - 40 CFR Part 68: Accidental Release Prevention Requirements: Risk Management Program under CAA, Section 112(r)(7)
- Commonly referred to as the "RMP Rule"

Individual States Legislation in the USA

1985:

- Hazardous Materials Management, California.

1986:

- Toxic Catastrophic Prevention Act, New Jersey.
- Air Control Board Permit Review Program, Texas.

1988:

- Extremely Hazardous Substances Risk Management Act

Occupational Safety and Health Administration (OSHA), Process Management of Highly Hazardous Chemicals and Blasting Substances Regulations – 29 CFR 1910.119

- Process Safety Management of Highly Hazardous Chemicals and Blasting Substances
- Driven by the Pasadena incident in Texas
- Amalgam of API 750, Community Awareness and Emergency Response (CAER) and 3 states legislations; Delaware, California & New Jersey.

Applies to:

- Specific hazardous chemicals (thresholds defined).
  - Flammable liquids and gases exceeding 10,000 lb. inventory

Excludes:

- Many storage-only type facilities.

Key Elements of OSHA 1910.119

- Employee Participation
- Process Safety Information
- Process Hazards Analysis
- Operating Procedures
- Training
- Contractors
- Pre-startup Safety Review
- Mechanical Integrity
- Hot Work Permit
- Management of Change
- Incident Investigations
- Emergency Planning & Response
- Compliance Audits
- Trade Secrets

Employee Participation

Employee Participation requires employers to involve employees at an elemental level of the PSM program. Minimum requirements for an Employee Participation Program for PSM must include a written plan of action for implementing employee consultation on the development of process hazard analyses and other elements of process hazard management contained within 1910.119. The employer must also provide ready access to all the information required to be developed under the standard.

Process Safety Information

With Process Safety Information the intent is to provide complete and accurate information concerning the process which is essential for an effective process safety management program and for conducting process hazard analyses. The employer is required to compile written process safety information on process chemicals, process technology, and process equipment before conducting any process hazard analysis.

Process Hazard Analysis

The intent of performing Process Hazards Analyses is to require the employer to develop a thorough, orderly, systematic approach for identifying, evaluating and controlling processes involving highly hazardous chemicals. Minimum requirements include:

(1) Setting a priority order and conducting analyses according to the required schedule;

(2) Using an appropriate methodology to determine and evaluate the process hazards;

(3) Addressing process hazards, previous incidents with catastrophic potential, engineering and administrative controls applicable to the hazards, consequences of failure of controls, facility siting, human factors, and a qualitative evaluation of possible safety and health effects of failure of controls on employees;

(4) Performing PHA by a team with expertise in engineering and process operations, the process being evaluated, and the PHA methodology used;

(5) Establishing a system to promptly address findings and recommendations, assure recommendations are resolved and documented, document action taken, develop a written schedule for completing actions, and communicate actions to operating,

maintenance and other employees who work in the process or might be affected by actions;

(6) Updating and revalidating PHA's at least every 5 years; and

(7) Retaining PHA's and updates for the life of the process.

Operating Procedures

For Operating Procedures the intent is to provide clear instruction for conducting activities involved in covered processes that are consistent with the process safety information. The operating procedures must address steps for each operating phase, operating limits, safety and health considerations, and safety systems and their functions.

Training

Training helps employees and contractor employees understand the nature and causes of problems arising from process operations, and increases employee awareness with respect to the hazards particular to a process. An effective training program significantly reduces the number and severity of incidents arising from process operations, and can be instrumental in preventing small problems from leading to a catastrophic release. Minimum requirements for an effective training program include: Initial Training, Refresher Training, and Documentation.

Contractors

The intent of addressing Contractors (including Subcontractors) is to require employers who use them to perform work in and around processes that involve highly hazardous chemicals to establish a screening process so that they hire and use contractors who accomplish the desired job tasks without compromising the safety and health of employees at a facility. The contractor must assure that contract employees are trained on performing the job safely, of the hazards related to the job, and applicable provisions of the emergency action plan.

Pre-startup Safety Review

The intent of Pre-Startup Safety Review is to make sure that, for new facilities and for modified facilities, when the modification necessitates a change to process safety

information, certain important considerations are addressed before any highly hazardous chemicals are introduced into the process. Minimum requirements include that the pre-startup safety review confirm the following: construction and equipment is in accordance with design specifications; safety, operating, maintenance, and emergency procedures are in place and adequate; for new facilities, a PHA has been performed and recommendations resolved or implemented; modified facilities meet the requirements of management of change; and training of each employee involved in the process has been completed.

Mechanical Integrity

Mechanical Integrity requirements mean that equipment used to process store, or handle highly hazardous chemicals is designed, constructed, installed, and maintained to minimize the risk of releases of such chemicals. A mechanical integrity program must be in place to assure the continued integrity of process equipment. The elements of a mechanical integrity program include the identification and categorization of equipment and instrumentation, development of written maintenance procedures, training for process maintenance activities, inspection and testing, correction of deficiencies in equipment that are outside acceptable limits defined by the process safety information, and development of a quality assurance program.

Hot Work Permit

The intent of Hot Work Permitting is to ensure that employers control, in a consistent manner, non-routine work conducted in process areas. Specifically, this is concerned with the permitting of hot work operations associated with welding and cutting in process areas.

Minimum requirements include: that the employer issue a hot work permit for hot work operations conducted on or near a covered process and that hot work permits shall document compliance with the fire prevention and protection requirements of 29 CFR 1910.252(a).

Management of Change

Management of Change requires management of all modifications to equipment, procedures, raw materials and processing conditions other than "replacement in kind" by identifying and reviewing them prior to the implementation of the change. Minimum requirements for management of change include: establishing written procedures to manage change; addressing the technical basis, impact on safety and health, modification to operating procedures, necessary time period, and authorizations required; informing and training employees affected; and updating process safety information and operating procedures or practices.

Incident Investigations

The employer is required to investigate each incident which resulted in, or could reasonably have resulted in a catastrophic release of highly hazardous chemical in the workplace. An investigation shall be initiated no later than 48 hours following the incident. An investigation team shall be established and a report prepared which includes: 1) Date of incident 2) Date investigation began 3) Description of incident 4) Factors that contributed to the incident 5) Recommendations from the investigation. The employer is required to establish a system to promptly address the incident report findings and recommendations, documenting all resolutions and corrective actions. Incident reports shall be reviewed with all affected personnel whose job tasks are relevant to the investigation and retained for five years.

Emergency Planning and Response

Emergency Planning and Response requires the employer to address what actions employees are to take when there is an unwanted release of highly hazardous chemicals. The employer must establish and implement an emergency action plan in accordance with the provisions of 29 CFR 1910.38(a) and include procedure for handling small releases. Certain provisions of the hazardous waste and emergency response standard, 29 CFR 1910.120(a) which addresses scope, application, and definitions for the entire standard, while (p), which addresses treatment, storage, and disposal (TSD) facilities under the Resource Conservation and Recovery Act (RCRA) and (q), which addresses requirements for facilities that are not RCRA TSD's, where there is the potential for an emergency incident involving hazardous substances may also apply.

Compliance Audits

Compliance Audits are required so that employers can self-evaluate the effectiveness of their PSM program by identifying deficiencies and assuring corrective actions. Minimum requirements include: audits at least every three years; maintenance of audit reports for at least the last two audits; audits conducted by at least one person knowledgeable in the process; documentation of an appropriate response to each finding; documentation that the deficiencies found have been corrected.

Trade Secrets

The intent with Trade Secrets is to require employers to provide all information necessary to comply with the standard to personnel developing Process Safety Information, Process Hazard Analysis, Operating Procedures, Engineering Planning and Response and Compliance Audits without regard to possible trade secrets. In addition, employees and their designated representatives shall have access to trade secret information contained within documents required to be developed by the standard.

Environmental Protection Agency (EPA), Risk Management Plan (RMP) Rule – 40 CFR Part 68

- Enacted on: June 20, 1996
- Final RMP Submission Deadline: June 21, 1999
- Chemical Safety Information, Site Security and Fuels Regulatory Relief Act, 1999:
  - o Parts of the 'RMP Info' that contain Offsite Consequence Analyses (OCA) information will not be accessible to the public over the Internet as was planned for June 21, 1999.
  - o OCA information is accessible in the form of paper copies of Sections 2 through 5 of Risk Management Plans at the eleven Federal Reading Rooms, open to public as of March 12, 2001.

Applies to:

- Specific hazardous substances with defined threshold
- Covered hazardous substances specified in List Rule of January 31, 1994 (40 CFR Parts 9 and 68)

Compared to OSHA 1910.119:

- Applies to all facilities containing greater than threshold quantity, including storage-type facilities for hazardous substances

  Risk Management Program requirements include implementation of:

  - o Hazard Assessment - Worst Case, Alternative Case Scenarios, 5-Year Accident History
  - o Prevention Programs - Level 1 to 3
    - Level 1 - No impact level
    - Level 2 - Streamlined Mini-OSHA PSM Requirements
    - Level 3 - Requirements very similar to OSHA PSM
  - o Emergency Response Programs

In addition, a Risk Management Plan must be submitted to EPA consisting of:

- o Executive Summary
- o Registration Information
- o Offsite Consequence Analysis
- o Five-year Accident History
- o Prevention Program Information - Level 2 and 3
- o Emergency Response Program Information
- o Certification Statement

List of Hazardous Substances

The list is composed of three categories:

- 77 toxic substances; threshold quantities established from 500 to 20,000 pounds.
  - o 63 flammable substances; threshold quantity is established at 10,000 pounds.
  - o Explosive substances with a mass explosion hazard by Department of Transportation (DOT). Threshold quantity is established at 5,000 pounds.

Amendments to the List Rule

On August 25, 1997

- o Changed the listed concentration of hydrochloric acid.

On January 6, 1998

- o Delisted Division 1.1 explosives (classified by DOT), to clarify certain provisions related to regulated flammable substances and the transportation exemption.

On March 13, 2000

- o In accordance with the *Chemical Safety Information, Site Security and Fuels Regulatory Relief Act,* the list of regulated flammable substances excludes those substances when used as a fuel or held for sale as a fuel at a retail facility.

Amendments to the RMP Rule

On January 6, 1999

- o Added several mandatory and optional RMP data elements

    - Established procedures for protecting confidential business information
    - Adopted a new industry classification system

On May 26, 1999

- o Modified the requirements for conducting Worst Case Release Scenario Analyses for flammable substances and to clarify its interpretation of CAA sections 112(1) and 112(r)(11) as they relate to DOT requirements under the Federal Hazardous Transportation Law.

United Kingdom

Health and Safety at Work Etc. Act (1974)

1974 - Health and Safety Executive (HSE)

Health and Safety Commission (HSC) - Advisory Committees

- Advisory Committee on Dangerous Substances (ACDS)
- Advisory Committee on Toxic Substances (ACTS)
- Chemical Industries Forum

HSE's Safety Policy Directorate

- Control of Major Accident Hazards (COMAH) regulations - 1999

HSE's Health Directorate

- Control of Substances Hazardous to Health (COSHH) regulations - 1999

HSE Guides for COMAH & COSHH

- A Guide to the Control of Major Accident Hazards (COMAH) Regulations, 1999; Guidance on Regulations, HSE
- COMAH Safety Report Assessment Manual, HSE
- Major Accident Prevention Policies for Lower-Tier COMAH Establishments, HSE
- COSHH Essentials: Easy Steps to Control Chemicals: Control of Substances Hazardous to Health Regulations, HSE
- A Step-By-Step Guide to COSHH Assessment, HSE
- Technical Basis for COSHH Essentials; Easy Steps to Control Chemicals, HSE

European Commission (EC)

Seveso I Directive (1982)

- Seveso I Directive (1982) was based on Article 174 of EC Treaty.
- Identification of installation concerned (based on substance and quantities handled).
- Operator provides safety report to authorities.
- Emergency Response Plan (ERP) must be established.
- Community Awareness of Risks and Emergency Response Plan.
- Accident notification procedures.

Seveso II Directive (1999)

- Seveso II Directive was proposed in December 1996 to include an extended scope and introduction of
    - Safety management systems,
    - Emergency planning,
    - Land-use planning,
    - Reinforcement of the provisions on inspections.
- Driven by the incident at Seveso, Italy. Amended twice, after accidents at
    - Bhopal, India (1984), Union Carbide
    - Basel, Switzerland (1986), Sandoz
- Seveso II has fully replaced the original Seveso Directive as of February 1999.

Seveso II Directive (Cont'd)

The Seveso II Directive is implemented in the UK as the COMAH Regulations. These came into force in February 1999 and it improves Seveso I Directive by

- Emphasizing management factors
- Introducing a Major Accident Prevention Plan (MAPP)
- Emphasizing that Safety Reports should
    1. Address potential hazards
    2. Be submitted to credible authorities
    3. Consider management and organizational issues

- Applying provisions to individual installations (plants) as well as whole plants
- Considering effects of an incident on surrounding plants
- Publishing the reports (after removing confidential information)
- Having Emergency plans
    1. With content defined explicitly in Directive
    2. That are tested regularly

Ongoing Revisions to Seveso II Directive

- Currently, revisions to Seveso II Directive are underway following accidents at
    - a mining facility in Baia Mare, Romania (Jan 2000), and
    - *storage* facility of fireworks in Enschede, Netherlands (May 2000).
- These events drive the need for Seveso II Directive to cover hazards from
    - storage and processing activities *in mining,* and
    - storage and manufacturing of *pyrotechnic substances*, specifically.

SUGGESTED READING (URLs current at time of publication)
OSHA Process Management of Highly Hazardous Chemicals & Blasting Substances – 29 CFR 1910.119 (Website) www.osha-slc.gov/FedReg_osha_pdf/FED19990323.pdf
EPA, Risk Management Plan (RMP) Rule – 40 CFR Part 68 (Website) www.access.gpo.gov/nara/cfr/cfrhtml_00/Title_40/40cfr68_00.html
Seveso II Directive Information (Website) www.ipk.ntnu.no/ross/Info/Law/Seveso2.htm
Control of Major Hazards (COMAH) Regulations (Website) www.hse.gov.uk/spd/noframes/spdcomah.htm
Control of Major Hazards (COMAH) Assessment Manual (Website) www.hse.gov.uk/hid/land/comah2/
API (American Petroleum Institute) Recommended Practice (RP) 750 : Management of Process Hazards http://api-ep.api.org/filelibrary/ACF4B.pdf
"Guidance on the Preparation of a Safety Report to Meet the Requirements of Council Directive 96/82/EC (SEVESO II)" by G.A.Papadakis & A.Mendola, published by the Institute for Systems Informatics and Safety (Website) www.ipk.ntnu.no/fag/SIO3043/Notater/Rapporter/safety-report-txt.RTF
"Model Risk Management Plan Guidance for Petroleum Refineries":API 760, 1997, American Petroleum Institute http://api-ep.api.org/filelibrary/ACF4B.pdf
"Model risk management program and plan for ammonia refrigeration", US EPA/CEPPO, 1996 (Website) www.epa.gov/swercepp/rules/ammon.pdf
"COMAH and the Environment – Lessons Learned from Major Accidents 1999 – 2000" by A.Whitfield, Process Safety and Environmental Protection pub. By IChemE, January 2002, pages 40 to 46 www.icheme.org/framesets/aboutusframeset.htm

CHAPTER 3

Risk Terminology

Administrative Controls

Procedural mechanisms, such as lockout/tagout procedures, used for directing and/or checking human performance on plant tasks.

Autoignition Temperature

The autoignition temperature of a substance, whether solid, liquid or gaseous, is the minimum temperature that is required to initiate or cause self-sustained combustion in air without a specific source of ignition. (It may also be noted that for paraffinic hydrocarbons the autoignition temperature decreases with increasing molecular weight).

BLEVE (Boiling-Liquid-Expanding-Vapor Explosion)

A type of rapid phase transition in which a liquid which is contained above its atmospheric boiling point is rapidly depressurized, causing a nearly instantaneous transition from liquid to vapor with a corresponding energy release. A BLEVE is often accompanied by a large fireball, if a flammable liquid is involved, since an external fire impinging on the vapor space of a pressure vessel is a common BLEVE scenario. However, it is not necessary for the liquid to be flammable to have a BLEVE occur.

Catastrophic Incident

An incident involving a major uncontrolled toxic emission, fire or explosion with an outcome effect in which the zone extends offsite into the surrounding community.

Combustible

A term used to classify certain liquids that will burn on the basis of flash points. Both the National Fire Protection Association (NFPA) and the Department of Transportation (DOT) define "combustible liquids" as having a flash point of 100°F (37.8°C) or higher

Importance: Combustible liquid vapors do not ignite as easily as flammable liquids; however, combustible vapors can be ignited when heated and must be handled with caution. Class II liquids have flash points at or above 100°F, but below 140°F. Class III liquids are subdivided into two subclasses.

Class IIIA: Those having flash points at or above 140°F but below 200°F.

Class IIIB: Those having flash points at or above 200°F.

Deflagration

The chemical reaction of a substance in which the reaction front advances into the unreacted substance at less than sonic velocity. Where a blast wave is produced that has the potential to cause damage, the term *explosive deflagration* may be used.

Detonation

A release of energy caused by the extremely rapid chemical reaction of a substance in which the reaction front advances into the unreacted substance at equal to or greater than sonic velocity.

DIERS (Design Institute for Emergency Relief Systems)

Institute under the auspices of the American Institute of Chemical Engineers founded to investigate design requirements for vent lines in case of two-phase venting.

DIPPR (Design Institute for Physical Property Data)

Institute under the auspices of the American Institute of Chemical Engineers, founded to compile a database of physical, thermodynamic, and transport property data for most common chemicals.

Dow's Fire and Explosion Index (F&EI)

A method (developed by Dow Chemical Company) for ranking the relative fire and explosion risk associated with a process. Analysts calculate various hazard and explosion indexes using material characteristics and process data.

Dow's Chemical Exposure Index (CEI)

A method (developed by Dow Chemical Company) for computing airborne releases from release scenarios, and distances pertaining to ERPG (Emergency Response Planning Guidelines) and EEPG (Dow Emergency Exposure Planning Guidelines, which are the Dow equivalent to the American Industrial Hygiene Association ERPGs) for a wide range of commonly manufactured industrial hazardous chemicals.

Emergency Response Planning Guidelines (ERPG)

Guidelines established by the American Industrial Hygiene Association (AIHA) which are intended to provide estimates of concentration ranges where one might reasonably anticipate observing adverse effects. (Thus, based upon methodologies, not specified by AIHA, distances from the source point of the release to the receptor may be computed or estimated: these distances can differ for different chemicals, different release scenarios and different meteorological conditions – refer to Chapter 20 on Quantitative Risk Assessment). The different ERPG levels are:

- **ERPG – 1:** The maximum airborne concentration below which it is believed that nearly all individuals could be exposed for one hour without experiencing other than mild transient adverse health effects or perceiving a clearly objectionable odor.

- **ERPG – 2:** The maximum airborne concentration below which it is believed that nearly all individuals could be exposed for one hour without experiencing irreversible or other serious health effects or symptoms that could impair their abilities to take protective action.

- **ERPG – 3:** The maximum airborne concentration below which it is believed that nearly all individuals could be exposed for one hour without experiencing or developing life-threatening health effects.

Explosion

A release of energy that causes a pressure discontinuity or blast wave.

Fire Point

The temperature at which a material continues to burn when the ignition source is removed.

Fireball

The atmospheric burning of a fuel-air in which the energy is mostly emitted in the form of radiant heat. The inner core of the fuel release consists of almost pure fuel whereas the outer layer in which ignition first occurs is a flammable fuel-air mixture. As buoyancy forces of the hot gases begin to dominate, the burning cloud rises and forms a more spherical shape.

Flammability Limits

The range of gas or vapor amounts in air that will burn or explode if a flame or other ignition source is present.

Note: The range represents an unsafe gas or vapor mixture with air that may ignite or explode. Generally, the wider the range, the greater the fire potential.

Flammable

A "Flammable Liquid" is defined by NFPA as a liquid with a flash point below 100°F (37.8°C).

Note: Flammable liquids provide ignitable vapor at room temperatures and must be handled with caution. Precautions such as bonding and grounding must be taken. Flammable liquids are: Class I liquids and may be subdivided as follows:

Class IA: Those having flash points below 73°F and having a boiling point below 100°F.

Class IB: Those having flash points below 73°F and having a boiling point at or above 100°F.

Flash Fire

The combustion of a flammable vapor and air mixture in which the flame passes through that mixture at less than sonic velocity, such that negligible damaging overpressure is generated.

Flash Point

The lowest temperature at which vapors above a liquid will ignite. The temperature at which vapor will burn while in contact with an ignition source, but which will not continue to burn after the ignition source is removed. There are several flash point test methods, and flash points may vary for the same material depending on the method used. Consequently, the test method is indicated when the flash point is given. A closed cup type test is used most frequently for regulatory purposes.

Note: The lower the flash point temperature of a liquid, the greater the chance of a fire hazard.

Hazard (or Hazardous Event or Incident)

An inherent chemical or physical characteristic that has the potential for causing damage to people, property, or the environment.

Hazards Identification

The process by which hazards are identified. Commonly known as Process Hazards Analysis (PHA). Structured analytical tools include:

- HAZard and OPerability Analysis (HAZOP)
- "What if" Analysis
- Failure Mode and Effects Analysis (FMEA)
- Checklist Analysis
- Preliminary Hazard Analysis (also known as PrHA or Screening Level Risk Analysis, SLRA)
- "What if" + Checklist

Note: Fault Tree and Event Tree analyses can be included, but they are most commonly used for Risk Quantification rather than Hazards Identification.

Hazards identification focuses attention on specific scenarios and examines:

- How these might occur, i.e., What are the ***causes***?
- What might happen, i.e., What are the ***consequences***?
- How is one currently protected either (or both) against the basic occurrence or the consequences, i.e., What are the ***safeguards***?
- What does one need to do if one is insufficiently protected, i.e., What are the ***actions*** required?

Individual Risk

Risk posed to an individual who is exposed to a hazardous activity.

***Example**: For smoking, the risk of death is around 1 per annum for 330 individuals who smoke, or 3×10^{-3}, deaths per year.*

Inert Gas

A noncombustible, nonreactive gas that renders the combustible material in a system incapable of supporting combustion.

Intrinsic and Extrinsic Safety (Passive and Active Methodologies)

Safety features that protect by virtue of their intrinsic nature and do not require activation or human intervention to be effective. Intrinsic Risk Control features are ***passive*** rather than ***active***.

***Example 1**: Reduced plant inventory or storage inventory of hazardous materials are inherent safety features since the consequences of hazardous events are reduced. This is a **passive** safety feature. Other examples include increased spacing of equipment and dyking around storage tanks.*

***Example 2:** Flammable gas detectors are an **active** safety feature since they depend upon automated components (some of which may fail). They would be classed as extrinsic safety features.*

Most facilities require a combination of both intrinsic and extrinsic safety features to meet acceptable risk standards.

LD50 and LC50

LD is an acronym for "Lethal Dose". LD50 is the amount of a material given, as a single dose, which causes the death of 50% (one half) of a group of test animals. The LD50 is one way to measure the short-term poisoning potential (acute toxicity) of a material.

Toxicologists often test using rats and mice. It is normally expressed as the amount of chemical administered (e.g., milligrams) per 100 grams (or kilogram) of the body weight of the test animal. The LD50 can be found for any route of entry or administration but dermal (applied to the skin) and oral (given by mouth) administration methods are the most common.

LC is an acronym for "Lethal Concentration". LC values usually refer to the concentration of a chemical in air and in environmental studies it can also mean the concentration of a chemical in water. With inhalation experiments, the concentration of the chemical in air that kills 50% of the test animals in a given time (usually four hours) is the LC50 value. Other common terms are:

LD01 - the lethal dose for 1% of the animal test population

LD100 - the lethal dose for 100% of the animal test population

LDLO - the lowest dose causing lethality

TDLO - the lowest dose causing a toxic effect.

Acute toxicity is the ability of a chemical to cause harm relatively soon after administering a dose or a 4-hour exposure to a chemical in air. "Relatively soon" is usually defined as a period of minutes, hours (up to 24) or days (up to about 2 weeks) but rarely longer

In general, if the immediate toxicity is similar in the different animals tested, the extent of immediate toxicity will likely be similar for humans. When the LD50 values are different for various animal species, one has to make approximations and assumptions when estimating the probable lethal dose for man. Special calculations are used when translating animal LD50 values to possible lethal dose values for humans. Safety factors of 10,000 or

1000 are usually included in such calculations to allow for the variability between individuals and how they react to a chemical, and for the uncertainties of experiment test results.

Likelihood

A measure of the expected frequency with which an event or incident occurs. This may be expressed as a frequency (e.g., events per year), a probability of occurrence during a time interval (e.g., annual probability), or a conditional probability (e.g., probability of occurrence, given that a precursor event has occurred).

Lower Explosive Limit (LEL) or Lower Flammable Limit (LFL)

The lowest concentration of a vapor or gas (the lowest percentage of the substance in air) that will produce a flash of fire when an ignition source (heat, arc, or flame) is present. (Also see Upper Explosive Limit or Upper Flammable Limit).

Note: At concentration lower than the LEL/LFL, the mixture is too “lean” to burn.

Oxidant

Any oxidizing agent that can react with a combustible material (either in the form of liquid, solid, gas, dust or mist) to produce combustion. Oxygen in air is the most common oxidant.

Pool Fire

The combustion of material evaporating from a layer of combustible liquid at the base of the fire.

Process Safety

A discipline that focuses on the prevention of fires, explosions, and accidental chemical releases at process plant facilities. It normally classic worker health and safety issues involving working surfaces, ladders, protective equipment, etc.

Purge Gas

A gas that is continuously or intermittently added to a system to render the atmosphere non-combustible by, typically, excluding air. The purge gas itself may be an inert gas (e.g., nitrogen) or a combustible gas (e.g., fuel gas).

QRA

QRA stands for **Q**uantitative **R**isk **A**ssessment, as opposed to Hazards Identification, which is qualitative in nature. Hazards Identification is a necessary prerequisite to QRA

Quenching

Rapid cooling from an elevated temperature such that the further decomposition is halted or severely reduced.

Risk

A measure of the consequence of a hazard and the frequency with which it is likely to occur. Risk is expressed mathematically as:

RISK = CONSEQUENCE × FREQUENCY OF OCCURRENCE

Note: Individual risks, for mutually independent events, can be added to provide overall risk.

Risk Analysis

The process of evaluating the consequences and frequencies of occurrence of hazardous activities.

Risk Appraisal

Judging the acceptability of risks.

Criteria are usually reached by consensus between Risk Analysts and are published through expert bodies, e.g., *UK Health & Safety Executive.*

Risk Assessment

Combination of Risk Analysis and Risk Appraisal.

Risk Contour (or Risk Isopleth)

Line drawn around a facility connecting all points having the same level of risk.

Risk Control (also called Risk Mitigation)

Method(s) existing or introduced for the express purpose of reducing the frequency or consequences of a hazardous event. Methods are often categorized as active or passive.

Risk Measurement

Usually measured in terms of:

- Death (Lethality)
- Property Damage ($)
- Lost Production ($)

- Environmental Damage

Note: In addition to the above, loss of market share and impact on community/public relations may also need consideration.

Risk Management

The process of acting upon information supplied on Hazards Identification, Risk Assessment and Risk Control for management decision-making purposes.

Risk Mitigation or Risk Control

Lessening the risk of an incident sequence by acting on the source in a preventive way by reducing the likelihood of occurrence of the event, or in a protective way by reducing the magnitude of the event and/or the exposure of local persons or property or the environment.

Runaway Reaction

A thermally unstable reaction system which shows a rapid escalation of temperature increase and reaction rate.

Safety

A judgment of the acceptability of risk.

An activity is deemed as "safe" if its risks are judged to be acceptable when compared with other common daily activities.

No activity is totally free from risk. Provided the activity is undertaken, risk can never be totally eliminated. However, it can usually be reduced to acceptable levels with the use of adequate safeguarding.

Societal Risk

Risk posed to a societal group who are exposed to a hazardous activity.

Example: *For smoking, the risk of death, per 100,000 persons who smoke, is about 300 deaths per year.*

Upper Explosive Limit (UEL) or Upper Flammable Limit (UFL)

The highest concentration of a vapor or gas (the highest percentage of the substance in air) that will burn when an ignition source (heat, arc, or flame etc.) is present.

Note: At concentrations higher then the UEL, the mixture is too "rich" to burn.

Vapor Cloud Explosion (VCE)

A vapor cloud explosion is the explosive oxidation of a vapor cloud. The flame speed may accelerate to high velocities and produce significant blast overpressure. Vapor cloud explosions in densely packed plant areas (pipelines, units, etc.) may show accelerations in flame speeds and intensification of blast.

Vapor Density

The weight of a vapor or gas compared to the weight of an equal volume of air; an expression of the density of the vapor or gas. The Molecular Weight (MW) of the gas is a measure of its density, relative to air, at the same pressure and temperature: those with MW greater than 28.8 are heavier than air and those less than 28.8 are lighter than air. Materials lighter than air have vapor densities less than 1.0 (example: acetylene, methane, hydrogen). Materials heavier than air (examples: propane, hydrogen sulfide, ethane, butane, chlorine, sulfur dioxide) have vapor densities greater than 1.0.

Note: All vapors and gases will mix with air, but the lighter materials will tend to rise and dissipate (unless confined). Heavier vapors and gases are likely to concentrate in low places – along or under floors, in sumps, sewers and manholes, in trenches and ditches – and can travel great distances undetected where they may catch fire (and flash back) or cause health hazards.

Vapor Pressure

The pressure exerted by a vapor which is in equilibrium with its own liquid.

Note: The higher the vapor pressure, the easier it is for a liquid to evaporate and fill the work area with vapors which can cause health or fire hazards.

Voluntary versus Involuntary Risk

Greater levels of risk may be accepted by people choosing to accept that risk activity *(e.g., mountain climbing)* versus risk they consider to have imposed on them *(e.g., toxic waste facility in the vicinity)*.

Worst Possible Scenario, (also known as Worst Case Scenario with EPA RMP)

Largest possible release and consequential damage (human, property, financial, environmental) without regard for its likelihood. May result in an overly pessimistic view.

Worst Credible Scenario (also referred to as Alternate Case Scenario with EPA RMP)

Largest credible release and consequential damage (human, etc.) taking into account its likelihood.

Specific Safety Terms

Availability

The percentage of the time that a protective system is available for operation (e.g., when a protective system is being tested it may cease to be available for that test period).

Common Mode Failure

An event having a single cause with multiple failure effects.

Demand Rate

The rate (occasions/year) at which a protective system is called upon to act.

Double (or Multiple) Jeopardy

The chance that two (or more) unrelated events or incidents will occur at the same time. (It is important to note that two (or more) events or incidents arising from a common cause do not qualify).

Specific double or multiple jeopardy events are frequently considered to be so rare that their consideration does not warrant further examination.[However non-specific multiple jeopardy events in general are not rare and frequently involve human error with multiple complex stages/interactions. Since their potential number are extremely high, although the probability of a specific multiple jeopardy event is extremely low, this makes non-specific (very-hard-to-predict) multiple jeopardy events fairly likely].

Emergency Shut Down System (ESD)

A safety control system which is installed in a facility and is capable of shutting the facility or unit in the event of an emergency. The ESD over-rides the action of the basic control system when predetermined conditions are violated. Usually triggered automatically or by human intervention.

Equipment Reliability

The probability that, when operating under stated environment conditions, process equipment will perform its intended function adequately for a specified exposure period.

Fail-Safe

Design features which provide for the maintenance of safe operating conditions in the event of a malfunction of control devices or an interruption of an energy source (e.g., direction of failure of a pneumatically actuated valve on loss of instrument air). Features incorporated for automatically counteracting the effect of an anticipated possible source of failure. A system is fail-safe if failure of a component, signal or utility initiates action that return the system to a safe condition.

Interlock System

A system that detects out-of-limits or abnormal conditions or improper sequences and either halts further action or starts corrective action.

Note: An interlock system is frequently connected to an Emergency Shutdown System.

Protective Device

Any device that alarms or trips a system, or part of a system, or relieves the condition in a safe manner *(e.g., a pressure relief valve)*.

Programmable Electronic System (PES)

A system based on a computer connected to sensors and/or actuators in a plant for the purpose of control, protection or monitoring (includes various types of computers, programmable logic controllers, peripherals, interconnect systems, instrument distributed control system controllers, and other associated equipment).

Programmable Logic Controller (PLC)

A microcomputer-based control device. A solid-state control system which receives inputs from user-supplied control devices such as switches and sensors, implements them in a precise pattern determined by instructions stored in the PLC memory, and provides outputs for control or user-supplied devices such as relays and motor starters.

Redundancy

Additional or spare protective devices that will operate in the event of first line devices failing.

SUGGESTED READING (Note: URLs current at date of publication)
"Guidelines for Chemical Process Quantitative Risk Analysis" by AIChE, CCPS, 2000. See Glossary, pages 725 to 737 www.aiche.org/pubcat/seadtl.asp?ACT=C&Category=Sect4
"Guidelines for Engineering Design for Process Safety" by AIChE, CCPS, 1993. See Glossary, pages xxi to xxvi www.aiche.org/pubcat/seadtl.asp?Act=C&Category=Sect4&Min=20
"Glossary of Terms", (Website) http://www.multiplan.co.ae/new_page_45.htm
"What is an LD50 and LC50", Canadian Centre for Occupational Health and Safety (CCOHS), (Website) http://www.ccohs.ca/oshanswers/chemicals/ld50.html
"Guidelines for Evaluating the Characteristics of Vapor Cloud Explosions, Flash Fires and BLEVE's" by AIChE, CCPS, 1994. See Glossary, pages x to xii www.aiche.org/pubcat/seadtl.asp?Act=C&Category=Sect4&Min=20

4

CHAPTER 4

Process Hazards & Risk Management Alternatives

Hazards that Concern us

Industrial substances that are stored and/or processed or created in sufficient quantities so as to exceed a specific threshold defined by the hazard level per unit mass of that substance.

Hazardous levels of substances due to their ability to:

- Spontaneously decompose releasing energy, toxins.

  Example: Explosives such as TNT, nitroglycerin, organic peroxides.

- Vaporize as a toxic gas causing harmful effects.

  Example: *Chlorine, anhydrous ammonia.*

- Combine with air and catch fire and/or form an explosive mixture.

  Example: *Liquefied petroleum gases, hydrogen, fuels.*

- Form an explosive dust when in the finely divided state.

  Example: *Coal dust, flour, cork dust.*

- Substances corrosive to the flesh.

  Example: *Caustic soda, sulfuric acid.*

- Toxic liquids that can enter water courses causing environmental damage and harm to humans.

***Example**: Dry cleaning fluids, aromatic petroleum compounds with appreciable solubility.*

Other hazards include:

- Hot surfaces
- Electrocution
- Radioactivity
- Falling objects and missiles
- Excessively high sound levels

Note: These are not examined in further detail other than to indicate their existence.

What Increases the Potential for Industrial Facilities to Become More Hazardous?

- Large inventories of hazardous materials, processed or stored.
- Trend towards processing materials at higher pressures and temperatures.
- Greater levels of complexity with process facilities allowing higher chances of failure.
- Older facilities that have extended their life span so that they are vulnerable to decay, excessive corrosion, loss of mechanical integrity "industrial geriatrics".
- Greater diversity than before of toxins and flammables currently manufactured.
- Economic downsizing of staff leading to less maintenance and reduced reliability.
- Encroachment of housing projects on industrial facilities without regard for potential hazards.

What Makes Transportation of Dangerous Goods More Hazardous?

- Larger quantities transferred.
- More pipelines and remote transfer stations.
- More toxic/flammable materials transported (greater diversity).
- More vehicular traffic on road and rail.
- More hazardous goods transfers along specific transport corridors.

How are Process Risks Analyzed ?

Risk is analyzed in three distinct stages as shown in Figure 4.1, namely:

- Stage 1: Hazard Identification
- Stage 2: Risk Assessment
- Stage 3: Risk Management

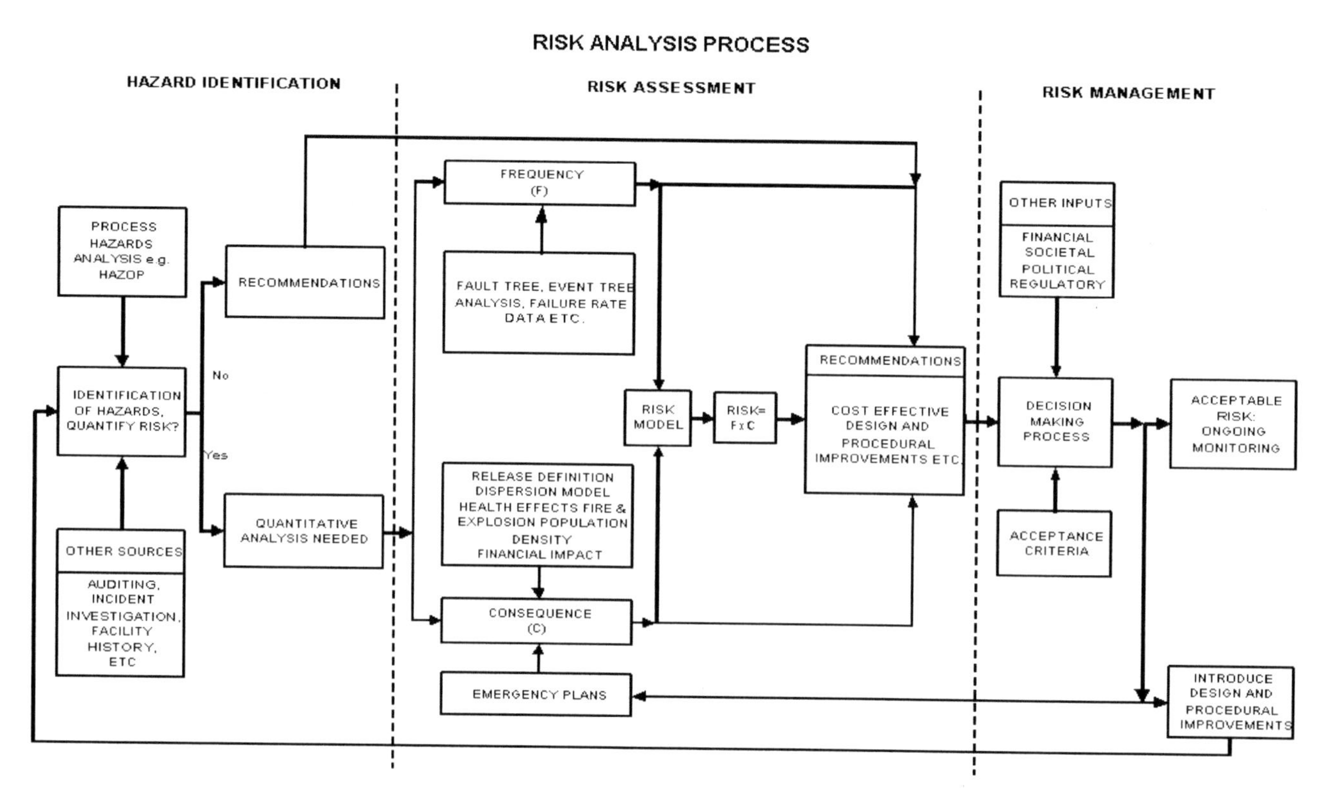

Figure 4.1: How Risk is Analyzed

Stage 1: Hazard Identification

Risk cannot be evaluated without first identifying the hazards involved. Many of the hazards will be identified by conducting a Process Hazards Analysis (PHA), e.g., such as HAZOP, What if/Checklist, FMEA .The hazards may arise from a wide range of sources such as fires, fireballs, BLEVEs, explosions, toxic releases and so forth. They have the potential to do harm to people, property and to the environment, but at the identification stage there is no clear or concise picture of what this harm might be or how often it might occur. At this stage it may be felt, as with a HAZOP, or other forms of PHA, that the use of a risk matrix of Severity versus Likelihood provides an adequate pseudo-measure or approximate gauging of risk so that a full quantification of the risk would not be necessary.

Other sources of hazards identification include results emerging from a plant safety audit, recommendations from the results of an incident investigation, from the results of near misses or from facility histories of similar or related process facilities.

If it is clear that one or more hazards pose significant risks that require further study then the decision to proceed with a Quantitative Risk Assessment (QRA) will have to be made. Typically where a plant incident could impact surrounding communities, or when the hazard is sufficiently great to the facility in question or may be seriously jeopardized, then a QRA may be justified.

Stage 2: Risk Assessment

If Recommendations from the Hazards Identification stage are not questioned via the QRA route then they will be reviewed from an economic standpoint for cost effectiveness and for implementation. For new facilities that are being designed this will be incorporated into the basic design. For existing facilities the recommendations may be processed through the Management of Change (MOC) route.

If QRA is the chosen route then the mechanism for calculating the basic components of the Risk Equation, namely, Consequence and Frequency, in the equation

Risk = (Consequence of Incident) x (Frequency at which Incident Occurs)

must be determined.

The **Consequence** is evaluated in a number of steps, these are:

(a) The Release Definition of HOW MUCH (e.g., lbs, kg, tons) of WHAT (i.e., what chemical, flammable or explosive material) is released over HOW LONG (i.e., seconds, minutes, hours).

(b) The Physical Effect, depending on the nature of the hazard, for example:

- Pool fires will emit thermal radiation depending on their size and configuration, rate of burning, flame emissivity etc.

- Fireballs will emit thermal radiation depending on the fireball diameter, emissivity, dwell time, configuration, rate of rise etc.

- Explosions will create overpressure (and under-pressure) waves together with momentum forces and generation of missiles

- Toxic vapors will create a toxic vapor cloud. Depending on the substance, the nature of the release, the atmospheric (including weather) conditions and wind force and direction, the vapor cloud will disperse and decrease in concentration with increasing distance from the source release point. Depending on the nature and temperature of the release the cloud may hug the ground, if the cloud is "heavy", or may be neutrally buoyant, if close to the density of air, or may rise, if lighter than air.

(c) The Impact on People, Flora and Fauna, Property and the Environment, for example:

- Toxic effects and Thermal Radiation impacts, in terms of probability of mortality, may be modeled from the dosages received resulting from the Physical Effects of the hazard.

The **Frequency** may be evaluated in a number of ways. Frequency may be evaluated from historical data of similar facilities or from fault or event tree modeling using failure rate data of system components.

Since Risk is the product of Consequence and Frequency, by knowing the probability of death for the Consequence and by knowing the potential rate of occurrence from the Frequency, the Risk may be determined. Since Risk is additive, all potential scenarios must be evaluated and the Risks summated, to calculate the Overall Risk.

Risk Mitigation (also known as Risk Control) Measures may need to be evaluated from an economic/design/procedural standpoint.

Stage 3: Risk Management

Risk may be managed once the hazards have been identified, and if the QRA route has been taken, when the Risks have been assessed. At this stage, if QRA has been done, then the calculated Overall Risk should be compared to accepted Risk Criteria. Depending on the level of Risk tolerable, the decision to accept the risk or take remedial action(s) must be made. If the level of risk is within accepted margins, then no further action may be necessary. If the level of risk is higher than desired, then actions requiring remediation and costing plant modifications, procedural changes, emergency response planning may be needed.

If the plant is an existing one, then remediation may likely require steps to reduce the frequency of potential incidents. With new facilities, prior to design, the consequences may be reduced since features, such as increased plant spacing, additional dikes for tanks etc, may be incorporated.

Principle and Practice of Risk Analysis via Quantitative Risk Assessment

Risk may be analyzed as indicated. It involves identifying hazards or examining what in a particular situation could cause harm or damage and then the assessment that the likelihood that harm will actually be experienced by an individual or specified population and what the consequences would be (i.e., the risk). The overall objective is to obtain a view on how to manage the risk or to compare the risk with other risks through the risk management process.

At a conceptual level, it has proved useful to make a distinction between an assessment of the risks (the evaluation of the likelihood of harm and its consequences for populations or individuals as described) and risk control (the prioritization of risks and the introduction of measures that might be put in place to reduce, if not prevent, the harm from occurring.

In practice it is often difficult to say where an assessment of risks ends and risk control begins or to assess risks without making a number of assumptions. In other words, a risk assessment is an order of magnitude estimate and is directional in nature. Unless inputs and assumptions are very similar, the repeatability of risk assessment results is rarely achieved.

As such, risk assessment is essentially a tool for extrapolating from statistical, engineering and scientific data, a value which people will accept as an estimate of the risk attached to a particular activity or event. Though there are many techniques for arriving at such a value or number, tailored to different applications and covering a wide range of sophistication, risk assessment is a composite of established disciplines, including toxicology, engineering, statistics, economics and demography.

The true value of risk assessment, through the QRA route, lies mainly in comparing Overall Risk Levels both before and after Risk Remediation is incorporated. For example, let us suppose that an Overall Risk Level of 10^{-4} deaths per annum is evaluated for a Facility and that post-remediation it would be 10^{-6} deaths per annum. This means that as a result of remediation, the facility would be 100 times safer. This improvement may be considered more important than the exact levels of risk both pre and post remediation.

Risk versus Safety: a Comparative View

It is often difficult to present the concept of "Acceptable Risk" since, no matter how low the risk is, certain levels of the population cannot or are not willing to accept such a concept. Therefore it may be easier to present a concept of safety, rather than risk. If we consider the following levels of individual risk as follows:

Risk Expressed as Individual Risk	Generally Perceived Level of Individual Risk
10^{-10} Deaths per Annum	Ultra Low Risk
10^{-9} Deaths per Annum	Extremely Low Risk
10^{-8} Deaths per Annum	Very Low Risk
10^{-7} Deaths per Annum	Medium Low Risk
10^{-6} Deaths per Annum	Low Risk
10^{-5} Deaths per Annum	Medium High Risk
10^{-4} Deaths per Annum	High Risk
10^{-3} Deaths per Annum	Very High Risk
10^{-2} Deaths per Annum	Extremely High Risk
10^{-1} Deaths per Annum	Ultra High Risk

By taking the negative value of the indices we can re-define risk in terms of levels of safety, namely:

Safety Expressed as an Indexed Level	**Generally Perceived Level of Individual Risk**
10	Ultra Safe
9	Extremely Safe
8	Very Safe
7	High Safe
6	Safe
5	Medium Safety
4	Limited Safety
3	Unsafe
2	Very Unsafe
1	Extremely Unsafe

Note: The exact risk and safety definitions shown above should be treated as relative, as opposed to being treated as absolute.

Risk Management Alternatives for New (Proposed) & Existing Hazardous Facilities

The following tables describe the advantages and disadvantages of various risk management alternatives for new and existing hazardous facilities.

Table 4-1: Advantages and Disadvantages of Various Risk Management Alternatives for New Facility

New facility		
Action	**Advantage**	**Disadvantage**
Ignore the risk	Financial benefits, provided no incidents	Incident could result in significant property damage, human health etc.
Build the facility elsewhere (NIMBY – Not in my backyard)	No risk to community being considered	Loss of potential employment, economic benefits
Provide adequate buffer zones (away from residential)	Reduces risk	Land is costly, may significantly increase facility cost
Build with adequate active & passive safeguards	Ensures incidents less likely	Not always effective unless properly managed
Incorporate emergency response plan (ERP)	Reduces risk to local community	Makes local community over anxious, creates worry, concern.

Table 4-2: Advantages and Disadvantages of Various Risk Management Alternatives for Existing Facility

Existing facility		
Action	**Advantage**	**Disadvantage**
Ignore the risk	Financial benefits, provided no incidents	Incident could result in significant property damage, human health etc.
Close the facility down	No risk to local community	Loss of potential employment, economic benefits
Introduce active & passive safeguards	Ensures incidents less likely	Not always effective unless properly managed
Introduce emergency response plan (ERP)	Reduces risk to local community	Makes local community over anxious, creates worry, concern.

SUGGESTED READING (Note: URLs current at date of publication)
"Loss Prevention in the Process Industries" by F.P.Lees, published by Butterworth-Heinemann, 1996. (Volume 1, pages 2/10 to 25) www.aiche.org/pubcat/seadtl.asp?Act=C&Category=Sect4&Min=50
"Probabilistic Risk Assessment in the CPI" by P.Guymer et al., Chemical Engineering Progress, January 1987, pages 37 to 45 www.che.com/
"Enhancing Safety through Risk Management" by G.A. Melhem & R.P. Stickles, Chemical Engineering, October 1997, pages 118 to 124 www.che.com/
"Quantified risk assessment: Its input to decision making" published by UK Health & Safety Executive, 1980 (Website) www.hse.gov.uk/dst/ilgra/minrpt1.htm#CONTENTS
"Process Safety Knowledge – The Route to Business Success" by B.D. Kelly, CCPS International Conference and Workshop MAKING PROCESS SAFETY PAY: THE BUSINESS CASE, 2001, pages 403 to 414 www.aiche.org/pubcat/seadtl.asp?Act=C&Category=Sect4&Min=50
"Inherently safer design principles are proven in expansions", by A.J.McCarthy and U.R.Miller, Hydrocarbon Processing, April, 1997, pages 122 to 125 www.hydrocarbonprocessing.com/contents/publications/hp/
"What is your corporate perspective on loss prevention?" by R.Scholing and P.Rieff, Hydrocarbon Processing, October 1997, pages 69 to 74 www.hydrocarbonprocessing.com/contents/publications/hp/

CHAPTER 5

Identification of Hazards and Structured Hazards Analysis Tools

How do we identify Hazards?

1. Identify potential loss of containment situations.
2. Identify causes that can result in loss of containment.
3. Identify potential consequences of loss of containment.
4. Identify potential safeguards that may:
    - Prevent loss of containment
    - Mitigate or reduce the consequences (such as fire, explosion, toxic release)

Depending on Step 1 to 4, actions may be introduced to reduce the hazard(s).

Widely Used Methodologies to Identify Hazards

- Preliminary Hazards Analysis (PrHA). Also known as Screening Level Risk Analysis (SLRA)
- Hazard and Operability Analysis (HAZOP)
- Failure Mode and Effects Analysis (FMEA)
- What If Analysis
- Checklist
- What If + Checklist

Preliminary Hazards Analysis (PrHA) - Also known as Screening Level Risk Analysis (SLRA)

When to Use Preliminary Hazards Analysis

PrHA is normally used on new or existing facilities to get an ***overall*** but not a detailed view of where the major areas of hazardous concerns exist.

The methodology can be used for new designs at the conceptual stage in order to assist with layouts, etc. and for existing facilities where some level of prioritization is needed prior to more detailed hazards analysis, *e.g., HAZOPs*. The method may also be considered as synonymous with *HAZID* Analysis (Hazards Identification Analysis) and Screening Level Risk Analysis (*SLRA*).

[See Chapter 10 on Screening Level Risk Analysis (SLRA) for further details].

Hazards And Operability Analysis (HAZOP)

When to Use HAZOP

HAZOP is a highly structured hazards identification tool.

HAZOP can be used at practically any stage. It is so widely used that almost any form of process hazards analysis is referred to as "HAZOP".

It is best used as late as possible with a new design, in order to be as complete as possible. With an existing facility it can be used at ***any*** time.

HAZOP can also be used for analyzing operating instructions and procedures so that sources of human error can be identified (and corrected).

It is ***extremely*** basic in its approach and makes practically ***no*** assumptions.

Advantage

HAZOP is very thorough, because you force yourself to painstakingly examine most aspects.

Disadvantage

HAZOP is very time consuming and costly. If not set up correctly and managed properly, it can be ineffective. Needs ***Leadership*** by an ***Expert*** in the field of HAZOP.

Basis

Simulates abnormal situations by using ***Guidewords*** applied to ***Parameters*** and ***Operations*** to create ***Deviations***.

HAZOP is the most widely used methodology used in the world today as a tool for hazards identification.

Methodology

1. Collect applicable documents and drawings, *e.g., process flow diagrams, piping and instrument diagrams, plot plans, etc.*

2. Break facility down into manageable sections (***"Nodes"***).

3. Prepare list of ***Parameters*** and ***Operations*** to be examined, composition, pressure, temperature, flow, etc. For batch operations list specific operations, *e.g., Transfer feed charge to reactor.*

4. Apply ***Guidewords*** to ***Parameters*** and ***Operations***.

Main Guidewords

- **More** or **High** or **Higher** or **Greater** (words that imply an ***excess***) than the design intent.
- **No**, **None**, **Less** or **Low** or **Lower** or **Reduced** (words that imply ***insufficiency***) than the design intent.
- **Part of** or **Not all of** or **Partially** (words that imply ***incompleteness***) than the design intent.
- **As well as** or **In addition to** (words that imply ***additional*** things occurring) the design intent.
- **Reverse** or **Opposite to** or **Instead of** (words that imply the ***reverse*** of something happening) the design intent.
- **Other than** or **What else** (words that imply something may have been ***overlooked***) the design intent.

Guidewords Applied to Time (Used for BATCH and Periodic Type transfer Operations)

- **Sooner** than intended
- **Later** than intended
- **Before** what is intended
- **After** what is intended
- **While** what is intended is occurring

Design Intent

The design intent reflects the specific purpose for an item of equipment, piping, etc. It does not necessarily imply its normal operating state. For example a section of line with a pressure relief valve may never, or very rarely, operate at conditions of elevated flows and pressure. Nonetheless the design intent is to meet such conditions as and when they do occur.

Parameters and Operations

Applicable **parameters** typically include:

- Pressure
- Temperature
- Flow
- Composition
- Level
- Reaction Rate
- Viscosity
- pH

Applicable **operations** typically include:

- Filling
- Transferring
- Purging
- Emptying
- Draining
- Venting
- Maintenance
- Start-up
- Shut-down

5. For each ***Node*** create ***Deviations***, *e.g., High pressure, High temperature, High flow, Low pressure, Low temperature, Low flow, Reverse flow, etc*.

6. List and record Causes for each Deviation.

7. List and record ***Consequences*** associated with each ***Cause***.

8. List and record ***Safeguards*** or ***Controls*** that may prevent the ***Cause*** and/or the ***Consequences***.

9. List any future ***Actions*** or ***Recommendations*** you think should be implemented.

Basically You Are

- Analyzing for potential hazards and deficiencies.
- Indicating the Cause mechanisms.
- Indicating potential Consequences.
- Identifying potential Safeguards & redeeming features.
- Providing Recommendations for any fix-it/remedial type solutions.

Note: A Risk Matrix may be applied to consequences. It is recommended to rate the Severity i.e. the Consequence based upon **no** Safeguards being present. The Likelihood should be evaluated **with** existing Safeguards present. If there are no Safeguards, then the Likelihood should be based upon the frequency, i.e. Likelihood of the Cause.

Failure Mode and Effects Analysis (FMEA)

5

When to Use FMEA

- Analyzing specific systems or items of equipment that are best handled as objects rather than by the use of parameters or operations.
- Analyzing pumps, compressors and items of equipment having interactive mechanical and/or electrical components.
- Splitting equipment into components and further splitting into sub-components
- Postulating failures, examine effects, record safeguards, and recommend modifications
- Consequence, severity and likelihood of failure can be used to indicate priority through use of risk matrix

Advantage

Very good for analyzing complex equipment items such as compressors, prime movers, etc. Widely used in the nuclear industry where failure of components in reactor circuits can have major consequences.

Disadvantage

Does not relate specific failures that have common causes. Needs to be used with ***Fault Tree Analysis*** to broaden scope.

Methodology

1. Select system or component and split into subsystems or subcomponents as required.
2. Postulate a failure mode of the subsystem or subcomponent.
3. List the effects of failure of that subsystem or subcomponent.
4. List safeguards or controls that might prevent or mitigate the effects of failure.
5. Recommend remedial actions (if needed) to prevent or mitigate the failure.

What If Analysis

When to Use What If Analysis

"What If" can be used at any time for new or existing facilities. Requires an experienced team and adequate preparation. Best results when used in conjunction with the Checklist method otherwise the team's imagination may prove inadequate at the time of analysis.

Advantage

Easy to learn and use. Powerful tool in hands of experienced personnel and when used in conjunction with Checklist Method.

Disadvantage

Much less structured than other methods and can give poor results unless personnel are experienced and well prepared.

Methodology

1. Divide the facility or unit into nodes that relate common functions (in a way very similar to HAZOP).
2. Postulate problems and failures by asking the question "What if..."
3. For each "What if" question record the Consequences.
4. For each "What if" question record any Safeguards present that may prevent the occurrence or may mitigate the consequences.
5. For each "What if" question, recommend any Actions needed to prevent the occurrence or mitigate the consequences.

Note: A Risk Matrix may also be used with What If, with similar considerations as for Guide Word HAZOP.

Checklist Analysis

When to Use Checklist Analysis

Checklist Analysis can be used at any time throughout a design or with an existing facility. Where there is a lack of experienced personnel the use of existing checklists is a valuable tool for identifying hazards. Useful where teams of personnel are not available and individuals are required to perform the analysis.

Advantage

Valuable method where less experienced personnel are involved.

Best used in conjunction with "What If" to get best results.

Disadvantage

Requires time up-front obtaining data and information.

Not thorough enough in many cases since it follows a non analytical, by rote, non interactive methodology.

Methodology

1. Obtain published and any available Checklists for analysis.
2. Where no Checklists are available consult whatever sources of information are available, such as MSDS sheets, textbook data, etc., in order to create Checklist.
3. Where Checklist items are not applicable record as N/A.
4. Where Checklist items are applicable, record Consequences, Safeguards present and any Actions needed.

Use of Risk Matrix With Hazards Identification *(also see Chapter 18 on Managing & Justifying Recommendations)*

A semi-quantitative methodology is often used with hazards identification tools. This permits a first order of magnitude identification of risk by addressing both frequency and consequence. This method can be very useful for prioritizing risk issues.

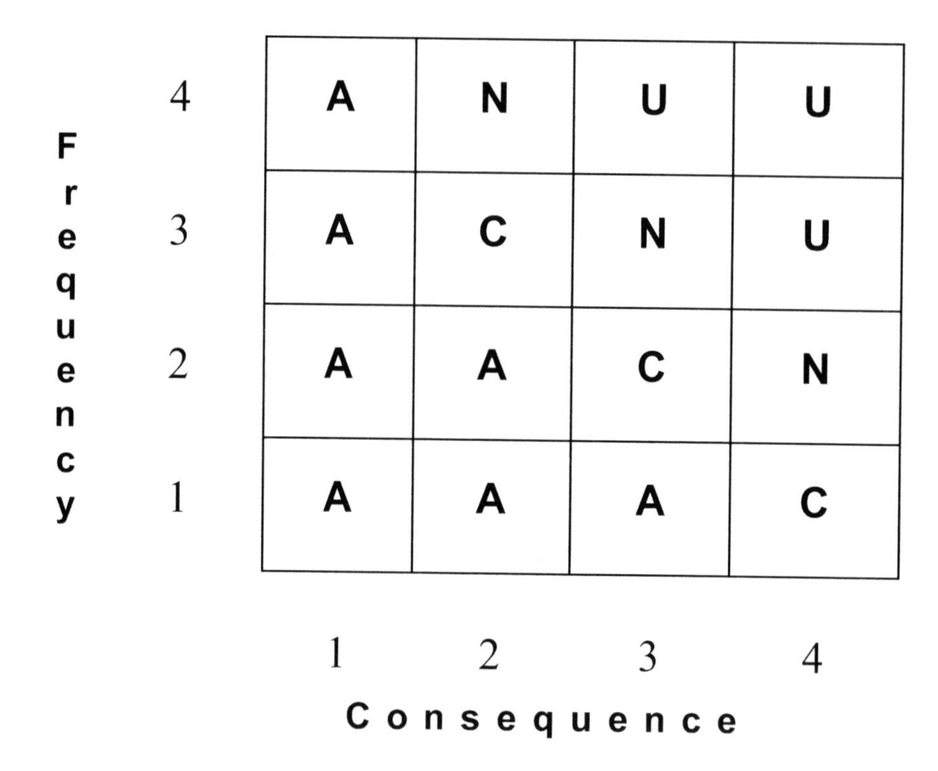

Figure 5-1: Typical Risk Matrix

A:	**Acceptable:**	No risk control measures are needed
C:	**Acceptable with Control:**	Risk control measures are in place
N:	**Not desirable:**	Risk control measures should be introduced within a specified time period
U:	**Unacceptable:**	Risk control measures should be introduced at the earliest opportunity

Example: Liquefied Petroleum Gas (LPG) Rail Car Loading Terminal

Location

In a rural area in North America, adjacent to a small urban community of around 50 people. Originally built in mid 1950's. Fed by gas plant 3 km away. Road runs adjacent to rail track (See Figure 5.2).

Capacity of Terminal

Terminal ships around 100,000 cubic meters of LPGs annually. Each rail car holds 128 cubic meters of LPG.

Terminal Description

- 2 loading racks with LPG piping, piping to ground flare, manual valves, loading hoses.
- Ground flare on SW perimeter.
- Operator's shack at middle of racks at one side.
- Propane fired vaporizer with its own small storage.
- Above ground shutdown/excess flow valves west of operator shack.
- Mercaptan stenching agent & dosing pump.
- Number of operators: One to two.

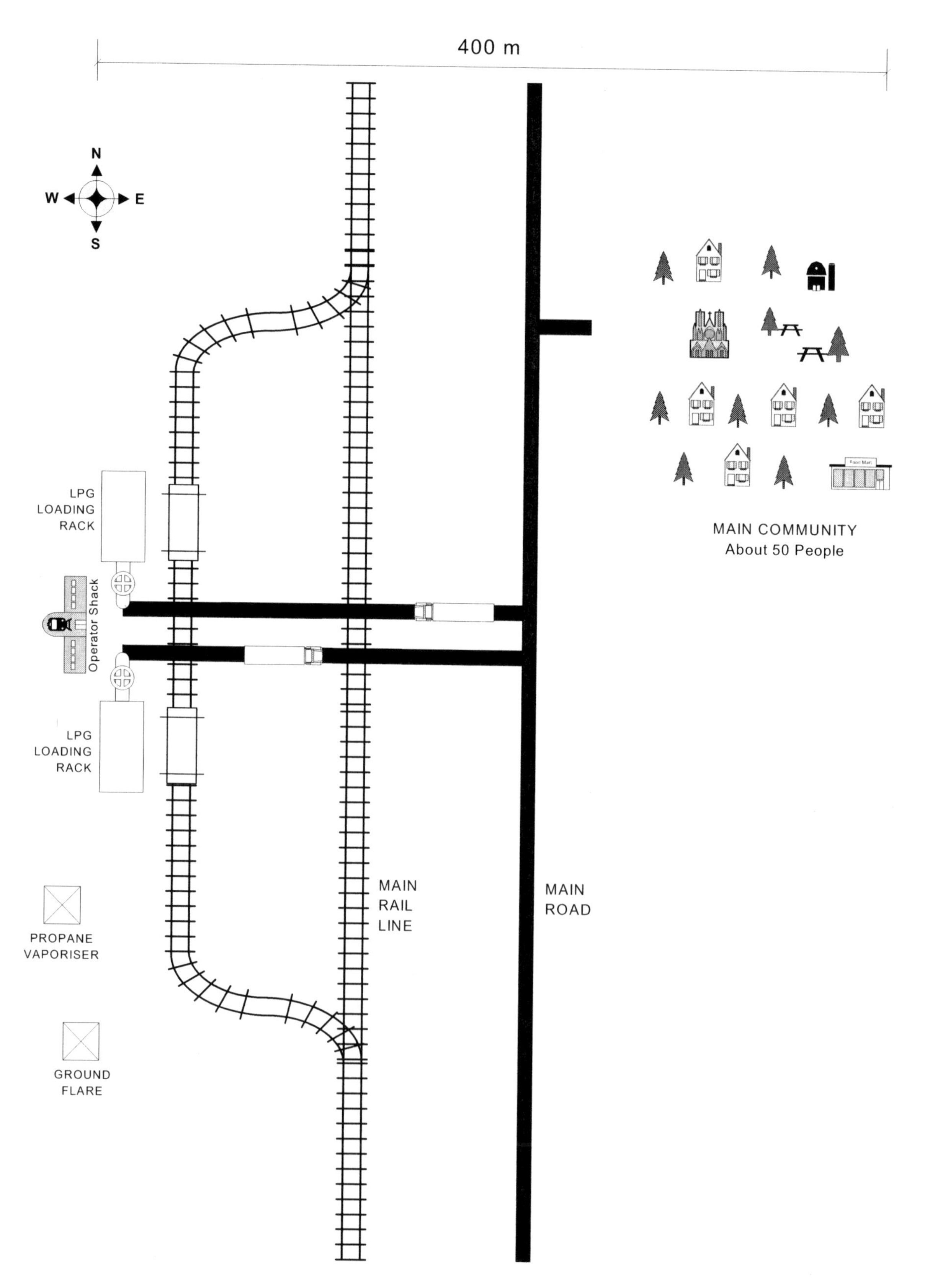

Figure 5-2: Schematic Representation of LPG Rail Car Loading Terminal

5

Table 5-1: FMEA Worksheet Report - LPG Railcar Example

Worksheet Report	5/28/01

Subsystems: 1.1. Tank car rupture	Drawing: A-234
Type: Vessel	

Failure Modes	Effects	Controls	Recommendations
1. Car rupture due to weld failure or corrosion.	1.1. Massive release of LPG, fireball, explosion on ignition of vapors.	1.1. Railcars are regularly inspected and tested.	5. No action required.

Subsystems: 1.2. Isolation valve failure on railcar	Drawing: A-234
Type: Line	

Failure Modes	Effects	Controls	Recommendations
1. Valve leaks causing cold vapor to crawl across ground.	1.1. Jet fire on vapor ignition. If fire goes undetected and impinges on railcar, it could lead to rupture, BLEVE, fireball, etc.	1.1. Maintenance procedures by railcar owner	7. Add flammable gas detectors in area, sufficient in number, and to warn at 10% of LEL.

Subsystems: 1.3. Overfilling of railcar	Drawing: A-234
Type: Vessel	

Failure Modes	Effects	Controls	Recommendations
1. Operator trips, has accident, is immobilized.	1.1. Flashing of LPG. Vapor could ignite causing severe fire which could escalate.	1.1. If second operator available.	1. Recommend need for second operator to be present at all times.
			2. Operators should carry intercoms for emergency use.
			3. Provide safety harnesses for operators to use during filling operations.
			4. Provide positive displacement flowmeters with cut-out switches.
2. Operator overestimates filling time of	2.1. Flashing of LPG. Vapor could ignite causing severe fire which could escalate.	2.1. If second operator available.	1. Recommend need for second operator to be present at all times.

Failure Modes	Effects	Controls	Recommendations
railcar.			4. Provide positive displacement flowmeters with cut-out switches.
3. Operator falls sick, faints, etc.	3.1. Flashing of LPG. Vapor could ignite causing severe fire which could escalate.	3.1. If second operator available.	1. Recommend need for second operator to be present at all times.
			2. Operators should carry intercoms for emergency use.
			4. Provide positive displacement flowmeters with cut-out switches.
4. Operator bored or inexperienced.	4.1. Flashing of LPG. Vapor could ignite causing severe fire which could escalate.	4.1. Operator training manual available.	5. No action required.
		4.2. Only experienced personnel tend to be used.	
5. Poor lighting conditions cause operator to misread gage.	5.1. Flashing of LPG. Vapor could ignite causing severe fire which could escalate.	5.1. None, except that most loading occurs during daylight hours. However, this could change if nightime operations are required.	6. Provide improved lighting for terminal.
6. Overfilling of railcar to top so that there is no room for expansion on heat-up due to summer sun.	6.1. Massive release of LPG, fireball, explosion on ignition of vapors.	6.1. Procedures and facilities exist for depressurizing and flaring fluids from overfilled railcars.	5. No action required.

Subsystems: 1.4. Railcar derailment	Drawing: A-234
Type: Vessel	

Failure Modes	Effects	Controls	Recommendations
1. Railcar derails during shunting operation.	1.1. Could lead to damaged railcar. However, at low speeds damage is not likely to result in LPG spill.	1.1. Low speeds during shunting.	5. No action required.
	1.2. Could lead to severing of LPG lines in vicinity of tracks with potential for escalating fire.	1.2. Excess flow valves would isolate LPG release.	

Subsystems: 1.5. Highway vehicle collides with railcar	Drawing: A-658
Type: N/A	

Failure Modes	Effects	Controls	Recommendations
1. Highway vehicle loses control, enters loading area and impacts railcar, possibly side-on.	1.1. Release and flashing of LPG. Vapor could ignite causing severe fire which could escalate.	1.1. None.	8. Provide local crash barriers to prevent highway traffic entering loading area.

Subsystems: 1.6. Collision with mainline train	Drawing: A-658
Type: N/A	

Failure Modes	Effects	Controls	Recommendations
1. If points are not set correctly, the mainline train could be diverted off onto rail loading spur causing major incident due to collision with terminal railcars.	1.1. Rupture of at least one LPG railcar, release, and flashing of LPG. Vapor would ignite causing fireball, BLEVE, etc.	1.1. None.	9. Provide rail signals to give early indication that the rail siding spur is open to oncoming freight trains.
			10. Impose a speed limit when rail spur is approached by freight trains.

Subsystems: 1.7. Brake failure on railcar during loading	Drawing: A-658
Type: N/A	

Failure Modes	Effects	Controls	Recommendations
1. Brake failure during LPG loading operation leading to severing of loading hosing.	1.1. Release and flashing of LPG. Vapor could ignite causing severe fire which could escalate.	1.1. Wooden wedges which are inserted behind railcar wheels during loading.	7. Add flammable gas detectors in area, sufficient in number, and to warn at 10% of LEL.
		1.2. Excess flow valves are integral with hoses. On severance and excessive flow they should close.	

Subsystems: 1.8. Loading hose not disconnected - severs on shunting	Drawing: A-658
Type: N/A	

Failure Modes	Effects	Controls	Recommendations
1. Brake failure during LPG loading operation leading to severing of loading hosing.	1.1. Release and flashing of LPG. Vapor could ignite causing severe fire which could escalate.	1.1. Wooden wedges which are inserted behind railcar wheels during loading.	7. Add flammable gas detectors in area, sufficient in number, and to warn at 10% of LEL.
		1.2. Excess flow valves are integral with hoses. On severance and excessive flow they should close.	

Subsystems: 1.9. Lightning strike during LPG loading operation	Drawing: A-658
Type: N/A	

Failure Modes	Effects	Controls	Recommendations
1. Lightning strike during LPG loading operation.	1.1. Damage to hosing, release and flashing of LPG. Vapor could ignite causing severe fire which could escalate	1.1. None.	11. Provide adequate grounding connections to minimize chances of ignition due to local sparking.

Subsystems: 1.10. Contaminants (air) in railcar prior to loading	Drawing: A-234
Type: Vessel	

Failure Modes	Effects	Controls	Recommendations
1. Air or oxygen atmosphere in railcar prior to filling.	1.1. Potential explosion on pressurization of railcar during filling with LPG.	1.1. Railcars normally under positive pressure with propane but since butane boils at -0.5C this could be a problem.	12. Consider nitrogen purging facility for purging railcars, in winter, prior to filling with butane.

Subsystems: 2.1. Block valve(s) jam open during filling operation	Drawing: A-234
Type: Line	

Failure Modes	Effects	Controls	Recommendations
1. Block valve(s) jam open during filling operation and cannot be closed.	1.1. Release adn flashing of LPG. Vapor could ignite causing severe fire which could escalate.	1.1. Upstream shutdown valves can isolate flow of LPG.	5. No action required.

Subsystems: 2.2. Contaminated hoses	Drawing: A-234
Type: Line	

Failure Modes	Effects	Controls	Recommendations
1. Hosing contaminated with water causing valves to jam open.	1.1. Release and flashing of LPG. Vapor could ignite causing severe fire which could escalate.	1.1. Upstream shutdown valves can isolate flow of LPG.	5. No action required.

Subsystems: 2.3. Gasket failure(s)	Drawing: A-234
Type: Line	

Failure Modes	Effects	Controls	Recommendations
1. Gasket leaks at flange.	1.1. Release and flashing of LPG. Vapor could ignite causing severe fire which could escalate.	1.1. Upstream shutdown valves can isolate flow of LPG.	13. Review viability of providing firewater to site to limit spread of fire.

Subsystems: 3.1. Pilot fails on ground flare	Drawing: A-234
Type: Line	

Failure Modes	Effects	Controls	Recommendations
1. Pilot fails to ignite flared vapors, releasing unignited vapors to atmosphere.	1.1. Flash fire, or fireball if sufficient vapor is released.	1.1. Twin pilots are used and intruments are regularly checked.	5. No action required.

Subsystems: 4.1. Heater fails to vaporize liquid LPG	Drawing: A-234
Type: Heater	

Failure Modes	Effects	Controls	Recommendations
1. Pilot or flame extinguished on propane vaporized.	1.1. Liquid passing to ground flare could extinguish ground flare pilots causing inignited vapor release causing flash fire and small fireball.	1.1. Redundancy on pilots to propane heater. Loss of flame is alarmed.	5. No action required.

Subsystems: 5.1. Vandalism	Drawing: A-658
Type: N/A	

Failure Modes	Effects	Controls	Recommendations
1. Vandals cracking open valves allowing LPG to escape.	1.1. Lines could depressurize or railcar inventory depressurized with potential for major incident.	1.1. None.	14. Provide limited access to facility. Review feasibility of fencing area and locking drawbridges on loading racks.
			15. Add 24 hour TV surveillance to terminal.

Table 5-2: FMEA Recommendations Report

Recommendations Report	6/13/01

#	Recommendation	Responsibility	Pri	Comments
1.	Recommend need for second operator to be present at all times.	John Phantom	8	
2.	Operators should carry intercoms for emergency use.	Bill Phoner	7	Provide all operators with intercoms plus back-up spares in hut.
3.	Provide safety harnesses for operators to use during filling operations.	Safety Coordinator	8	Ensure designs of harness do not hamper operators.
4.	Provide positive displacement flowmeters with cut-out switches.	Instrument Section	5	
5.	No action required.		0	
6.	Provide improved lighting for terminal.	Electrical	7	
7.	Add flammable gas detectors in area, sufficient in number, and to warn at 10% of LEL.	Instrument Section	8	Also investigate regular maintenance of gas detectors.
8.	Provide local crash barriers to prevent highway traffic entering loading area.	Civil Eng Gp.	8	
9.	Provide rail signals to give early indication that the rail siding spur is open to oncoming freight trains.	Rail Authority	7	
10.	Impose a speed limit when rail spur is approached by freight trains.	Rail Authority	8	Study must address stopping distance with respect to speed of train.
11.	Provide adequate grounding connections to minimize chances of ignition due to local sparking.	Electrical	4	
12.	Consider nitrogen purging facility for purging railcars, in winter, prior to filling with butane.	Process Eng.	8	
13.	Review viability of providing firewater to site to limit spread of fire.	Civil Eng Gp.	5	
14.	Provide limited access to facility. Review feasibility of fencing area and locking drawbridges on loading racks.	Civil Eng Gp.	8	Also address problems of access and escape with fenced systems in event of incident.
15.	Add 24 hour TV surveillance to terminal.	Instrument Section	6	Confirm vendor cont'd recommendation regarding robustness. Relay surveillance back to control room of main gas plant.

SUGGESTED READING (URLs current at time of publication)
"Guidelines for Hazard Evaluation Procedures" by AIChE, CCPS, 2nd edition, 1992 plus "Guidelines for Hazard Evaluation Procedures" by AIChE, CCPS, 1st edition, 1985 www.aiche.org/pubcat/seadtl.asp?Act=C&Category=Sect4&Min=20
"HAZOP and HAZAN" by T.Kletz, pub. by IChemE, 1992 www.icheme.org/framesets/aboutusframeset.htm
"Hazard Identification, Analysis and Control" by H.Ozog, Chemical Engineering, February 18, 1985, pages 161 to 170 www.che.com/
"Design plants for safety" by R.Roberts, Hydrocarbon Processing, September 1989, pages 92-98 www.hydrocarbonprocessing.com/contents/publications/hp/
"Guidelines for Process Safety Documentation" by AIChE, CCPS, 1995, pages 73 to 105 www.aiche.org/pubcat/seadtl.asp?Act=C&Category=Sect4&Min=30
"Process Safety Analysis, An Introduction" by Bob Skelton, IChemE, 1997 www.icheme.org/framesets/aboutusframeset.htm
"Process Hazards Analysis" by I.Sutton, published by SW/Sutton & Associates, 2002 http://www.swbooks.com/books/book_prha.shtml
"Hazard identification and management – an overview" by M.P.Sukumaran Nair, Hydrocarbon Processing, July 2002, pages 63 to 67 www.hydrocarbonprocessing.com/contents/publications/hp/
"Hazard Analysis", (Website) http://www-compsci.swan.ac.uk/~csetzer/lectures/critsys/critsysfinalhrefa2.pdf
"Introduction to process safety management", (Website) http://www.nd.edu/~ed/Cheg448/lecture17_web.pdf
"A Code of Practice for Risk Assessment in the Department of Physics –Advanced Techniques", (Website) http://www.phy.cam.ac.uk/cavendish/hands/cops/RAcopAdv.pdf
"Risk based decision-making guidelines – Vol 3" (Website) www.uscg.mil/hq/g-m/risk/e-guidelines/html/vol3/00/v3-00.htm
"A Review of Hazard Identification Techniques and Their Application to Major Accident Hazards", by S.T.Parry, March 1986, SRD R 379, United Kingdom Atomic Energy Authority www.ukaea.org.uk/contact/mail.htm

CHAPTER 6

Basics of HAZOP

What Did we Do Before HAZOP Came Along?

We relied upon:

- Good engineering practices
- Codes of practice such as ASME, API, NFPA, etc.
- Informal safety reviews

So what slips through the cracks?

- Interface problems between equipment and systems
- Abnormal conditions not envisioned during design
- Human error in design, operation, maintenance

How Do We Know If a Plant Is Safe?

- Historical record of plant incidents and near misses
- History of incidents on similar plants
- Record of onstream time
- Can apply some risk ranking, *e.g., Dow/Mond indices*

Ask: *What guarantee do we have that there isn't a serious accident about to happen?*

If there is **NO guarantee**:

What can we do about it?

⇨ **Increase Hazards & Risk Awareness**

by:

⇨ **Process Safety Management Program.**

HAZOP Methodology

HAZOP

Acronym for HAZards and Operability Analysis.

Originated by Imperial Chemical Industries (ICI) in Mond Division.

Basic Concept

Simulate abnormal behavior by considering ***deviations*** and disturbances due to ***causes*** likely to impact immediate and surrounding plant resulting in ***consequences***. Then decide whether the design has adequate features (i.e., ***safeguards***) that can prevent occurrence or limit the consequential effects. If no such safeguards exist, then consider what ***actions*** are needed to remedy the situation.

High Deviation
Normal (Design Intention)
Low Deviation

Other Deviations typically include:

- Reverse of what was intended.
- What else can happen?
- System only partially functions.
- What additional things can occur?

Methodology for Generating Deviations

Guide Word + Property = Deviation

For example:

- **When Property = Parameter:**

High	+	Flow	=	High Flow
Low	+	Pressure	=	Low Pressure
More	+	Reaction	=	Greater Reactivity

- **When Property = Operation:**

No	+	Transfer	=	No Transfer
Less	+	Empty	=	Residue Remaining

- **When Property = Material:**

No	+	Steam	=	No Steam
More	+	Diluent	=	More Diluent

What Type of HAZOP Should You Use?

Parametric Deviation (e.g., High pressure, Low temperature, etc.)

- Good for continuous processes.
- Most widely used in world today.

Critical Examination

The approach examines:

- Material
- Activities
- Sources and Destinations

Good for batch operations, start-up, shut down.

Procedural Methodology

Useful for HAZOPing:

- Operating manuals, procedures
- Batch operations
- Start-up, shut down

Knowledge Based HAZOP (more like 'What if' with established Checklist)

Mainly applicable to:

- Well established (continuous) processes
- Organizations with very high quality engineering practices & standards

Steps in the HAZOP Process

1. Preparation

Assemble:

- P&IDs (Full size and reduced copies for the team)
- PFDs plus material and energy balances
- Equipment specifications
- Layout drawings

2. Facilitator and Process Engineer

Break P&IDs down into Nodes.

- Nodes are equipment items (or numbers of items).
- If nodes are too small you can loose sense of analysis and incur excessive repetition.
- If nodes are too large, hard to handle, becomes confusing.

Question: How do you size a node?

Answer: Based on system function.

Example: *Reactor feed system may consist of Pump + Line + Exchanger.*

3. Prepare HAZOP Outline with List of Deviations

4. Assemble HAZOP Team

5. *Facilitator Explains*

- The facilitator or one of the team members explains the purpose and scope of the HAZOP and sets the rules for the study.

6. *Process Engineer Explains*

- Process in general
- Immediate Node being HAZOPed

7. *HAZOP Each Node Using Deviations Listed in Outline Working Through the P&ID.*

Produce HAZOP worksheet recording the following:

- Causes
- Consequences
- Safeguards
- Actions/Recommendations
- Remarks

8. *At the End of HAZOP, the Facilitator Issues Preliminary HAZOP Report (issuance is optional) consisting of*

- Attendance
- Outline
- Detail Report
- Action/Recommendations Register

9. *Issue Final Report Giving Full Details*

- A sample of table of contents is given in page 6-11.

Variations in HAZOP Types

Three basic types:

- Guide Word HAZOP
- Knowledge Based HAZOP
- "Creative Checklist"

The Guide Word method is the most accepted method. There are five main variations:

- Cause-by-cause
- Consequence-by-consequence
- Deviation-by-deviation
- Exception only
- Action/Recommendation item only

Cause-By-Cause Methodology

Correlates Consequences, Safeguards and Actions to each particular Cause of a Deviation.

- Precise method
- Reduces ambiguity
- Detail print-out can be followed, is fully auditable

Example**: Deviation: **Line Rupture

Cause #1	***Consequence***	***Safeguard***	***Action***
Line overstressed	*Flammable release, fire*	*Pipe stress analysis*	*Check fire protection*

Cause #2	***Consequence***	***Safeguard***	***Action***
Brittle fracture	*Flammable release, fire*	*Charpee tested steel*	*Check fire protection*

Cause #3	***Consequence***	***Safeguard***	***Action***
Vehicular impact	*Flammable release, fire*	*None*	*Provide crash barrier*

Cause-By-Cause provides ***full*** cross-referencing.

Consequence-By-Consequence Methodology

Correlates Consequences, Safeguards and Actions to each particular Consequence of a Deviation.

- Precise method
- Reduces ambiguity
- Detail print-out can be followed, is fully auditable

Example**: Deviation: **Line Rupture

Cause #1	***Consequence***	***Safeguard***	***Action***
Line overstressed	*Flammable release*	*Pipe stress analysis*	
	Fire		*Check fire protection*

Cause #2	***Consequence***	***Safeguard***	***Action***
Brittle fracture	*Flammable release*	*Charpee tested steel*	
	Fire		*Check fire protection*

Cause #3	***Consequence***	***Safeguard***	***Action***
Vehicular impact	*Flammable release*	*None*	*Provide crash barrier*
	Fire		*Check fire protection*

Consequence-By-Consequence provides ***full*** cross-referencing.

Deviation-By-Deviation Methodology

All Causes, Consequences, Safeguards and Actions are related only to a particular Deviation.

- Fairly simple to execute
- Some ambiguity
- Fairly rapid
- Detail print-out hard to follow

Example**: Deviation: **Line Rupture

Causes	***Consequences***	***Safeguards***	***Actions***
Line overstressed	*Flammable release, fire*	*Pipe stress analysis*	*Check fire protection*
Brittle fracture		*Charpee tested steel*	*Provide crash barrier*
Vehicular impact			

Deviation-By-Deviation provides ***no*** cross-referencing.

Exception Only Methodology (Not Recommended)

Includes only those deviations for which team believes there are credible causes.

- Reduces time
- Cannot be audited

Citations have been issued by OSHA in the USA against covered facilities using this method.

Action/Recommendations Item Only Methodology (Not Recommended)

Only suggestions that team makes for action items are recorded. (No proper analysis).

- Not auditable

Preparation of HAZOP Reports

Basic Report Should Consist of

1. HAZOP Outline
   - Nodes
   - Deviations
   - Guide Words
   - Parameters
   - Design Intent
   - Design Conditions

2. Detail Report

   Lists output of sessions.

   For each Node and Deviation lists:
   - Causes
   - Consequences (+ Risk Ranking)
   - Safeguards
   - Actions/Recommendations
   - Remarks

3. Attendance Register
   - Facility, Unit
   - Location
   - Team members and expertise
   - Attendance (Present/Absent/Not required/Part-time)

4. Action/Recommendations Register

- Action/Recommendations Item
- Person(s) responsible for follow-up
- Prioritization
- Status
- Target date for completion (Resolution)

Other

- Risk ranking
- Categorization

Final Report

1. Executive Summary
2. Introduction
3. Process Description
4. Hazards of the process
5. Hazard and Operability Methodology
6. Conclusions and Recommendations

Appendices

- o Outline of Hazard and Operability Study
- o Drawings
- o Project Information Report
- o Drawing Report
- o Team Members Report
- o Risk Matrix Report
- o Worksheet Report
- o Action/Recommendations Report
- o Computer files

HAZOP Example

Table 6-1: Scope of HAZOP and Process Description

Company: XYZ Processing Corporation

Location: Anyplace

Unit: Light Ends Separation

Project ID: ABC 001

Start Date: 1/21/2002

End Date: 1/22/2002

Comment: LIGHT ENDS RECOVERY UNIT (See Figures 6-1 and 6-2)

Purpose:

To recover the light ends portion of a liquid feed stream containing 50%, by weight, of light ends using a continuous fractionator. The feed rate is 100,000 lb/hour and the intent is to recover 81% of the light ends fed to the stripper in order to produce a 90%, by weight, light ends distillate.

Process Description:

The unit is fed from an upstream feed drum, V-101. Liquid feed is supplied at a battery limits pressure of around 90 psig and at 220 F. The feed is pre-heated by heat exchange with the light ends stripper bottoms bottoms, at 300 F, in the feed/bottoms exchanger, EX-101, before it enters the light ends stripper, C-101.

The light ends stripper, C-101, is a 22 plate fractionation column, using valve-type trays with the feed point located on plate 12.

Overhead vapors from the light ends stripper, C-101, pass to the light ends condenser, EX-102, which is water cooled. Fluids from this condenser flow to the reflux drum, V-102. Non condensible vapors entering the reflux drum, V-102, are vented to the flare system while the condensed liquids pass to the reflux pump, P-101 or spare. The total distillate is split so that product distillate is sent directly to storage while the main portion is refluxed back to the top of the light ends stripper, C-101.

At the base of the light ends stripper, C-101, there is a vertical thermosiphon reboiler, EX-103, which is heated by 300 psig steam on the shell side. Bottoms liquid from the base of the light ends stripper, C-101, is pumped by a bottoms pump, P-102 or spare, to the tube side of the feed/bottoms exchanger, EX-101, in which it is cooled by feed before passing to storage.

Process Controls:

The feed to column is under flow control via loop FRC-101.

The reflux flow is under flow control via loop FRC-116.

Distillate withdrawal is under level control via loop LIC-107, from the reflux drum level.

Non-condensibles bleed off under pressure control via loop PIC-106 based on the overheads column pressure.

The rate of bottoms withdrawal is under level control via loop LIC-119 based on the column bottom level.

The steam flow to the reboiler is under composition control via temperature control loop TRC-126 based on the process side of the reboiler outlet.

Protective Devices:

Relief valve PSV-105, protects against overpressuring of the light ends stripper and connected components.

Relief valve, PSV-106, protects against thermal expansion on the cooling water side of the light ends condenser.

High and low level conditions, LAH-120, LAL-121 and LAH & LAL-107, are alarmed respectively on the light ends stripper and the reflux drum. The low level condition, LSL-121, on the stripper is also interlocked to stop the bottoms pump. The low low level condition on the reflux drum stops the reflux pump.

In event of failure of the bottoms pump, the spare pump is started by a low low pressure switch, PSLL-125. The same arrangement, for the reflux pump, is also supplied by a low low pressure switch, PSLL-109.

High or low column pressures are alarmed by PAH & PAL-106 respectively.

Loss of reflux is alarmed by FAL-116.

Loss of steam to the reboiler is alarmed by TAL-126.

A minimum flow bypass on the bottoms pump protects against the no flow condition.

Remotely operable motor operated valve, MOV-122, can be manually initiated in an emergency, such as bottoms line leak/fracture, to prevent significant flammables inventory loss and fire.

The instrument air failure positions of the control valves are indicated as F.C. (fail close) or F.O. (fail open).

Car seal open (CSO) valves are as indicated.

Assumptions:

During a normal HAZOP you would normally have access to full equipment specifications, plant layout drawings, piping specifications, line lists, tie points and other pertinent documents. As this sample demonstrates PHA-Pro, rather than being an exercise in design, such documents are not included. Therefore make whatever assumptions you think reasonable if you wish to modify or extend the HAZOP as shown.

Normal Operating Conditions:

Stream #1, Feed @ 220 F, 90 PSIG, 100,000 Lb/Hour, 50% Light Ends
Stream #2, Overhead @ 200 F, 75 PSIG, 135,000 Lb/Hour, 90.2% Light Ends
Stream #3, Bottoms @ 300 F, 120 PSIG, 50,000 Lb/Hour, 9.5% Light Ends
Stream #4, Reflux @ 200 F, 75 PSIG, 85,000 Lb/Hour, 90% Light Ends
Stream #5, Non Condensibles @ 200 F, 75 PSIG, 5,000 Lb/Hour, 95% Light Ends
Stream #6, Distillate @ 200 F, 150 PSIG, 45,000 Lb/Hour, 90% Light Ends
Stream #7, Reboiler Feed @ 300 F, 80 PSIG, 185,000 Lb/Hour, 9.5% Light Ends
Stream #8, Steam Flow @ 420 F, 300 PSIG, 25,000 Lb/Hour

Heat Exchanger Duties:

Condenser, EX-102: 19.5 MMBTU/HR
Feed/Bottoms Exchanger, EX-101: 1.9 MMBTU/HR
Reboiler, EX-103: 19.9 MMBTU/HR

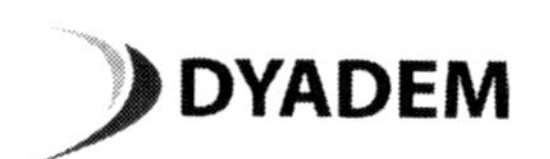

Figure 6-1: P&ID of Light Ends Process

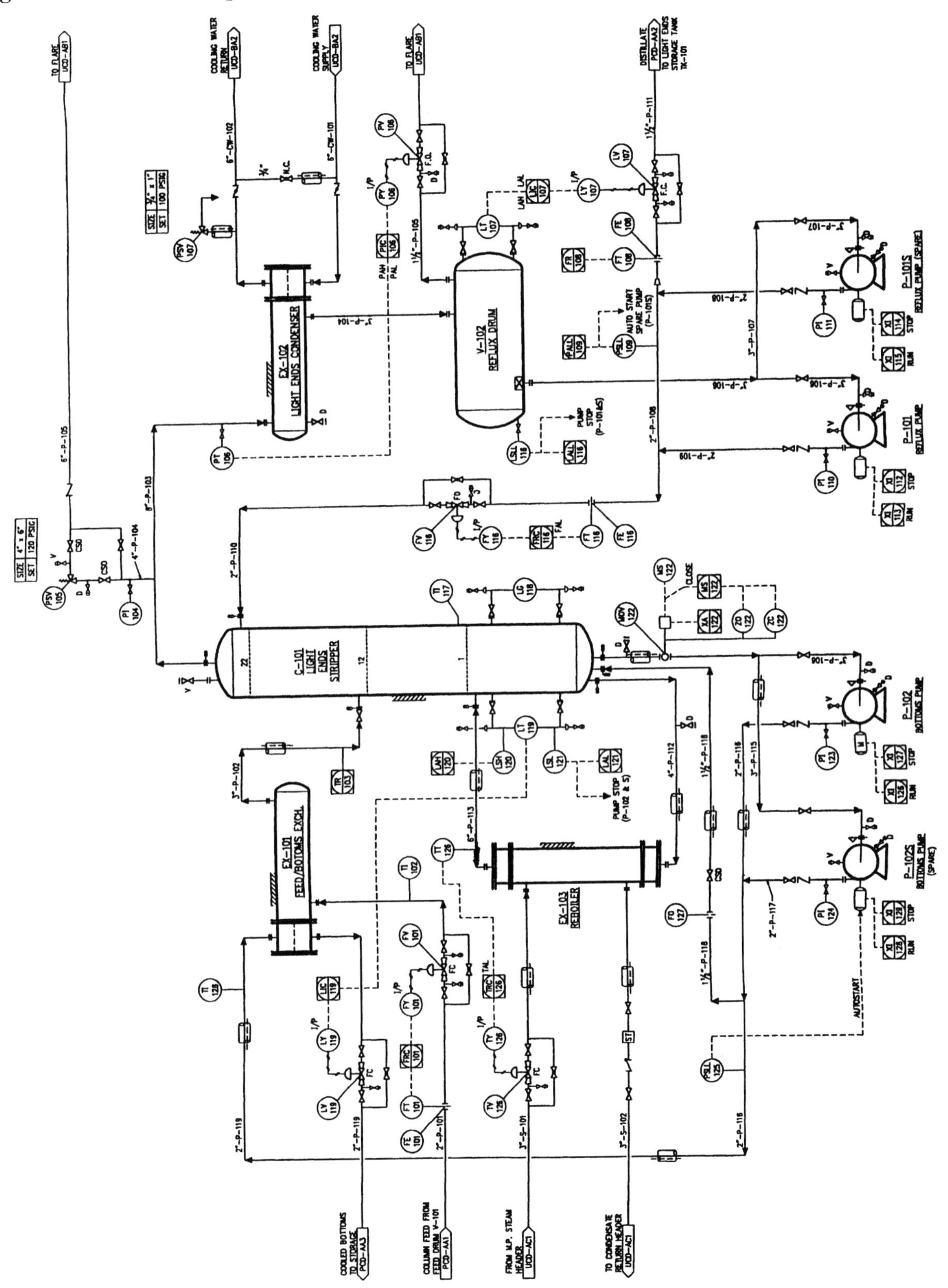

Figure 6-2: Process Flow Diagram

PROCESS FLOW SKETCH FOR LIGHT ENDS RECOVERY UNIT
XYZ Processing Corporation

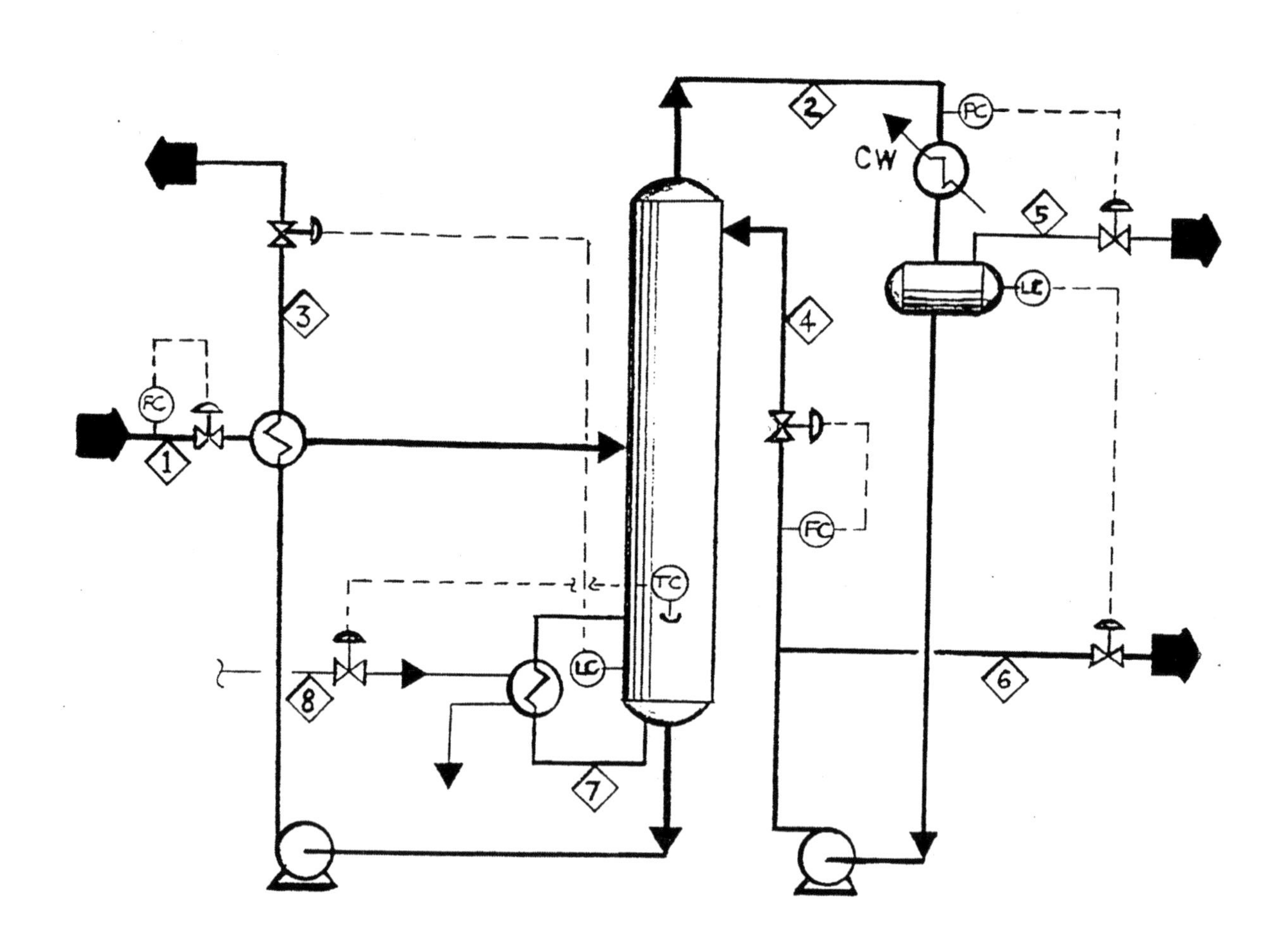

Condenser (EX-102)

Feed/Bottoms Exch (EX-101) — Light Ends Stripper (C-101) — Reflux Drum (V-102)

Reboiler (EX-103)

Bottoms Pump (P-102/S) — Reflux Pump (P-101/S)

Stream	1	2	3	4	5	6	7	8
Description	Feed	Over-head	Bottoms	Reflux	Non Conden-sibles	Distillate	Reboiler Feed	Steam Flow
Degrees F	220	200	300	200	200	200	300	420
P.S.I.G.	90	75	120	75	75	150	80	300
Lb/Hour	100,000	135,000	50,000	85,000	5,000	45,000	185,000	25,000
Light Ends	50%	90.2%	9.5%	90%	95%	90%	9.5%	N/A

HEAT EXCHANGER DUTIES: CONDENSER: 19.5 MMBTU/HR
FEED/BOTTOMS EXCHANGER: 1.9 MMBTU/HR
REBOILER: 19.9 MMBTU/HR

Table 6-2: List of Deviations

Node: 1. Light Ends Recovery Unit	Drawings: 1. Process Flow Sketch of Light Ends Recovery Unit; 2. Piping & Instrumentation Diagram # PCD-A1
Types: Centrifugal Pump, Column, Heat Exchanger, Line, Maintenance problems, Vessel	
Equipment ID: Distillation unit with heat recovery from bottoms heating feedstream	
Design Conditions/Parameters: Design conditions are listed in Process Flow Sketch of Light Ends Recovery Unit provided	

Deviation	Guide Word	Parameter	Session	Revision #	Design Intent
1.1. High Flow	High	Flow	1	0	As per Process Flow Diagram
1.2. Low/No Flow	Low/No	Flow	1	0	As per Process Flow Diagram
1.3. Reverse/Misdirected Flow	Reverse/Misdirected	Flow	1	0	As per Process Flow Diagram
1.4. Other than Flow	Other than	Flow	1	0	As per Process Flow Diagram
1.5. High Temperature	High	Temperature	1	0	As per Process Flow Diagram
1.6. Low Temperature	Low	Temperature	1	0	As per Process Flow Diagram
1.7. High Pressure	High	Pressure	2	0	As per Process Flow Diagram
1.8. Low Pressure	Low	Pressure	2	0	As per Process Flow Diagram
1.9. High Level	High	Level	2	0	As per Process Flow Diagram
1.10. Low Level	Low	Level	2	0	As per Process Flow Diagram
1.11. Cavitation	As well as	Performance	2	0	As per Process Flow Diagram
1.12. Column Flooding	Part of	Performance	2	0	As per Process Flow Diagram
1.13. Low Tray Level	Low	Level	2	0	As per Process Flow Diagram
1.14. High Concentration of Impurities	As well as	Composition	2	0	As per Process Flow Diagram
1.15. Leak	As well as	Flow	2	0	As per Process Flow Diagram

Deviation	Guide Word	Parameter	Session	Revision #	Design Intent
1.16. Rupture	Other than	Flow	2	0	As per Process Flow Diagram
1.17. Start-up/Shutdown Hazards	Other than	Start-up/Shutdown	2	0	As per Operating Instructions
1.18. Maintenance Hazards	Other than	Maintenance	2	0	As per Maintenance Data

Table 6-3: Sample Worksheet

Node: 1. Light Ends Recovery Unit	Drawings: 1. Process Flow Sketch of Light Ends Recovery Unit, 2. Piping & Instrumentation Diagram # PCD-A1
Types: Centrifugal Pump, Column, Heat Exchanger, Line, Maintenance problems, Vessel	
Equipment ID: Distillation unit with heat recovery from bottoms heating feedstream	
Design Conditions/Parameters: Design conditions are listed in Process Flow Sketch of Light Ends Recovery Unit provided	

1.1. High Flow

Causes	Consequences	Safeguards	S	L	RR	Recommendations	Responsible	Remarks
1. Control valve or controller fails CV FV-101 open or bypass left open	1.1. Off-spec products	1.1. PSV-105	1	3	3	1. Add high flow alarm to FT-101 which is independent of FRC-101	Tom Volke	
						2. Investigate recycling line and/or additional storage for handling off-spec products	Phillip Smith	
	1.2. Possible gas blow-through from upstream feed drum V-101 leading to overpressuring of column and possible tray damage					3. Provide maintenance log of bypass valves around control valves (typical)	Milo Rimer	
						4. Provide low low level switch from V-101 (dwg. no. PCD-AA1) to trip FV-101 closed using solenoid valve to cut instrument air to CV	Tom Volke	
2. FE-101 reads falsely low	2.1. Off-spec products	2.1. PSV-105	1	4	4	5. Provide regular maintenance inspection of flow devices, such as FE-101	Milo Rimer	
	2.2. Possible gas blow-through from upstream feed drum V-101 leading to overpressuring of column and possible tray damage							
3. 1. Control valve or controller fails CV LV-119 open or bypass left open	3.1. Loss of products	3.1. LSL-121 trips P-102/S						3.1. Safeguards are adequate
	3.2. Low level in column could cavitate/damage bottoms pumps P-102/S							
	3.3. Off-spec products							

Causes	Consequences	Safeguards	S	L	RR	Recommendations	Responsible	Remarks
4. Control valve or controller fails CV TV-126 open or bypass left open	4.1. High bottoms temperature 4.2. Off-spec products	4.1. PSV-105	1	3	3	6. Make TI-117 control room monitored as opposed to local device and add high temperature switch and alarm	Tom Volke	
						7. Check sizing of control valve TV-126 so that on full opening column will be unlikely to flood due to excess vapor flow	Phillip Smith	
	4.3. Over-pressuring of column 4.4. Possible flooding of column	4.2. TI-117				8. Add high temperature alarm, TAH-126	Tom Volke	
5. Control valve or controller fails CV PV-106 open or bypass left open	5.1. Column will depressure to flare 5.2. Loss of products 5.3. Off-spec products	5.1. PAL-106 (provided PIC-106 is functional)	1	3	3	9. Add independent pressure monitor on column overheads with high and low pressure switches and alarms	Tom Volke	
6. Control valve or controller fails CV LV-107 open or bypass left open	6.1. Loss of products 6.2. Off-spec products	6.1. LAL-121	1	4	4			6.1. Safeguards are adequate
	6.3. Low level in column could cavitate/damage bottoms pumps P-102/S	6.2. LSL-121 trips P-102/S						
7. Control valve or controller fails CV FV-116 open or bypass left open	7.1. Excess reflux to column	7.1. None	1	4	4	10. Add high flow alarm to FRC-116	Tom Volke	
	7.2. Uneconomical performance					11. Consider monitoring steam flow to column by adding flow indicator on 3"-S-101 as check on energy consumption	Tom Volke & Carl Hanks	
8. PSV-105 fails open due to spring failure	8.1. Column will depressure to flare 8.2. Loss of products 8.3. Off-spec products	8.1. Block and bypass valves around PSV plus PI-104	1	1	1			8.1. Safeguards are adequate

Causes	Consequences	Safeguards	S	L	RR	Recommendations	Responsible	Remarks
	8.4. [Failure too infrequent to raise significant concerns]	for manual pressure relief when PSV-105 is removed for repair						
9. Steam trap on 3"-S-102 sticks open	9.1. Steam wastage	9.1. None	1	3	3	11. Consider monitoring steam flow to column by adding flow indicator on 3"-S-101 as check on energy consumption	Tom Volke & Carl Hanks	
	9.2. Uneconomical performance							

6

1.2. Low/No Flow

Causes	Consequences	Safeguards	S	L	RR	Recommendations	Responsible	Remarks
1. Control valve or controller fails CV LV-119 closed	1.1. No withdrawal of bottoms product	1.1. LAH-120	1	3	3			1.1. Safeguards are adequate
	1.2. Loss of products	1.2. LG-118						
	1.3. High level in base of column	1.3. Minimum flow bypass on bottoms pumps						
	1.4. Damage to bottoms pumps							
2. Control valve or controller fails CV FV-101 closed	2.1. Loss of products	2.1. None	1	4	4	12. Add low flow alarm to FT-101 which is independent of FRC-101	Tom Volke	
3. FE-101 reads falsely high	3.1. Loss of products	3.1. None	1	4	4	5. Provide regular maintenance inspection of flow devices, such as FE-101	Milo Rimer	
4. Control valve or controller fails CV TV-126	4.1. Loss of products	4.1. TI-117	1	4	4	13. Consider adding nitrogen pressure balancing system to make up in the event of sudden loss of steam to reboiler	Phillip Smith	

Causes	Consequences	Safeguards	S	L	RR	Recommendations	Responsible	Remarks
closed	4.2. Possible vacuum in column causing tray damage					14. Interlock reflux return, FV-116, and feed, FV-101 to close when PAL-106 is actuated	Tom Volke	
5. Control valve or controller fails CV PV-106 closed	5.1. Column will overpressure	5.1. PAH-106 (provided PIC-106 is functional)	1	3	3	9. Add independent pressure monitor on column overheads with high and low pressure switches and alarms	Tom Volke	
	5.2. Loss of products	5.2. PSV-105						
	5.3. Off-spec products							
6. Control valve or controller fails CV LV-107 closed	6.1. Loss of products	6.1. LAH-107 (provided LIC-107 is functional)	1	4	4	15. Add high level switch and alarm on reflux drum V-102	Tom Volke	
	6.2. High level in reflux drum							
7. Control valve or controller fails CV FV-116 closed	7.1. Loss of reflux to column	7.1. FAL-116 (provided FIC-116 is functional)	1	4	4	11. Consider monitoring steam flow to column by adding flow indicator on 3"-S-101 as check on energy consumption	Tom Volke & Carl Hanks	
	7.2. Off-spec products	7.2. PSV-105				16. Add independent pressure monitor on column overheads with high and low pressure switches and alarms	Tom Volke	
	7.3. Over-pressuring of column							
8. Steam trap on 3"-S-102 sticks closed	8.1. Reboiler will flood	8.1. None	1	3	3	11. Consider monitoring steam flow to column by adding flow indicator on 3"-S-101 as check on energy consumption	Tom Volke & Carl Hanks	
	8.2. Loss of products							
9. Bottoms pump P-102/S stops	9.1. No withdrawal of bottoms product	9.1. LAH-120	1	3	3			9.1. Safeguards are adequate
	9.2. Loss of products	9.2. Spare pump						
	9.3. High level in base of column	9.3. LG-118						
10. Reflux pump P-101/S stops	10.1. Loss of reflux to column	10.1. FAL-116	1	4	4			10.1. Safeguards are adequate
	10.2. Off-spec products	10.2. PSV-105						

Causes	Consequences	Safeguards	S	L	RR	Recommendations	Responsible	Remarks
	10.3. Over-pressuring of column	10.3. Spare pump						
11. MOV-122 fails closed	11.1. No flow would cavitate/damage bottoms pumps P-102/S	11.1. Interlock on MOV-122 positioner stops bottoms pumps when valve closes	2	3	6	17. Provide interlock on MOV-122 positioner ZC-122 to stop bottoms pumps when MOV-122 valve closes	Tom Volke	
12. Temporary strainers on P-101/S plugged	12.1. No flow would cavitate/damage bottoms pumps P-1021S	12.1. None	2	4	8	18. Ensure temporary strainers on P-101/S are cleaned and removed when no longer required	Carl Hanks	
13. Loss of overhead condenser.	13.1. Overpressuring of column to relief condition.	13.1. Low flow alarm FAL-116 on loss of reflux.	2	3	6	19. Check PSV-105 for controlling case for sizing valve. Must handle fire case, tube rupture in reboiler,total loss of reflux, loss of cooling medium, instrument or controller failure, instrument air failure, power failure, etc.	Phillip Smith	
		13.2. Pressure relief valve PSV-105.				20. PV-106 to be checked for maximum discharge flow in event of cooling water failure to EX-102.	Phillip Smith	
		13.3. PAH-106 & PV-106 opening to flare.						

1.3. Reverse/Misdirected Flow								
Causes	Consequences	Safeguards	S	L	RR	Recommendations	Responsible	Remarks
1. Backflow from flare via PV-106 at start-up	1.1. Possible explosive /combustible mixture in columns	1.1. None	3	3	9	21. Provide check valve downstream of PV-106	Allen Brown	

Table 6-4: List of Recommendations

#	Recommendation	Resp	Status	Pri	Place(s) Used
1.	Add high flow alarm to FT-101 which is independent of FRC-101	Tom Volke	Incomplete	5	1.1.1
2.	Investigate recycling line and/or additional storage for handling off-spec products	Phillip Smith	Study	7	1.1.1
3.	Provide maintenance log of bypass valves around control valves (typical)	Milo Rimer	Incomplete	7	1.1.1
4.	Provide low low level switch from V-101 (dwg. no. PCD-AA1) to trip FV-101 closed using solenoid valve to cut instrument air to CV	Tom Volke	Incomplete	7	1.1.1
5.	Provide regular maintenance inspection of flow devices, such as FE-101	Milo Rimer	Incomplete	6	1.1.2; 1.2.3
6.	Make TI-117 control room monitored as opposed to local device and add high temperature switch and alarm	Tom Volke	Incomplete	6	1.1.4
7.	Check sizing of control valve TV-126 so that on full opening column will be unlikely to flood due to excess vapor flow	Phillip Smith	Study	7	1.1.4
8.	Add high temperature alarm, TAH-126	Tom Volke	Incomplete	6	1.1.4
9.	Add independent pressure monitor on column overheads with high and low pressure switches and alarms	Tom Volke	Incomplete	8	1.1.5; 1.2.5
10.	Add high flow alarm to FRC-116	Tom Volke	Incomplete	7	1.1.7
11.	Consider monitoring steam flow to column by adding flow indicator on 3"-S-101 as check on energy consumption	Tom Volke & Carl Hanks	Study	5	1.1.7,9; 1.2.7,8
12.	Add low flow alarm to FT-101 which is independent of FRC-101	Tom Volke	Incomplete	7	1.2.2
13.	Consider adding nitrogen pressure balancing system to make up in the event of sudden loss of steam to reboiler	Phillip Smith	Study	8	1.2.4
14.	Interlock reflux return, FV-116, and feed, FV-101 to close when PAL-106 is actuated	Tom Volke	Incomplete	8	1.2.4
15.	Add high level switch and alarm on reflux drum V-102	Tom Volke	Incomplete	6	1.2.6
16.	Add independent pressure monitor on column overheads with high and low pressure switches and alarms	Tom Volke	Incomplete	8	1.2.7
17.	Provide interlock on MOV-122 positioner ZC-122 to stop bottoms pumps when MOV-122 valve closes	Tom Volke	Incomplete	9	1.2.11
18.	Ensure temporary strainers on P-101/S are cleaned and removed when no longer required	Carl Hanks	Incomplete	7	1.2.12
19.	Check PSV-105 for controlling case for sizing valve. Must handle fire case, tube rupture in reboiler,total loss of reflux, loss of cooling medium, instrument or controller failure, instrument air failure, power failure, etc.	Phillip Smith	Study	10	1.2.13; 1.14.2
20.	PV-106 to be checked for maximum discharge flow in event of cooling water failure to EX-102.	Phillip Smith	Study	8	1.2.13
21.	Provide check valve downstream of PV-106	Allen Brown	Incomplete	9	1.3.1

#	Recommendation	Resp	Status	Pri	Place(s) Used
22.	Check on flow regime in 6"-P-113 to confirm that regime is non-slugging	Phillip Smith	Study	7	1.4.1
23.	Check that line 3"-P-104 is both self-venting and is not pocketed	Phillip Smith	Study	7	1.4.2
24.	Evaluate need for emergency depressuring to prevent BLEVE in event of fire	Phillip Smith	Study	9	1.5.4
25.	Provide sample point on inlet feed. Also consider need for on-line analyzer for column feed.	Tom Volke	Incomplete	6	1.5.5
26.	Provide quality control check on feed stream to column	Carl Hanks	Study	5	1.5.5
27.	Add high temperature alarm on overheads to indicate trend towards off-spec distillate	Tom Volke	Incomplete	5	1.5.5
28.	Modify TRC-103 to show TR-103 only	Phillip Smith	Incomplete	5	1.5.6; 1.6.3
29.	Add low temperature alarm to TR-103.	Tom Volke	Incomplete	6	1.6.3
30.	Consider adding independent high high level switch and alarm on reflux drum	Tom Volke	Study	7	1.9.2
31.	Check sizing of control valve TV-126 so that CV is not oversized and could cause column flooding when fully open. If necessary consider adding upper limit stop on control valve.	Phillip Smith & Tom Volke	Study	8	1.12.1
32.	Check as to whether upstream water separation precautions are met. Reconvene meeting if not met.	Phillip Smith	Incomplete	9	1.14.2
33.	Provide bolt torquing procedure as part of regular maintenance	Milo Rimer	Incomplete	9	1.15.1
34.	Consider need for environmental monitors.	Mary Patterson	Incomplete	6	1.15.3
35.	Add isolation valve immediately upstream of stripper on reflux line 2"-P-110.	Allen Brown	Incomplete	6	1.15.4
36.	Make valve on 3"-P-102 feed to column car seal open.	Phillip Smith	Incomplete	7	1.16.3
37.	Add check valve to 3"-P-102, close to stripper feed inlet.	Allen Brown	Incomplete	4	1.16.4
38.	Provide procedure for nitrogen purging during column start-up that ensures oxygen content of system is well below lower flammable limit.	Carl Hanks	Incomplete	9	1.17.2
39.	Confirm that C-101 and all vessels & heat exchangers are designed for full vacuum.	Phillip Smith	Incomplete	6	1.18.2
40.	(Action on hold relating to check whether there are maintenance vents & drains around EX-101).	Milo Rimer	Incomplete	8	1.18.3

SUGGESTED READING (URLs current at time of publication)
"Guidelines for Hazard Evaluation Procedures" by AIChE, CCPS, 2nd edition, 1992 plus "Guidelines for Hazard Evaluation Procedures" by AIChE, CCPS, 1st edition, 1985 www.aiche.org/pubcat/seadtl.asp?Act=C&Category=Sect4&Min=20
"HAZOP and HAZAN" by T.Kletz, published by IChemE, 1992 www.icheme.org/framesets/aboutusframeset.htm
"Size up plant hazards this way" by H.G.Lawley, Hydrocarbon Processing, April 1976, pages 247 to 258 www.hydrocarbonprocessing.com/contents/publications/hp/
"Eliminating Potential Process Hazards" by T.Kletz, Chemical Engineering, April1, 1985, pages 48 to 59 www.che.com/
"An Introduction to Hazard and Operability Studies – The Guide Word Approach" by R.E.Knowlton, published by Chemetics International, 1981 www.kvaerner.com/companies/companiesdetail.asp?id=796
"A Manual of Hazard & Operability Studies – The Creative Identification of Deviations and Disturbances", published by Chemetics International, 1992 www.kvaerner.com/companies/companiesdetail.asp?id=796
"Some Features of and Activities in Hazard and Operability (Hazop) Studies", by J.R.Roach and F.P.Lees, The Chemical Engineer,October, 1981, pages 456 to 462 www.icheme.org/framesets/aboutusframeset.htm
"HAZOP: Guide to best practice" by F.Crawley, M.Preston, B.Tyler, IChemE, 2000 www.icheme.org/framesets/aboutusframeset.htm
"The HAZOP (Hazard and Operability) Method" (Website) www.acusafe.com/Hazard_Analysis/HAZOP_Technique.pdf
"Hazard and Operability Studies", by M.Lihou (Website) www.lihoutech.com/hzp1frm.htm
"Hazard and Operability Studies", University of Florida, (Website) http://pie.che.ufl.edu/guides/hazop/index.html
"Process Hazards Analysis" by I.Sutton, published by SW/Sutton & Associates, 2002 http://www.swbooks.com/books/book_prha.shtml

CHAPTER 7

Pitfalls with HAZOP, Optimization of PHAs & Sizing of Nodes

Pitfalls with HAZOP

Inadequate Preparation

- Slows team down
- Excessive man-hour consumption

Do Not Assume Everyone Understands HAZOP

- Prepare team. This objective can be achieved through the services of risk management consultants.

Wrong Team Players Can Damage HAZOP

Typically you need:

- Facilitator/Scribe (Facilitator can usually function as scribe)
- Process Engineer
- Plant Operator
- Plant Maintenance
- Instrument Engineer
- Mechanical Engineer (Part-time)

Do Not Have Too Many or Too Few People

- More than 10 persons are hard to control; wasted man-hours.
- 3 persons or less: input too limited.
- Optimal number: 4 to 6.

Avoid Getting Sidetracked

- Avoid getting off topic.
- Avoid "hobby horses".
- Avoid redesigning during HAZOP. Identify Action Items for further study.

Do Not Run HAZOP Sessions for Excessively Long Periods

Use the Right Type of PHA Methodology in Relation to the Risk of the Unit

- Guide Word HAZOP for high risk units, *e.g., where pressure is above 1,000psi.*
- What if/Checklist for medium risk units.
- Checklist for low risk units.

If You Decide not to Evaluate a Specific Deviation for a Specific Node, Make Sure You Fully Understand the Ramifications

- The criterion for rejection is, ***primarily***, that ***no cause exists*** for the deviation coming from either within or outside of the node. Under these conditions record ***no cause for deviation***.

- It is irrelevant whether the consequences of the deviation impact the node in question or other nodes. These are ***not*** criteria for rejection of the deviation.
- Build-up of node deviation data is time consuming in the first place. However, once this data bank of information is established the HAZOP will proceed much faster.
- The secret of saving time and maintaining efficiency is ***not*** by rejecting valid deviations but by ensuring that each node is not undersized, thereby avoiding unnecessary repetition.

Address Group Participation

- Avoid team sessions being dominated totally by one or two people.
- Ensure everyone is encouraged to input. Use "round table" techniques. Share the responsibility of the HAZOP.

Make Sure You A*lways* Address *Hazards* and *Risk* Items.

Some HAZOPs have **addressed** everything but these items!

Are you protected against major hazards such as:

- Overpressure?
- Overtemperature & BLEVE potential?
- Loss of containment?
- Toxic releases, *e.g., Hydrogen Sulfide*?
- Fire?
- Explosions, especially with respect to buildings such as control centers?

When Listing Action Items

- Record the drawing number(s) with the Action Item so that it can be easily traced.
- Record the Action Item so that it can be acted upon by the responsible person designated to execute it. Avoid indecisive instructions such as "Consider studying..."
- Do not propose Actions that are just "wish list". Excessive numbers of Actions tend to devalue their worth. Be critical, but not over or under zealous.

Prioritize Your Analyses

- Analyze the most hazardous units first, *e.g., hydrocracker.*
- Some operations need early HAZOPs, *e.g., furnace start-ups.*
- Some equipment needs early analysis, *e.g., sour gas compressors, hydrogen compressors.*

Avoid Using HAZOP as a Design Tool

- HAZOP is an ***audit*** tool.
- Be wary of the expression "Leave it until we do the HAZOP, we will consider it then".

Just Doing HAZOP Isn't the End of the Story, It's Just the Beginning

- Follow Management Programs that specifically address the full spectrum, *e.g.,* API 750, OSHA 1910.119.

Optimization: When to Do What

Grass Root Design (or New Unit)

- **Phase 1:** Conceptual hazard review of process.

  ***Example**: Preliminary Hazards Analysis: process concerns, maximum inventories, effects on layout of worst credible scenario.*

- **Phase 2:** Use of Checklists during preparation of P&IDs.
- **Phase 3:** What if/Checklist on client approval issue of P&IDs.
- **Phase 4:** . Guide Word HAZOP on P&IDs issued for construction.

Note: Often Phase 3 is final & PHA is either Guide Word HAZOP or What if/Checklist on detailed issue of P&IDs.

Revamp Projects

- **Phase 1:** What if/Checklist on client approval issue of P&IDs.
- **Phase 2:** Guide Word HAZOP on P&IDs issued for construction. (Or single Guide Word HAZOP or What if/Checklist on detailed issue of P&IDs).

Existing Units

- **Step #1:** Establish priority for units, Risk ranking, *e.g., Dow F&EI.*
- **Step #2:** Perform What if/Checklist or Guide Word HAZOP.

Choosing & Sizing of Nodes for HAZOP

The Concept of Nodes

In HAZOP, the term "node" is used to describe the selection of one or more items of equipment as a focal point of study. A node could be as small as a line, a pump, a vessel or a heat exchanger, or as large as an entire process plant. The practicality of not only selecting nodes, but also of sizing nodes, is of critical importance.

Early Method of Assigning Nodes

Let us examine the early method for assigning nodes for Guide Word HAZOP.

Consider a vessel where there are a number of lines entering the vessel and a number of lines leaving the vessel. The early method was to take each of the lines entering the vessel, in turn, and treat them as separate nodes, applying deviations, such as High Flow, Low/No Flow, Reverse/Misdirected Flow, High Pressure, Low Pressure and so forth. Each line leaving the vessel was also treated as a separate node. The vessel itself was not treated as a separate node because it was considered to be adequately addressed by applying deviations to the entry and exit lines.

Later Methods of Assigning Nodes

Following on from the early method of line-by-line assignment of nodes, the concept of compound nodes was devised. With compound nodes, a section of routing, say, involving feed piping from a feed vessel, a centrifugal pump, a control valve set and a heat exchanger supplying a reactor vessel would be considered as a single node.

In time, compound nodes were expanded considerably to typically include all of the equipment shown on one or more piping and instrument diagrams (P&IDs).

Experience Gained in Choosing & Sizing Nodes

The early methodology of choosing single lines as nodes, although comprehensive, proved to be extremely time consuming and resulted in extensive repetition of recorded data. This led to extreme fatigue and loss of interest by HAZOP teams, resulting in low-efficiency HAZOPs.

Increasing the size of nodes to take into account more equipment items was found to result in less repetition, greater progress and more efficient HAZOPs.

Maximizing Node Sizes

For the relative newcomer to HAZOP, small node sizes, even those confined to single lines, can have the benefit of familiarization with the HAZOP methodology. Thus, as greater familiarity and confidence are gained with the HAZOP methodology, the node size can be increased to include more equipment.

What therefore is the practical and optimized limit to node size?

Given that small node sizes are inefficient, very large node sizes may also be inefficient when they become unwieldy and hard to handle.

In general, the optimum node size can include multiple items of equipment, provided that they share a common function.

When there is a discrete change in functionality, this becomes a demarcation point, and one or more additional nodes need to be designated to reflect the different functional groupings.

By way of example, where a complete P&ID, several P&IDs or several sections of one or more P&IDs have a common functionality, this can be deemed as a discrete node. For example, a fired furnace oil heater may show the following main components:

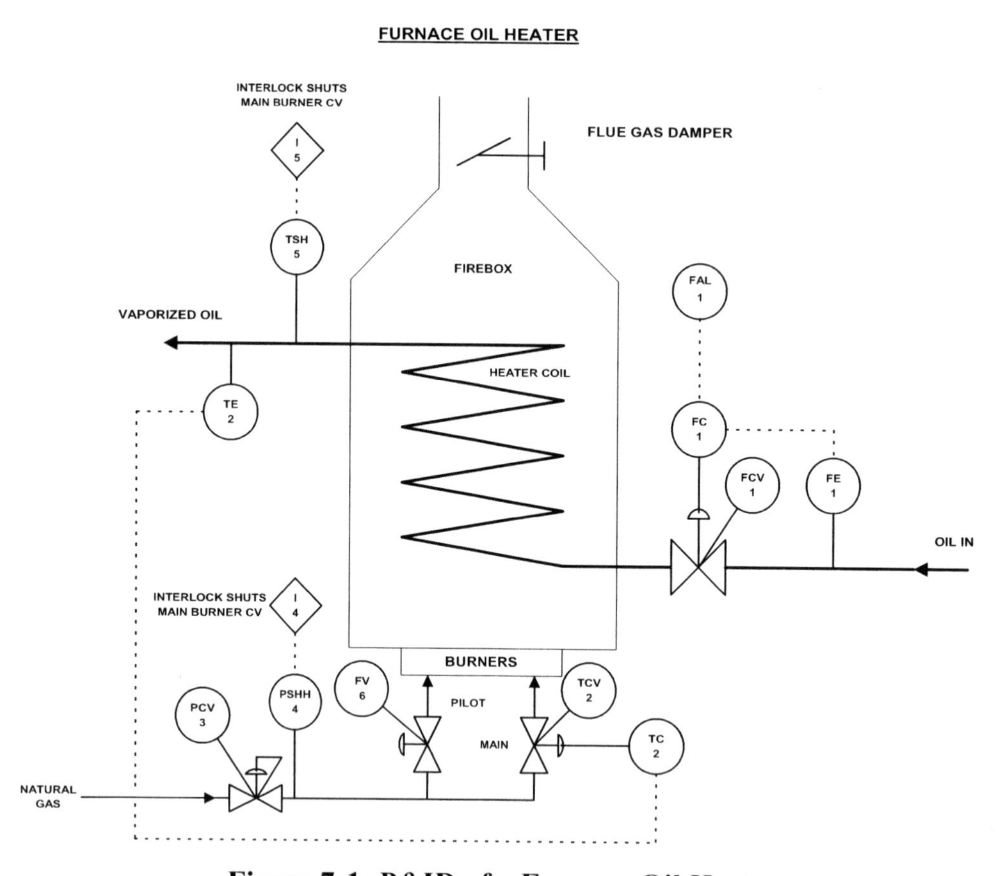

Figure 7-1: P&ID of a Furnace Oil Heater

The furnace oil heater may be designated as a complete node. Alternatively, the process flow (oil side, including the heating coil) could be designated as a node, the burner management system (fed by natural gas) could be counted as another node and the firebox itself as a third node.

One of the questions frequently asked is "If I create a large node, won't I perform a less-thorough HAZOP than if I break it down into multiple smaller nodes?" The answer to this question is that a number of experienced HAZOPers have tried both methods and have found that relatively little, if anything, is lost by choosing large nodes. In fact, with large nodes there is usually a better overview of systems. As well, important synergies and interactions are maintained, while repetition is minimized.

This speeds up the entire HAZOP process, making it more interesting for the HAZOP team as a whole, and overall gains usually exceed potential losses.

SUGGESTED READING (URLs current at time of publication)
"Guidelines for Hazard Evaluation Procedures" by AIChE, CCPS, 2nd edition, 1992 plus "Guidelines for Hazard Evaluation Procedures" by AIChE, CCPS, 1st edition, 1985 www.aiche.org/pubcat/seadtl.asp?Act=C&Category=Sect4&Min=20
"HAZOP and HAZAN" by T.Kletz, pub by IChemE, 1992 www.icheme.org/framesets/aboutusframeset.htm
"Oversights and mythology in a HAZOP program" by W.J. Kelly, Hydrocarbon Processing, October 1991, pages 114 to 112 www.hydrocarbonprocessing.com/contents/publications/hp/
"Hazard and Operability Studies", by M.Lihou (Website) www.lihoutech.com/hzp1frm.htm
"Process Hazards Analysis" by I.Sutton, published by SW/Sutton & Associates, 2002 http://www.swbooks.com/books/book_prha.shtml

CHAPTER 8

What If/Checklist

What If

Preparation

- Assemble drawings.
- Assign nodes.
- Prepare checklist.

Assemble Team

Explain process.

Commence Analysis

Add additional checklist items at the end of review of individual nodes.

Proceed on Same Basis as HAZOP

Advantages and Pitfalls

Advantages

- Interesting for participants.
- Usually very productive with experienced team.
- Versatile: No limiting formats or constraints.

Pitfalls

- May not cover all cases.
- Very dependent on experience of team members.
- Dependent on obtaining/creating/using good checklists.

Checklist

Available checklists, other data and your own experience may be used to create Checklists. Failing that, or in addition, you can use the following to assist with the preparation:

1. What does the equipment actually do? In what ways can the equipment actually fail?
2. What are the major hazards associated with the material being handled by the equipment?
3. What potential interactions between upstream or downstream equipment or conditions could lead to problems?
4. Could an external event give problems?
5. Could supporting utility failure(s) give problems?
6. Could environmental conditions give problems, *e.g., low temperatures*?
7. Could individual component failures, *e.g., control valves, level switches*, give problems?
8. Any problems with start-up or shut down?
9. Any problems maintaining equipment or individual components?
10. Sparing philosophy, equipment reliability?
11. Instrumentation & control system failures: what will happen?
12. Are there adequate protective systems? If so, how about redundancy?

13. Have you considered:

    - Power failure?
    - Instrument air failure?
    - Cooling water failure?
    - Steam failure?
    - Have effects of all of these been considered in relation to flare system sizing?

14. Do system components, *e.g., control valves*, fail safe?

15. Have you considered:

    - Equipment isolation?
    - Drainage?
    - Venting?
    - Blinding?
    - Emergency interlocks?

16. Have you considered any special operations, *e.g., presulfiding, on-site catalyst loading/unloading, on-site regeneration, etc.?*

17. Have you looked at common problems, such as:

    - High pressure/low pressure interfaces?
    - Possibility of reverse flow?
    - Chances of seal ruptures?
    - Equipment plugging?
    - Gas breakthrough on level control failure?
    - Bypasses being left open around control valves?
    - Tube ruptures in furnaces and heat exchangers?
    - Water hammer/two phase flow damaging lines?
    - Stress corrosion cracking e.g., stainless steel in presence of chlorides?

Checklist Applied to a Furnace Oil Heater

1. The Functionality of the Furnace Oil Heater

   - Furnace heats the oil and vaporizes it
   - Uses natural gas to heat oil
   - Controls the flow of oil
   - Containment of oil in tubes
   - Controls the temperature of oil
   - Maintains negative pressure in the furnace
   - Controls combustion air through grating
   - Controls natural gas pressure
   - Pilot flame ensures combustion

   The following are some general what if questions for the Furnace Oil Heater:

   a. What if heating is lost?
   b. What if the oil gets over heated?
   c. What if there is a disruption in the natural gas supply?
   d. What if the flow of oil is too low?
   e. What if the flow of oil is too high?
   f. What if the tubes containing the oil, ruptures?
   g. What if the temperature of the oil is too high?
   h. What if the temperature of the oil is too low?
   i. What if there is too little combustion air?
   j. What if there is too much combustion air?
   k. What if the gas pressure is too low?
   l. What if the gas pressure is too high?
   m. What if the pilot is extinguished?
   n. What if the Pilot doesn't initiate the flame?

2. Major Hazards

    a. What if the tube containing the oil ruptures?
    b. What if there is insufficient purging? (fire box explosion a possibility)

3. Flammable Release (Upstream /Down stream conditions)

    a. What if the tubes containing the oil rupture?
    b. What if there is insufficient purging?
    c. Is there a firewall protection?

4. Control

    a. Is the temperature of coil adequately controlled?
    b. Is the Firebox pressure controlled /regulated?

5. Worst event/Worst Credible Scenarios

    a. What are the mitigation steps taken to reduce the effects of fire box explosion?
    b. What are the mitigation steps taken to reduce the effects of tube rupture?
    c. What are the mitigation steps taken to reduce the effects of vapor cloud explosion?

6. Supporting Utilities

    a. Is the combustion air regulated?
    b. Is the natural gas supply continuous? Chances of interruption?

7. Process Side

    a. Is the temperature in the process area monitored and maintained at acceptable level?

8. Individual Component Failures

    a. Are there any controls to regulate/detect high pressure on tube side? Are they fail safe?
    b. Could heater be source of ignition for the vapor cloud release of flammable?

9. Startup/Shutdown

    a. Are the CVs or controllers open?
    b. Are there extra temperature monitors to detect temperature fluctuations during emergency shut down?
    c. Is the cooling water system linked to the emergency shutdown?

10. Are there sufficient monitors/alarm switches to detect:

    a. local tube overheating
    b. high firebox pressure
    c. loss/escape of pilot flame

11. Emergency Shut Down

    a. Will furnace trip on, on power failure?
    b. Will furnace trip on, on loss of instrument air? Are controls fail safe if the instrument air is disrupted?
    c. Will the furnace trip on, on loss of process/utility flows?
    d. Will furnace trip on, on loss of steam?

a. Are flow control valves fail safe on loss of actuating medium (e.g., instrument air failure)?

b. Is the PSV sized to take the vaporization load?

c. Is the heater located upwind/down wind?

d. Has an integrity check been made of the burner?

e. If relief valve could plug or coke up, is a purge stream (e.g., steam) included?

f. Have decoking provisions been reviewed?

g. Are there emergency shut off valves?

h. Are coil drains provided?

i. Is there adequate venting in place?

j. Are there spectacle blinds provided at all process inlets and outlets?

k. If process flow is lost will burner shut down?

l. Are there adequate high/low pressure alarms on fuel gas & pilot gas provided?

m. What if reverse flow into the firebox occurs in the event of tube failure?

n. Are emergency shutoff valves installed on fuel lines?

o. If burner tip is plugged, could atomizing stream cause higher pressure than fuel?

p. Are emergency shut down valves separate from control valves?

q. What if bypasses are left open around control valves?

r. What mitigation steps are taken to reduce the effects of tube rupture in the furnace?

s. Are valve closure times low enough to prevent water hammer?

t. Are piping materials suitable for maximum possible operating temperatures?

What If Example

Figure 8-1: P&ID of Ammonia Refrigeration Unit

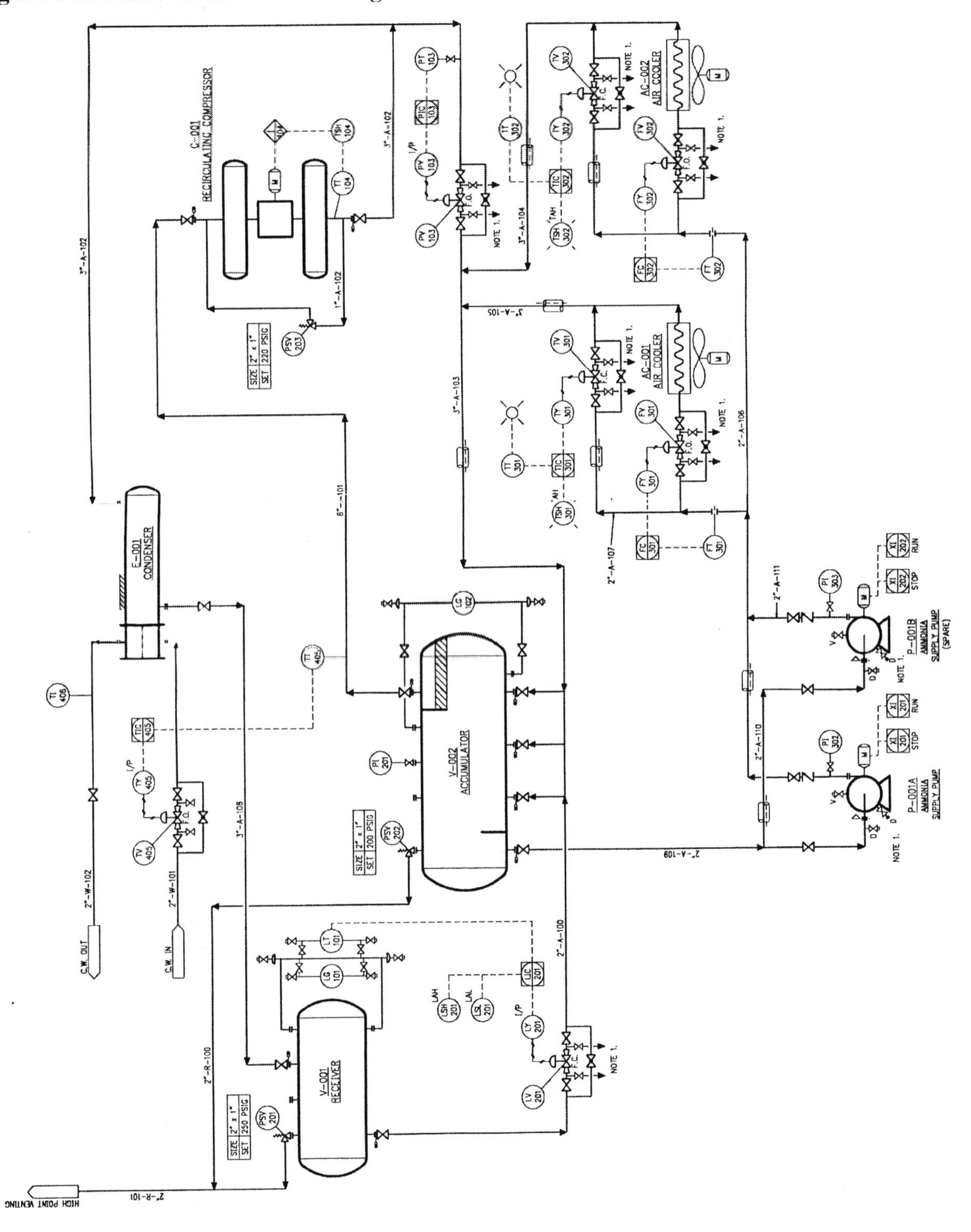

Figure 8-2: Process Flow Diagram

PROCESS FLOW SKETCH FOR ANHYDROUS AMMONIA REFRIGERATION UNIT
XYZ Cold Storage Corporation

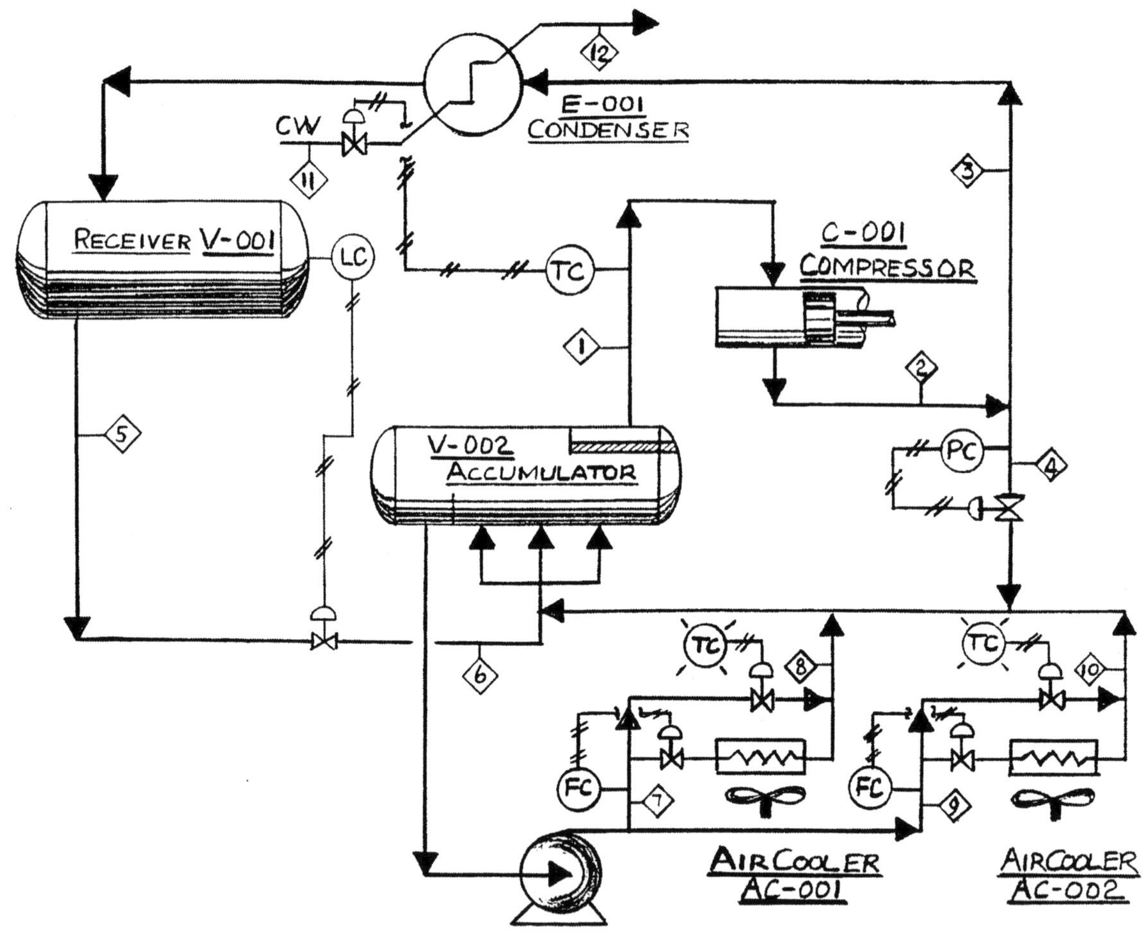

Stream	1	2	3	4	5	6
Description	Flow to Comp.	Comp. Disch.	Flow to Cond.	Recycle Gas	Cond. Liquid	Flash Return
Degrees F	21.7	84	84	84	76	21.7
P.S.I.G.	35	140	140	140	138	36
Lb/Hour	3,930	3,930	0 to 3,930	3,930 to 0	0 to 3,930	0 to 430/3500
State	Gas	Gas	Gas	Gas	Liquid	Gas/Liquid

Stream	7	8	9	10	11	12
Description	Ferm. Cooling	Ferm. Return	Stor. Cooling	Stor. Return	C. Water In	C.Water Out
Degrees F	21.7	21.7	21.7	21.7	60	84
P.S.I.G.	55	36	55	36	50	40
Lb/Hour	973	973	2,529	2,529	82,635	82,635
State	Liquid	Gas/Liquid	Liquid	Gas/Liquid	Liquid	Liquid

E-001, CONDENSER DUTY:	1.983 MMBTU/HR
C-001, REFRIGERATION COMPRESSOR:	907 BRAKE HP
AC-001, FERMENTER AREA AIR COOLER DUTY:	0.537 MMBTU/HR
AC-002, STORAGE ROOM AIR COOLER DUTY:	1.395 MMBTU/HR

Table 8-1: Thermodynamic Properties of Saturated Ammonia by Temperature

Temper-ature F	Pressure Lbs. Per Sq. In. Absolute	Specific Heat BTU/Lb. F	Specific Volume Cu. Ft. Per Lb.		Enthalpy of Liquid BTU/Lb.	Latent Heat of Evaporation BTU/Lb.	Enthalpy of Vapor BTU/Lb.	Entropy BTU/Lb. F		Density Pounds Per Cu. Ft.	
			LIQUID	VAPOR				LIQUID	VAPOR	LIQUID	VAPOR
−60	5.55	1.054	0.02278	44.73	−99.08	610.8	511.7	−0.2225	1.3061	43.91	0.02235
−50	7.67	1.058	0.02299	33.08	−88.51	604.3	515.8	−0.1964	1.2789	43.49	0.03023
−40	10.41	1.062	0.02322	24.86	−77.90	597.6	519.7	−0.1708	1.2534	43.08	0.04022
−30	13.90	1.066	0.02345	18.97	−67.24	590.7	523.5	−0.1457	1.2293	42.65	0.05271
−20	18.30	1.070	0.02369	14.68	−56.54	583.6	527.1	−0.1210	1.2066	42.22	0.06813
−10	23.74	1.075	0.02393	11.50	−45.79	576.4	530.6	−0.0969	1.1850	41.78	0.08695
0	30.42	1.080	0.02419	9.116	−34.98	568.9	533.9	−0.0732	1.1644	41.34	0.1097
10	38.51	1.085	0.02446	7.304	−24.11	561.1	537.0	−0.0499	1.1449	40.89	0.1369
20	48.21	1.091	0.02474	5.910	−13.19	553.1	539.9	−0.0270	1.1261	40.43	0.1692
30	59.74	1.097	0.02503	4.825	− 2.19	544.8	542.6	−0.0044	1.1082	39.96	0.2073
40	73.32	1.104	0.02533	3.971	8.87	536.2	545.1	0.0177	1.0910	39.49	0.2518
50	89.19	1.112	0.02564	3.294	20.04	527.3	547.3	0.0397	1.0745	39.00	0.3036
60	107.6	1.120	0.02597	2.751	31.28	518.1	549.4	0.0614	1.0586	38.50	0.3635
70	128.8	1.129	0.02632	2.312	42.64	508.6	551.2	0.0829	1.0432	38.00	0.4325
80	153.0	1.138	0.02668	1.955	54.09	498.7	552.8	0.1041	1.0283	37.48	0.5115
90	180.6	1.147	0.02707	1.661	65.65	488.5	554.1	0.1250	1.0138	36.95	0.6019
100	211.9	1.156	0.02747	1.419	77.31	477.8	555.1	0.1458	0.9997	36.40	0.7048
110	247.0	1.168	0.02790	1.217	89.11	466.7	555.8	0.1664	0.9858	35.84	0.8219
120	286.4	1.183	0.02836	1.047	101.08	455.0	556.1	0.1868	0.9719	35.26	0.9549
130	330.3	(1.197)	0.02885	.....	(113)	(443)	....	.....	.....	34.66	......
140	379.1	(1.213)	0.02938	.....	(125)	(430)	....	.....	.....	34.04	......
150	433.2	(1.23)	0.02995	.....	(138)	(416)	....	.....	.....	33.39	......
160	492.8	(1.25)	0.03056	.....	(151)	(401)	....	.....	.....	32.72	......
170	558.4	(1.27)	0.03124	.....	(163)	(386)	....	.....	.....	32.01	......
180	630.3	(1.30)	0.03198	.....	(177)	(369)	....	.....	.....	31.27	......
190	708.9	(1.34)	0.03281	.....	(191)	(351)	....	.....	.....	30.48	......
200	794.7	(1.38)	0.03375	.....	(205)	(332)	....	.....	.....	29.63	......
210	888.1	(1.43)	0.03482	.....	(219)	(310)	....	.....	.....	28.72	......
220	989.5	(1.49)	0.0361	.....	(235)	(287)	....	.....	.....	27.7	......
230	1099.5	(1.57)	0.0376	.....	(251)	(260)	....	.....	.....	26.6	......
240	1218.5	(1.70)	0.0395	.....	(268)	(229)	....	.....	.....	25.3	......
250	1347	(1.90)	0.0422	.....	(287)	(192)	....	.....	.....	23.7	......
260	1486	(2.33)	0.0463	.....	(309)	(142)	....	.....	.....	21.6	......
270	1635	(5.30)	0.0577	.....	(341)	(52)	....	.....	.....	17.3	......
271.4	1657	.....	0.0686	.....	(355)	0	....	.....	.....	14.6	......

*Data from Bureau of Standards Circular No. 142

Base Temperature: 32F.

Note: The figures in parentheses were calculated from empirical equations given in Bureau of Standards Scientific papers Nos. 313 and 315 and represent values obtained by extrapolation beyond the range covered in the experimental work.

Table 8-2: Worksheet for What If Analysis Example

Subsystems: 1.1. Receiver V-001	Drawing: ARU-A1
Type: Vessel	
Operating Conditions/Parameters: 138 psig, 76°F	

What If	Causes	Consequences	Risk Matrix			Safeguards	Recommendations	Responsibilities
			S	L	RR			
1. Overfilling?	1.1. LV-201 or controller LC-201 malfunctioning	1.1. Reverse flow backs up into condenser and reduces performance	1	3	3	1.1. Level Gauge (LG-101)	1. Establish operational procedures for inventory control of ammonia	Anna D
	1.2. Overfilling on initial fill of ammonia					1.2. Level Alarm High (LAH-201)	2. Design to ensure that inventories of ammonia in receiver and accumulator plus piping are compatible and do not result in flooding.	Nigel W
2. Level too low?	2.1. LV-201 or controller LC-201 malfunctioning	2.1. Accumulator tends to fill-up	1	3	3	2.1. Level Gauge (LG-101)	3. Accumulator should be large enough to ensure the full contents of the receiver.	Geoff B
	2.2. Compressor cuts out					2.2. Level Alarm Low (LAL- 201)		
3. Pressure exceeds design specifications?	3.1. Fire case	3.1. Release of ammonia through relief system				3.1. High point vent	4. Ensure fire monitors and extinguishers are nearby	Steve L
		3.2. Some fire hazard in vicinity					5. Locate relief valve vents in a safe location.	Anna D
		3.3. Hazard to personnel in vicinity	4	2	8			
4. Level gauge (LG-101) breaks?	4.1. Physical impact from wrench	4.1. Hazard to personnel in vicinity	4	3	12	4.1. None	6. Provide armor plated level gauges	Nigel W
5. Level transmitter (LT-101) fails?	5.1. Faulty manufacturing	5.1. Loss of control on refrig. unit	1	3	3	5.1. Level Gauge (LG-101)	7. No action required.	

What If	Causes	Consequences	Risk Matrix			Safeguards	Recommendations	Responsibilities
			S	L	RR			
6. Low temperature embrittlement?	6.1. Depressuring to atmosphere on pressure let down during maintenance	6.1. Steel can shatter if impacted by wrench				6.1. None	8. Check need for Charpee Tested Carbon Steel	Geoff B
		6.2. Hazard to personnel in vicinity	4	1	4			
7. Toxic or hazardous service?	7.1. Release from joints or flanges during maintenance	7.1. Hazard to personnel in vicinity	4	3	12	7.1. None	9. Self contained breathing apparatus	Nigel W
							10. Buddy system	Nigel W
							11. Develop operational procedures	Anna D
8. Vortex on liquid discharge?	8.1. Low liquid level in V-001	8.1. Loss of performance	1	4	4	8.1. Level Gauge (LG-101)	12. Install Vortex breaker	Roy S
						8.2. Level Alarm Low (LAL- 201)		

Subsystems: 1.2. Accumulator V-002	Drawing: ARU-A1
Type: Vessel	
Operating Conditions/Parameters: 35 psig, 21.7°F	

What If	Causes	Consequences	Risk Matrix			Safeguards	Recommendations	Responsibilities
			S	L	RR			
1. Overfilling?	1.1. LV-201 or controller LC-201 malfunctioning	1.1. Reverse flow backs up into condenser and reduces performance	1	3	3	1.1. Level Gauge (LG-102)	1. Establish operational procedures for inventory control of ammonia	Anna D
	1.2. Overfilling on initial fill of ammonia						2. Design to ensure that inventories of ammonia in receiver and accumulator plus piping are compatible and do not result in flooding.	Nigel W
	1.3. Compressor cuts out						15. Install High Level Alarm V-002	Geoff B

What If	Causes	Consequences	Risk Matrix			Safeguards	Recommendations	Responsibilitie s
			S	L	RR			
2. Level too low?	2.1. LV-201 or controller LC-201 malfunctioning	2.1. Accumulator tends to empty	1	3	3	2.1. Level Gauge (LG-102)	3. Accumulator should be large enough to ensure the full contents of the receiver.	Geoff B
		2.2. Damage to pump					16. Level Alarm Low (LAL- 201)	Geoff B
3. Pressure exceeds design specifications?	3.1. Fire case	3.1. Release of ammonia through relief system				3.1. High point vent	4. Ensure fire monitors and extinguishers are nearby	Steve L
		3.2. Some fire hazard in vicinity					5. Locate relief valve vents in a safe location.	Anna D
		3.3. Hazard to personnel in vicinity	4	2	8			
4. Level gauge (LG-102)breaks?	4.1. Physical impact from wrench	4.1. Hazard to personnel in vicinity	4	3	12	4.1. None	6. Provide armor plated level gauges	Nigel W
5. Low temperature embrittlement?	5.1. Depressuring to atmosphere on pressure let down during maintenance	5.1. Steel can shatter if impacted by wrench				5.1. None	8. Check need for Charpee Tested Carbon Steel	Geoff B
		5.2. Hazard to personnel in vicinity	4	1	4			
6. Toxic or hazardous service?	6.1. Release from joints or flanges during maintenance	6.1. Hazard to personnel in vicinity	4	3	12	6.1. None	9. Self contained breathing apparatus	Nigel W
							10. Buddy system	Nigel W
							11. Develop operational procedures	Anna D
7. Vortex on liquid discharge?	7.1. Low liquid level in V-002	7.1. Loss of performance	1	4	4	7.1. Level Gauge (LG-102)	12. Install Vortex breaker	Roy S
							16. Level Alarm Low (LAL- 201)	Geoff B

Table 8-3: Recommendations Report for a What If Analysis Example

#	Recommendation	Drawings
1.	Establish operational procedures for inventory control of ammonia	ARU-A1
2.	Design to ensure that inventories of ammonia in receiver and accumulator plus piping are compatible and do not result in flooding.	ARU-A1
3.	Accumulator should be large enough to ensure the full contents of the receiver.	ARU-A1
4.	Ensure fire monitors and extinguishers are nearby	ARU-A1
5.	Locate relief valve vents in a safe location.	ARU-A1
6.	Provide armor plated level gauges	ARU-A1
7.	No action required.	
8.	Check need for Charpee Tested Carbon Steel	ARU-A1
9.	Self contained breathing apparatus	ARU-A1
10.	Buddy system	ARU-A1
11.	Develop operational procedures	ARU-A1
12.	Install Vortex breaker	ARU-A1
13.	Install high-limit stops on all control valves: TV-301, 302, FV 301,302	ARU-A1
14.	Operating procedure for setting temperature controls TIC-301, 302	ARU-A1
15.	Install High Level Alarm V-002	ARU-A1
16.	Level Alarm Low (LAL- 201)	ARU-A1

8

SUGGESTED READING (URLs current at time of publication)
"Guidelines for Hazard Evaluation Procedures" by AIChE, CCPS, 2nd edition, 1992 plus "Guidelines for Hazard Evaluation Procedures" by AIChE, CCPS, 1st edition, 1985 www.aiche.org/pubcat/seadtl.asp?Act=C&Category=Sect4&Min=20
"Use 'What if' method for process hazard analysis" by L.Zoller, J.P.Esping, Hydrocarbon Processing, January 1993, pages 132-B to 132-G www.hydrocarbonprocessing.com/contents/publications/hp/
"Speed Your Hazard Analysis with the Focused What If?" by L.Goodman, Chemical Engineering Progress, July 1996, pages 75 to 79 www.cepmagazine.org/
"Process Safety Review Checklist", University of Florida, (Website) http://pie.che.ufl.edu/guides/safety_health/process_checklist.html
"Process Hazards Analysis" by I.Sutton, published by SW/Sutton & Associates, 2002 http://www.swbooks.com/books/book_prha.shtml

CHAPTER 9

Failure Mode and Effects Analysis

What Is FMEA?

FMEA is a failure mode and effects analysis tool that is used in various industries to

- Identify failures,
- Evaluate the effects of the failures, and
- Prioritize the failures according to severity of effects.

Prioritization or risk ranking is done mainly using

- Risk Matrix (Risk Priority Number)
- Criticality Analysis (FMECA)

Reasons for Using FMEA

- To identify specific accident situations
- To consider alternative safety improvements
- To obtain data for quantitative risk analysis (QRA)
- To evaluate hazards from preliminary designs and operating procedures
- To improve reliability of the process
- To meet regulatory requirements
- To document a systematic process hazard evaluation

- To evaluate complex processes where perceived risks are significant
- To identify single-point failures

When and Where to Use It?

- Implementing it as soon as the preliminary designs are ready ensures that the necessary design changes can be made at the earliest possible time.
- Its usefulness lies in preventing failures from occurring in the future. So, it is usually done in the design phase when the failure modes have not yet been built-in to the process.
- A good FMEA is an ongoing process whereby it is continuously updated and revised over the life of the process.

It is performed on

- Mechanical equipment such as pumps, compressors, etc. where there is a history of component failures.
- Systems for which there are few drawings or details but where individual components are readily identifiable.
- Reliability studies or for input into quantitative risk assessment studies.

Regulatory Compliance

Regulations generally recommend FMEAs to deal with

- Complaints
- Corrective actions
- Documentation
- Health and Safety
- Management of change

- Misuse or unintended use
- Operability problems
- Prevention
- Process hazards
- Regulatory compliance
- Risk management

Standards that specifically address FMEA methodologies

- MIL STD 1629
- SAE ARP5580
- SAE J1739

Others have it as a part of their mandate along with other PHAs.

- AIAG, APQP Manual
- FDA, GMP, QS Regulation Title 21, CFR Part 820
- ISO 9001 2000
- IATF, ISO/TS 16949
- PSM CFR 1910.119
- QS 9000

Different Types of FMEAs

- The nature of the study and the stage of the process life cycle it's conducted at, determines the type of the FMEA to be used.
- There are 6 types of FMEAs namely, machinery-FMEA, design-FMEA, system-FMEA, process-FMEA, application-FMEA, and product-FMEA.
- Each FMEA follows the same approach. The nature, purpose, and the scope of the study dictates which type of FMEA is used and to what extent of detail.
- Most processes, equipment, and designs can be broken into levels of systems, sub-systems, assemblies, sub-assemblies, components, parts, etc. Such a breakdown of the subject study helps to define the scope.

Methodology

1. Collect pertinent information, e.g., P&IDs, PFDs, site plans, charts, operations information, procedures, relevant data, design plans, etc.
2. Establish the purpose, scope, depth of the study, associated costs, expertise, experience available, and so on.
3. Break the system into logical and manageable items by function (cooling system, braking system, pumping, heat exchangers) or area location (bottom of the distillation tower, top of the tower, feed-line system, products-line system). Record this information in the tabular format of FMEA.
4. Identify all potential failure modes for each item.
5. Determine the causes of each failure mode.
6. Identify and list the current controls.
7. Assign a rating for severity, occurrence and detection for each failure.
8. Determine appropriate corrective actions.
9. Carry out the recommended actions.

Risk Analysis (prioritizing risks)

Since resources are usually scarce, prioritizing the recommendations helps to focus the efforts where they are most necessary. The severity of the risk posed and the magnitude of the risk reduction possible are two common criteria for prioritization.

Risk Ranking (using Risk Matrix)

- Severity is arbitrarily assigned values

Table 9-1: Sample of Severity Ranking

Rank	Description
1	No injury or health effects
2	Minor injury or minor health effects
3	Injury or moderate health effects
4	Death or severe health effects

- Likelihood is also arbitrarily assigned values

Table 9-2: Sample of Likelihood Ranking

Rank	Description
1	Not expected to occur during the facility lifetime
2	Expected to occur no more than once during facility lifetime
3	Expected to occur several times during the facility lifetime
4	Expected to occur more than once in a year

- Risk matrix is developed using these parameters

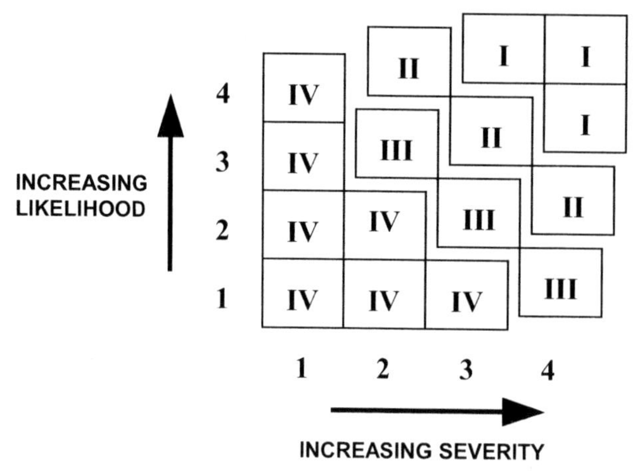

Figure 9-1: Sample Risk Matrix

- The risk ranking categories are predefined

Table 9-3: Sample of Risk Ranking Categories

Number	Category	Description
I	Unacceptable	Should be mitigated with engineering and/or administrative controls to a risk ranking of III or less within a specified time period such as six months
II	Undesirable	Should be mitigated with engineering and/or administrative controls to a risk ranking of III or less within a specified time period such as 12 months
III	Acceptable with controls	Should be verified that procedures or controls are in place
IV	Acceptable as is	No mitigation required

Risk Priority Number (RPN)

RPN is calculated by a simple multiplication of the S (Severity), O (Occurrence), and D (Detection) values. Companies usually establish minimum RPN values as their standard. Any RPN above minimum (say 100) would warrant further study and anything below is considered either safe, or acceptable risk, or very low priority for further analysis. Examples of Severity, Occurrence and Detection values for determining RPN are shown in Tables 9-4, 9-5 and 9-6 respectively.

Table 9-4: Sample of Severity Values used in Risk Priority Number Calculation

Effect	Rank	Criteria
None	1	Might be noticeable by the operator (Process). Improbable / not noticeable by the user (Product).
Very slight	2	No downstream effect (Process). Insignificant / negligible effect (Product).
Slight	3	User will probably notice the effect but the effect is slight (Process & Product).
Minor	4	Local and/or downstream processes might be affected (Process). User will experience minor negative impact on the product (Product).
Moderate	5	Impacts will be noticeable throughout operations (Process). Reduced performance with gradual performance degradation. User dissatisfied (Product).
Severe	6	Disruption to downstream process (Process). Product operable and safe but performance degraded. User dissatisfied (Product).
High Severity	7	Significant downtime (Process). Product performance severely affected. User very dissatisfied (Product).
Very High Severity	8	Significant downtime and major financial impacts (Process). Product inoperable but safe. User very dissatisfied (Product).
Extreme Severity	9	Failure resulting in hazardous effects highly probable. Safety and regulatory concerns (Process and Product).
Maximum Severity	10	Injury or harm to operating personnel (Process). Failure resulting in hazardous effects almost certain. Non-compliance with government regulations (Product).

Table 9-5: Sample of Occurrence Ranking used in Risk Priority Number Calculation

Occurrence	Rank	Criteria
Extremely Unlikely	1	Failure highly unlikely.
Remote Likelihood	2	Rare number of failures likely.
Very Low Likelihood	3	Very few failures likely.
Low Likelihood	4	Few failures likely.
Moderately Low Likelihood	5	Occasional failures likely.
Medium Likelihood	6	Medium number of failures likely.
Moderately High Likelihood	7	Moderately high number of failures likely.
High Likelihood	8	High number of failures likely.
Very High Likelihood	9	Very high number of failures likely.
Extremely Likely	10	Failure almost certain.

Table 9-6: Sample of Detection Ranking used in Risk Priority Number Calculation

Detection	Rank	Criteria
Extremely Likely	1	Controls will almost certainly detect the existence of the defect.
Very High Likelihood	2	Controls have very high probability of detecting existence of failure.
High Likelihood	3	Has high effectiveness for detection.
Moderately High Likelihood	4	Has moderately high effectiveness for detection.
Medium Likelihood	5	Has medium effectiveness for detection.
Moderately Low Likelihood	6	Has moderately low effectiveness for detection.
Low Likelihood	7	Has low effectiveness for detection.
Very Low Likelihood	8	Has lowest effectiveness in each applicable category.
Remote Likelihood	9	Controls have very low probability of detecting existence of defect.
Extremely Unlikely	10	Controls will almost certainly not detect the existence of a defect.

FMEA Worksheet Format

- Potential Failure Modes
- Potential Causes of Failure Modes
- Potential Effects of Failure Modes
- Current Controls (Existing Safeguards)
- Severity
- Occurrence/Likelihood
- Detection (RPN), or α, β, λ_p, and t (C_m and C_r for criticality analysis)
- Risk Ranking, Risk Priority Number (RPN) or Criticality Analysis
- Recommendations/Corrective Actions
- Responsibility
- Target Completion Date
- Actions Taken
- New Risk Ranking, Risk Priority Number or Criticality Analysis results
- Comments

FMECA

- The MIL STD 1629A standard titled "Procedures for Performing a Failure Mode, Effects, and Criticality Analysis" describes the FMECA exclusively.
- The classification of failure modes is based on severity (in the general FMEA) combined with the probability of occurrence.
- The criticality analysis requires first calculating failure mode criticality number (C_m): $C_m = \beta \alpha \lambda_p t$

  Where:

  β = conditional probability of mission loss

  α = failure mode ratio

  λ_p = part failure rate

  t = duration of operating time (hours) or number of operating cycles

- Then, C_r is calculated using C_m: $\sum_{n=1}^{j} (C_m)_n$

  Where:

  n = 1, 2, 3…j

  C_r = criticality number for the item (C_m is criticality of the failure mode)

- A criticality matrix of C_r versus Severity Categories is constructed that helps to determine what actions should be taken first in a manner similar to the Risk Matrix method.

9

General FMECA Methodology

1. Define the process or system to be analyzed.
2. Identify all potential failure modes, and assign effects to the failure modes and severity to effects.
3. Enter failure mode data such as failure detection methods, and failure rates.
4. Use severity and criticality to rank failure modes.
5. Highlight and report critical failures.
6. Reduce critical failures by implementing corrective actions.

Benefits of FMEA and FMECA

- Better company image and competitiveness
- Compliance with regulations, standards, and specifications
- Continuous improvement of product quality, reliability, and safety
- Defining corrective action.
- Documentation of the reasons for changes
- Improved reliability, productivity, quality, safety, and cost efficiency
- Increased liability prevention
- Increasing customer satisfaction
- Recognition and evaluation of potential failures and their effects
- Reduction of downtime
- Reduction of manufacturing process deviations
- Selection of alternative materials, parts, devices, components and tasks.
- Selection of optimal system design

Pitfalls with FMEA and FMECA

- Human errors and environmental influences are easily overlooked.
- Exclusively for single-point failures. Multiple-point failures are overlooked.
- Extremely tedious for large and complex processes.
- Successful completion requires expertise, experience, and good team skills.
- Can be costly and time consuming.
- Past failure rates are difficult to obtain (for criticality analysis).

FMEA Terminology

Causes

These are the root cause/s of the potential failure mode. Some examples are

- Over-stressing
- Uncontrollable rise in temperature
- Improper wall thickness

Criticality

It is a measure of the consequences of a failure mode determined from its severity and its probability of occurrence (actual *failure rate data from the past*). Criticality Analysis is the procedure in which this is achieved. Instead of risk ranking (using risk matrix) or method of risk priority number, certain parameters called criticality for item and failure mode (C_r and C_m, respectively) are calculated and used for prioritization.

Current Controls

This refers to existing safeguards or mitigation controls that are already implemented for prevention.

Detection

It is the ability to detect the failure before it affects the target. The levels of detection are assigned an arbitrary value, as are the levels for occurrence and severity, for calculating RPN.

Effects

These are consequences of the failure modes on different targets such as function, personnel, environment, etc.

Failure Mode

It is the manner in which the reviewed item can fail to perform its intent design function. Failure modes are verbs such as

- Fail to open/close
- Cracked
- Undersized/Oversized

Occurrence

It is the frequency of the failure (obtained over the past years) for the process or the part of a process that is being studied.

Risk Priority Number (RPN)

The Risk Matrix quantifies qualitative information by defining a Risk Priority Number. RPN is obtained by multiplying severity, occurrence and detection. Values for severity, occurrence, and detection are arbitrarily defined and must always be greater than zero. RPN by itself does not mean anything; it is only used to prioritize the actions to be taken.

Risk Priority Number = Severity x Occurrence x Detection

Single-Point Failure

An item or element whose failure would result in the failure of the system is termed a single-point failure.

Severity

It is a measure of the degree of damage a failure mode inflicts on the various targets. Severity can be reduced only through a change in the design.

Table 9-7: Sample of FMEA Report Using Software

Component: 2. Combustion Turbine

Item: 3. Four Stage Turbine

Item ID:

Item Function: To provide power to the generator set

Potential Failure Mode	Potential Effect of Failure	S	Potential Cause of Failure	O	Current Design Controls	D	RPN	Recom.	Resp.	Action Taken	Est. Start Date	After Actions Taken			
												S	O	D	RPN
1. Carbon deposits on flow surfaces during start-up	1. Degradation of performance	4	1. Removable deposits as a consequence of normal operation	2	1. Washing system provided as part of package	3	24	2. Implement regular washing schedule	Maintenance Dept.	Scheduled as per manufacturers instructions	3/13/01	4	1	3	12
2. Roughing of flow surfaces	1. Degradation of performance	4	1. Roughing caused as a consequence of normal operation - not removable by washing	2	1. Major overhaul at regular intervals.	3	24	1. Carry warehouse spares for turbine blade replacement	Inventory Control Section	Ordered spare turbine blades	3/13/01	4	2	3	24
3. Foreign object damage	1. Catastrophic failure/damage of turbine blades	8	1. Complete loss of filter function	1	1. Maintenance schedule	3	24	3. No action required							
4. Distortion of parts	1. Degradation of performance	4	1. Localized overheating from cooling system maldistribution	3	1. Maintenance schedule	8	96	4. Check for distortion during maintenance	Maintenance Dept.	Added distortion check to maintenance procedure	4/9/01	4	3	3	36
	2. Catastrophic damage and inoperability	7	1. Cooling system failure	1	1. Maintenance schedule	3	21	10. Adhere to quarterly inspection schedule	Maintenance Dept.			7	1	3	21
								2. Implement regular washing schedule	Maintenance Dept.	Scheduled as per manufacturers instructions	3/13/01				
			2. Metal fatigue of turbine blade leading to fracture	2	1. Quality control by manufacturer	2	28	8. Adhere to quarterly inspection schedule	Maintenance Dept.	Schedule as determined	3/13/01	7	1	3	21
					2. Maintenance schedule	3	42	9. Manufacturer to submit quality control information	Maintenance Dept.	Schedule as determined	3/13/01				
5. Cooling system failure	1. Catastrophic damage and inoperability	8	1. Defect within air cooling system distribution.	4	1. Quality control by manufacturer	3	96	8. Adhere to quarterly inspection schedule	Maintenance Dept.	Schedule as determined	3/13/01	4	2	3	24
			2. Pluggage in the air cooling system filter	2	1. Maintenance schedule	3	48	11. FMEA of the cooling system	Engineering Dept.	Outstanding	4/20/01	8	1	1	8

SUGGESTED READING (Note: URLs current at date of publication)
"Failure Mode and Effect Analysis" by D.H.Stamatis, published by ASQ Quality Press, 1995 http://qualitypress.asq.org/perl/catalog.cgi?item=H0856
MIL STD 1629A : Procedure for Performing a Failure Mode, Effects and Criticality Analysis, 1980 http://jcs.mil/htdocs/teinfo/software/ms18.html
MIL STD 1472D : Human Engineering Design Criteria, 1989 http://store.mil-standards.com/eproducts/doclist/MIL%20CD%20Power%20User.pdf
"Guidelines for Hazard Evaluation Procedures" by AIChE, CCPS, 2nd edition, 1992 plus "Guidelines for Hazard Evaluation Procedures" by AIChE, CCPS, 1st edition, 1985 www.aiche.org/pubcat/seadtl.asp?Act=C&Category=Sect4&Min=20
"Equipment health management program improves plant reliability" by G.Goacone and R.Hall, Hydrocarbon Processing, October 1997, pages 61, 62 www.hydrocarbonprocessing.com/contents/publications/hp/
"Failure Mode and Effects Analysis (FMEA)" by Chrysler Corporation (Website) http://tdserver1.fnal.gov/users/mc/blowers/Quality_resources-misc/FMEA-N.pdf
"Process Hazards Analysis" by I.Sutton, published by SW/Sutton & Associates, 2002 http://www.swbooks.com/books/book_prha.shtml

CHAPTER 10

Screening Level Risk Analysis (SLRA)

Basis

Formulates a list of hazards and generic hazardous situations by considering characteristics such as materials processed, operating environment (high pressure, etc.), equipment, inventories, and plant layout.

Purpose

Identification of hazards – provide ranking of hazards

When to Use SLRA

- Anytime in plant life or design phase.
- Often early in the development of a process.
- When there is limited information available.
- To assist with preliminary layout and siting studies.

SLRA Methodology

This methodology can be used for new designs at the conceptual stage in order to assist with layouts, etc. and for existing facilities where some level of prioritization is needed prior to more detailed hazards analysis. The SLRA methodology may include:

1. A list all substances that are both stored as well as processed at the facility together with hazardous properties of those materials including:
   - Toxicities
   - Flash points
   - Upper and lower explosive limits
   - Vapor pressures
   - Corrosive nature
   - Interactive properties with other substances
   - Auto-ignition temperatures
   - Any tendencies for auto decomposition

   Any other harmful properties: refer to MSDS sheets (if available)

2. List inventories of materials and where they are contained, by location.

3. List interactive points such as sources of ignition, *e.g., furnaces, boilers.*

4. List vulnerable locations, such as adjacent office blocks, housing, main highways.

5. List possible release type scenarios. These should typically include:
   - Leaks or ruptures of vessels
   - Leaks or ruptures of storage tanks
   - Leaks or ruptures of critical lines, fittings, vents, drains, blowdown, etc.

   Leaks or ruptures from seals of pumps and compressors, other prime movers

6. Estimate consequences in broad terms, from minor to severe and describe the nature of the consequences and qualitatively estimate their severity in terms of impact on employees, environment, capital equipment, and production.

7. Qualitatively estimate likelihood (frequency) of the hazardous events based on best judgment or historical records (if available).

8. Based on severity of consequences and likelihood, make a qualitative risk ranking of each hazardous event on employees, environment, capital equipment, and production

9. List existing safeguards that are present which can prevent or control the potential hazardous events. These should include:
   a. Safeguards against the cause or failure in the first place, e.g., interlocks and trips
   b. Detection and remedial action, e.g., pressure safety valves, instrumentation
   c. Mitigation of the consequences, e.g., flammable gas detectors, fire suppressions systems
   d. Post-incident response, e.g., emergency response plans, evacuation equipment and procedures

10. If existing safeguards are found to be inadequate, develop recommendations for further measures to prevent or control potential hazardous events

11. If there are areas of vulnerability identified, how best can these be handled so that risk is minimized?

12. All the hazardous events will be risk ranked using a risk matrix, which assigns risk levels, from highest to lowest

Results

- Ranking of hazards on a plant by plant basis
- Allows identification and resolution of high risk events
- Consequence and frequency
- Quantitative

Table 10-1: Example of SLRA Worksheet

Facility: XYZ Chemicals Inc.						
Chemical: 1.1. Chlorine Storage and Handling						Drawing: A - 135
Type: Chlorine Storage and Handling						
Design Conditions/Parameters: Chlorine stored as a liquefied gas under pressure in 1 ton containers						
Hazards & Source	Consequences	Existing Safeguards	S	L	RR	Recommendation
1. Accident during delivery and offloading of 1 ton chlorine containers.	1.1. Release of toxic chlorine vapor cloud from 1 ton container	1.1. High wall plus ditch would trap chlorine vapors, especially under "F" stability conditions. Some mitigation under other, more turbulent, weather conditions.	3	2	6	1. Provide an enclosed offloading area for trucks supplying 1 ton chlorine containers. Both offloading and storage area should be connected to a proposed new chlorine scrubbing system .
	1.2. Onsite & offsite toxic gas cloud hazard	1.2. Onsite personnel have SCBA (Self Contained Breathing Apparatus).				2. Provide a wet scrubbing system that can absorb a major chlorine release and neutralize using caustic soda solution. The scrubbing system should be switched on prior to offloading of chlorine and also scrubber to be interlocked with chlorine detectors to switch on with chlorine release.
2. Internal or external fire raises temperature above 160F causing fusible plugs to melt on 1 ton container(s) causing release of chlorine.	2.1. Release of toxic chlorine vapor cloud from one or more 1 ton containers	2.1. Chlorine detectors inside building	3	2	6	3. Provide an emergency deluge systems in the chlorine handling area. Also both inside and outside of the building first aid fire fighting should be provided.
	2.2. Onsite & offsite toxic gas cloud hazard	2.2. High wall plus ditch would trap chlorine vapors, especially under "F" stability conditions. Some mitigation under other, more turbulent, weather conditions.				
		2.3. Onsite personnel have SCBA (Self Contained Breathing Apparatus).				
3. Rupture of line from chlorine cylinder due to excessive force/human error, causing release of chlorine.	3.1. Release of toxic chlorine vapor cloud from 1 ton container	3.1. Chlorine detectors inside building	3	2	6	2.Provide a wet scrubbing system that can absorb a major chlorine release and neutralize using caustic soda solution. The scrubbing system should be switched on prior to offloading of chlorine and also scrubber to be interlocked with chlorine detectors to switch on with chlorine release.
		3.2. Onsite personnel have				4. Confirm rigorous hook-up and

Hazards & Source	Consequences	Existing Safeguards	S	L	RR	Recommendation
	3.2. Onsite & offsite toxic gas cloud hazard	SCBA (Self Contained Breathing Apparatus).				inspection procedures in place when connecting up chlorine containers.
		3.3. High wall plus ditch would trap chlorine vapors, especially under "F" stability conditions. Some mitigation under other, more turbulent, weather conditions.				5. The lighting in the chlorine and drum handling areas must be adequate at all times for unloading and hooking up of chlorine containers and other chemicals (so as to minimize the potential for human error).
4. Fractured manifold leads to release of chlorine.	4.1. Release of toxic chlorine vapor cloud from one or more 1 ton containers through manifold	4.1. Onsite personnel have SCBA (Self Contained Breathing Apparatus).	2	2	4	2. Provide a wet scrubbing system that can absorb a major chlorine release and neutralize using caustic soda solution. The scrubbing system should be switched on prior to offloading of chlorine and also scrubber to be interlocked with chlorine detectors to switch on with chlorine release.
		4.2. High wall plus ditch section across rail track section would trap chlorine vapors, especially under "F" stability conditions. Some mitigation under other, more turbulent, weather conditions.				
	4.2. Onsite & offsite toxic gas cloud hazard	4.3. Chlorine detectors inside building				
		4.4. Emergency shut-off valve.				

Chemical: 1.2. 100 drums of 98% red fuming nitric acid						Drawing: A - 136
Type: 100 drums of 98% red fuming nitric acid						
Design Conditions/Parameters: Fuming 98% red nitric acid stored under atmospheric pressure/temperature conditions						
Hazards & Source	**Consequences**	**Existing Safeguards**	**S**	**L**	**RR**	**Recommendation**
1. Spillage of a drum of nitric acid.	1.1. Release of nitrogen dioxide and nitric oxide fumes. Both NO2 and NO are very toxic and NO can accelerate burning in the fire situation. (Release is through diffusion into the atmosphere, as opposed to flashing off).	1.1. High wall plus ditch would trap NO2 and NO vapors, especially under "F" stability conditions. Some mitigation under other, more turbulent, weather conditions.	2	3	6	6. Provide diking to contain spillage of 98% fuming nitric acid.
		1.2. Onsite personnel have SCBA (Self Contained Breathing Apparatus).				

<table>
<tr><td rowspan="4">2. Vehicle or truck impacts a stack of drums (e.g., on pallets) of nitric acid.</td><td rowspan="4">2.1. Release of nitrogen dioxide and nitric oxide fumes. Both NO2 and NO are very toxic and NO can accelerate burning in the fire situation. (Release is through diffusion into the atmosphere, as opposed to flashing off).</td><td rowspan="2">2.1. High wall plus ditch section would trap NO2 and NO vapors, especially under "F" stability conditions. Some mitigation under other, more turbulent, weather conditions.</td><td rowspan="2">2</td><td rowspan="2">2</td><td rowspan="2">4</td><td>5. The lighting in the chlorine and drum handling areas must be adequate at all times for unloading and hooking up of chlorine containers and other chemicals (so as to minimize the potential for human error).</td></tr>
<tr><td>6. Provide diking to contain spillage of 98% fuming nitric acid.</td></tr>
<tr><td rowspan="2">2.2. Onsite personnel have SCBA (Self Contained Breathing Apparatus).</td><td rowspan="2"></td><td rowspan="2"></td><td rowspan="2"></td><td>7. Provide bollards, curbing etc. to prevent vehicles impacting drums of fuming nitric acid, thereby causing spillage</td></tr>
<tr><td>8. Provide adequate water spray/dilution available in the vicinity to reduce fuming nitric acid hazard, in event of spillage.</td></tr>
<tr><td rowspan="2">3. Gasoline leak and other flammable solvents causing fire to impact the nitric acid drum storage area.</td><td rowspan="2">3.1. Copious release of nitrogen dioxide and nitric oxide fumes. Both NO2 and NO are very toxic and NO will accelerate burning. Nitric acid fumes (NO2 & NO will boil off).</td><td>3.1. Onsite personnel have SCBA (Self Contained Breathing Apparatus).</td><td>4</td><td>2</td><td>8</td><td>9. Provide segregation of flammable solvents away from storage area for fuming nitric acid.</td></tr>
<tr><td>3.2. High wall plus ditch section would trap NO2 and NO vapors, especially under "F" stability conditions. However, hot vapors could rise and bridge this High wall. Some mitigation under other, more turbulent, weather conditions.</td><td></td><td></td><td></td><td>10. Provide fire fighting capability to mitigate fire involving fuming nitric acid, i.e., alcohol foam, carbon dioxide , dry chemical or water spray.</td></tr>
</table>

Chemical: 1.3. 40 drums of 50% caustic soda solution.						Drawing: A - 137
Type: 40 drums of 50% caustic soda solution.						
Design Conditions/Parameters: 50% caustic soda stored in warm environment to prevent freezing						
Hazards & Source	Consequences	Existing Safeguards	S	L	RR	Recommendation
1. 50% caustic inadvertently comes into contact with other chemicals, such as fuming nitic acid.	1.1. Violent boil-off reaction. Also heat of dilution could ignite combustible materials, e.g., flammable solvents.	1.1. High wall plus ditch section would trap NO2 and NO vapors released as a secondary effect of boil-off through caustic contacting nitric acid, especially under "F" stability conditions. Some mitigation under other, more turbulent, weather conditions.	2	1	2	5. The lighting in the chlorine and drum handling areas must be adequate at all times for unloading and hooking up of chlorine containers and other chemicals (so as to minimize the potential for human error).
		1.2. Onsite personnel have SCBA (Self Contained Breathing Apparatus).				11. Store the 50% caustic away from the fuming nitric acid and also from any flammable solvents or other combustible substances.

Chemical: 1.4. 50 drums of 35% Hydrochloric Acid						Drawing: A - 138
Type: 50 drums of 35% Hydrochloric Acid						
Design Conditions/Parameters: 35% hydrochloric acid stored under atmospheric pressure/temperature conditions						
Hazards & Source	Consequences	Existing Safeguards	S	L	RR	Recommendation
1. Hydrochloric acid coming into contact with 50% caustic.	1.1. Hydrochloric acid will be neutralized and boil off some hydrogen chloride.	1.1. High wall plus ditch section, with water in it, would likely absorb most of any residual hydrogen chloride vapors..	2	2	4	15. Provide segregation of hydrochloric acid storage/handling from caustic soda storage/handling.
		1.2. Onsite personnel have SCBA (Self Contained Breathing Apparatus).				

Chemical: 1.5. External 10 ton propane bullet feeding fuel to furnace inside building						Drawing: A - 139
Type: 5. External 10 ton propane bullet feeding fuel to furnace inside building						
Design Conditions/Parameters: Propane stored as liquefied gas under pressure						
Hazards & Source	**Consequences**	**Existing Safeguards**	**S**	**L**	**RR**	**Recommendation**
1. Propane leakage inside building	1.1. Potential explosion within building due to ignition by furnace	1.1. Regular predictive maintenance schedule	2	2	4	16. Provide flammable gas detectors inside building with automated fuel shutoff/venting
	1.2. Onsite explosion hazard					
2. If there is a gasoline or brush fire this could seriously impact the integrity of the propane bullet.	2.1. Could result in possible BLEVE explosion of propane bullet and missile generation.	2.1. None	4	2	8	12. Confirm that propane bullet is located on a sloped concrete pad (so that spills of flammables will not pond beneath bullet). Surround should be cleared of grass/brush that has any potential for catching fire in summer.
						13. Fence propane bullet in compound, with video surveillance, that provides security and limits access.

SUGGESTED READING (Note: URLs current at date of publication)
"Guidelines for Hazard Evaluation Procedures" by AIChE, CCPS, 1st edition, 1985 www.aiche.org/publications/reprints.htm
MIL-STD-882, Military Standard System Safety Program Requirements, January 2002 http://store.mil-standards.com/eproducts/doclist/MIL%20CD%20Power%20User.pdf
"Preliminary safety analysis" by G.Wells, J.Loss Prev. Process Ind., 1993, Vol 6, No. 1, pages 47 to 60 www.elsevier.nl/locate/jlp

CHAPTER 11

PHA Revalidation

Overview

The Occupational Safety and Health Administration's (OSHA) process safety management (PSM) regulation, 29 CFR 1910.119, requires companies to update or revalidate their Process Hazards Analyses (PHAs) at least every 5 years. In addition, the U.S. Environmental Protection Agency's (EPA) risk management program rule, 40 CFR Part 68, requires companies to perform quantitative off-site consequence analysis. Consequently, many facilities are attempting to establish an effective revalidation strategy to incorporate both OSHA and EPA requirements.

This chapter provides the considerations of PHA revalidation, an approach to determine the scope of a PHA revalidation study and a suggested procedure for conducting a PHA revalidation study.

Objectives of PHA Revalidation

Revalidate previous PHAs to ensure the following:

- Consistent with the current process;
- Known process hazards are identified;
- Adequate controls are in place to manage the hazards;
- Completeness of previous PHAs;
- Consistency with risk rankings;
- Compliance with applicable regulations (e.g. OSHA and EPA).

Considerations of PHA Revalidation

In order to determine the appropriate revalidation approach, the following issues should be considered:

Internal and/or External Influences

Some of the key internal and external issues are listed below:

- New regulatory requirements or new interpretations of the existing requirements;
- Major process modifications occurring (or planned) since the previous analysis;
- A full PHA not being performed for an extended period.

Initial/Previous PHA Quality

Some of the major factors relating to initial/previous PHA quality are listed below:

- Quality of process safety information (PSI) used for previous PHA;
- Documentation of PHA;
- Team composition;
- Analysis technique used for PHA;
- Adequacy of initial PHA scope.

Operating Experience

Some of the major issues relating to the facility history are given below:

- The number of years of process unit operating experience;
- The maturity of the MOC program;
- The number and/or significance of management of change (MOC) since the previous PHA;
- The number and/or significance of incidents or "near misses" that have occurred since the previous PHA.

Determination of the Scope of PHA Revalidation Study – 6-Step Approach (Ref: Philley, J. and Moosemiller, M., PHA Revalidation: A Six-step Approach, Chemical Process Safety Report, February 1997)

Philley and Moosemiller (1997) proposed a six-step approach for determining the scope of a PHA revalidation study. This includes deciding whether the original PHA needs to be redone entirely. The six-step approach is listed below:

1. Verify the quality and resolution of previous PHA study.
2. Verify the existence and rigor of MOC system and all changes carried out since the original PHA study.
3. Confirm current process safety information (PSI) package is up-to-date.
4. Consider the applicability of RMP regulations – evaluation of off-site impacts.
5. Review the changes in facility standards and protocols relating to the PHA.
6. Conduct PHA revalidation study.

Step 1 – Verify the quality and resolution of previous PHA study

As the foundation of revalidation is the quality of previous PHA study, it is essential to evaluate the quality of the study. Six crucial quality tests are proposed below:

1. Scope of the study

The adequacy of the initial PHA scope should be evaluated to determine if all critical equipment items and activities are covered. In addition, the following factors should be considered in the process of evaluation:

- Any MOC cases?
- Any new technology introduced?
- Any incidents since original PHA?
- Human factors considered?

If the initial scope is not adequate, redo the PHA; otherwise, proceed to the next test.

2. **Quality (competency) of the PHA team**

The quality of a PHA is directly related to the competency of the PHA team. Consequently, it is essential to determine if the PHA team of the previous study had the right composition and knowledge to perform the study. Some of the typical questions regarding the quality of the team are listed below:

- Did the previous team have the correct range of expertise, e.g., process design, instrumentation and control, operations, maintenance, etc.?
- Did the previous team have a person knowledgeable in the PHA method that was used?
- Did the previous team have a person knowledgeable in the process that was assessed (especially a technology new to the company)?

If the team is not qualified, redo the PHA; otherwise, proceed to the next test.

3. **Methodology appropriateness**

Considering the nature of the process, the potential hazards that exist, the actual operating experience and the type of design, was the correct PHA methodology used?

If the methodology used is not appropriate, redo the PHA; otherwise, proceed to the next test.

4. **Time and level of detail**

- Was sufficient time spent on the previous PHA so that all hazardous issues were addressed? The time spent is the actual session time and does not include time devoted to preparing for the sessions, report writing or following up on recommendations.
- Were issues addressed in sufficient detail?

Redo the PHA if insufficient time was spent on the PHA study or the issues had been inadequately addressed; otherwise, proceed to the next test.

5. PHA study reflects logical conclusions

The objectives of reviewing the PHA documentation are to identify potentially catastrophic events that were inadequately addressed or excluded from the previous study, as well as issues leading to inconsistent or illogical conclusions, and then incorporate them in the scope of the revalidation study. A complete redo of the PHA would be necessary if there were repeated and consistent failures to document a logical conclusion to an identified hazard. The following questions are typically asked during the review:

- Were actions or recommendations that were developed the logical conclusions of a thorough analysis?
- Was the previous PHA well documented and could it be easily understood?
- Were all concerns that were identified, and for which no actions or recommendations were deemed necessary, adequately safeguarded?

Examples of possible problems with the documentation of the PHA include these situations:

- ***Open-ended discussion*** – e.g. the discussion identifies an explosion hazard, lists the consequences as being severe, and then makes no further statement;
- ***Illogical conclusion*** – e.g. the discussion identifies safeguards or recommendations that do not address the identified hazard;
- ***Inconsistencies*** – e.g. the discussion makes a non-safety-related recommendation without a clear distinction from other safety-related recommendations.

If the conclusions drawn in the previous PHA are satisfactory, proceed to the next test.

6. Resolution of original concerns or action items

It is essential to verify if all the safety-related issues from the previous PHA have been resolved and documented. Any incomplete items should fall within the scope of the revalidation. The following are some typical questions that help to determine if the original concerns have been satisfactorily resolved:

- Were all items resolved during the previous PHA?
- Were all action items or recommendations that were developed during the PHA addressed and satisfactorily resolved?

Any incomplete items should fall within the scope of the revalidation.

Step 2 – Verify the existence and rigor of MOC system and all changes carried out since the original PHA study

All the changes made to the process and equipment since the previous PHA study should be evaluated to determine if they are adequately addressed either inside or outside the MOC system. Typically consider the following:

- Are known changes adequately incorporated under current competent MOC procedures and PHAs?
- Are there known changes that are not adequately addressed?
- Are there changes not known to the PHA revalidation team?
- Is there adequate documentation of MOCs?

Changes that were not adequately reviewed when they were implemented, or for which MOC documentation was inadequate, should be included in the scope of the PHA revalidation. When significant changes were made to the process or equipment, even if they were adequately addressed in the MOC system, it is essential to determine if the changes are so significant that a redo of the PHA study is required.

Step 3 – Confirm current process safety information (PSI) package is up-to-date

Before revalidation of the PHA, it is essential to ensure that the process safety information is updated:

- Were original PSIs adequate?
- With MOCs, is there a need for updating PSIs?
- Have PFDs and P&IDs been adequately updated (field verification)?
- Do operating procedures, manuals, alarm lists, emergency procedures, contractor orientation materials, etc. reflect changes?

Step 4 – Consider the applicability of RMP regulations – evaluation of off-site impacts

The RMP regulations, established by the EPA in 1996, are similar to the requirements in OSHA's PSM standard. The focus of the RMP regulations is on potential off-site consequences, as opposed to OSHA's focus on on-site effects on employees and contractors. For facilities required to conform to both OSHA and RMP requirements, the PHA should include:

- On-site scenarios as per current OSHA 1910.119;
- Off-site scenarios as per 40 CFR Part 68.

However, because OSHA's and the EPA's lists of regulated chemicals and threshold quantities are not identical, it may not be possible to combine the analysis in certain cases.

If the revalidated PHA is related to the RMP regulations, perform off-site quantitative consequence analysis, complete parts 68.20 to 60.42 in the RMP regulations on hazard assessment, and incorporate into the RMP document for submission.

The Off-site Consequence Analysis (OCA) requirements for PHAs in RMP are to evaluate the following scenarios for both toxic and flammable substances:

- Worst case scenarios – highly unlikely;
- Alternative case scenarios – for emergency planning purposes.

In addition, the PHA must document the designated "endpoint" for each release scenario's potential off-site consequences.

Worst Case Scenarios

- Rapid loss of major process or storage inventory in short period (10 minutes or 60 minutes);
- Examples – vessel or tank or line ruptures, leading to massive release rates.

Alternative Case Scenarios

- Transfer hose and coupling releases, split bellows joints;
- Flange and pipe fitting and valve seal leaks, joint failures;
- Seal, drain and plug bleeds and cracks, and other similar failures;
- Process vessel, tank, pump and compressor releases due to cracks, seal failures, etc.

The PHA should document the current and recommended mitigation measures. The following are typical examples for passive and active mitigations:

Passive Mitigation

- Dikes, berms;
- Enclosures;
- Drains, sumps;
- Fire walls, blast walls.

Active Mitigation

- Sprinkler and deluge systems;
- Water curtains;
- Neutralization;
- Excess flow valves;
- Flare systems;
- Scrubbers;
- Emergency shutdown systems.

Step 5 – Review the changes in facility standards and protocols relating to the PHA

During the process of reviewing the previous PHA, one should identify any significant changes to the facility's internal standards and protocols for PHA-related issues and incorporate the changes into the PHA revalidation study. The following questions would assist in the reviewing process:

- Relevant siting, layout and spacing standards used (e.g., API RP # 752 for process buildings)?
- PHA content and scope (e.g. CCPS Guidelines)?
- Acceptable operating practices (e.g. hot taps, purging, etc.)?
- Acceptable environmental criteria?
- Worker health/industrial hygiene exposure tolerances?
- Upgrades and changes to PSM systems such as mechanical integrity standards and permit standards?

Step 6 – Conduct PHA revalidation study

A suggested procedure for conducting a PHA revalidation study is proposed below:

1. Confirm scope.
2. Select leader.
3. Select PHA methodology.
4. Assemble the following information:
    - Previous PHA data;
    - Updated PFDs and P&IDs;
    - Updated equipment data;
    - A summary of process safety incidents since previous PHA;
    - Management of Change data;
    - Process Safety Information;
    - Revised operating procedures.
5. Review status since previous PHA (with reference to above).
6. Prepare list of factors to be reviewed (based on the above list).
7. Select revalidation PHA team.
8. Team to pre-study items of concern and verify assumptions.
9. Team Leader prepares basis for updated outline for PHA.
10. Team member sessions to revalidate previous PHAs to include:
    - MOC's incident reviews/near-misses;
    - Human factors;
    - Status of previous recommendations;
    - Facility siting issues.
11. Establish new recommendations and action plan.
12. Issue revalidated PHA.

PHA REVALIDATION CHECKLIST OF SUGGESTED ITEMS

The following checklist can be used with the 6-step approach proposed by Philley and Moosemiller (1997).

Table 11-1: Questions Relating to Initial Quality of PHA

1. QUALITY OF INITIAL PHA
▪ Was the scope of the previous PHA adequate, with all critical equipment items and activities covered? ▪ Did the previous team have the correct range of expertise, e.g., process design, instrumentation and control, operations, maintenance, etc.? ▪ Did the previous team have a person knowledgeable in the PHA method that was used? ▪ Did the previous team have a person knowledgeable in the process that was assessed? ▪ Considering the nature of the process, the potential hazards that exist, the actual operating experience and the type of design, was the correct PHA methodology used? ▪ Was sufficient time spent on the previous PHA so that all hazardous issues were addressed? ▪ Were issues addressed in sufficient detail? ▪ Were actions or recommendations that were developed the logical conclusions of a thorough analysis? ▪ Was the previous PHA well documented and could it be easily understood? ▪ Were all items resolved during the previous PHA? ▪ Were all concerns that were identified, and for which no actions or recommendations were deemed necessary, adequately safeguarded? ▪ Were all action items or recommendations that were developed during the PHA addressed and satisfactorily resolved?

Table 11-2: Questions Relating to MOCs

2. MANAGEMENT OF CHANGE (MOC) ISSUES
▪ Are you familiar with all the MOCs that have been issued since the previous PHA? ▪ Do you have a procedure for identifying which MOC issues need PHA? ▪ Have there been process changes since the previous PHA and have PHA reviews been augmented through the MOC system? ▪ Is the MOC documentation adequate? Is the update PHA documentation adequate? ▪ Is there a PHA update available for every applicable MOC? ▪ Was the correct PHA methodology used for each MOC? ▪ Did the PHA team assessing the MOC have adequate knowledge and expertise?

Table 11-3: Questions Relating to PSI

3. PROCESS SAFETY INFORMATION (PSI) PACKAGE
▪ Have any new substances been introduced into the process with potential for hazardous consequences? ▪ Has any new equipment been introduced into the process with potential for hazardous consequences? ▪ Have any new procedures been introduced into the process with potential for hazardous consequences? ▪ Have changes to the process been included on the process flow diagrams (PFDs)? ▪ Have changes to the process been included on the piping and instrumentation diagrams (P&IDs)? ▪ Are the PFDs and P&IDs up-to-date? ▪ Have changes to the process been included on equipment specification and data sheets?

3. PROCESS SAFETY INFORMATION (PSI) PACKAGE (Cont.)
▪ Have changes to the process instrumentation and control systems been included on appropriate documents (e.g. P&IDs, instrument data sheets, etc.)?
▪ Have changes to the process been included in the operating procedures, where required?
▪ Have changes to the process been included in the instrument alarm lists, where required?
▪ Have changes to the process been included in the training manuals, where required?
▪ Have changes to the process been included in the emergency procedures, where required?
▪ Have changes to the process been included in the contractor orientation materials, where required?

Table 11-4: Questions Relating to RMP Regulations

4. RMP REGULATIONS
▪ Have all cases with potential for off-site consequences been identified in the PHA?
▪ Where RMP Program levels have changed, e.g., from Program 2 to 3, does the previous PHA include sufficient data?
▪ If there has been an accident history since the previous PHA, has this update been included?
▪ Does the previous PHA reference the Emergency Response Plan, where appropriate?

Table 11-5: Questions Relating to PHA Standards and Protocol

5. PHA STANDARDS AND PROTOCOL
▪ Has there been any change for siting, layout and spacing issues that could require updating of the previous PHA? ▪ Have the requirements of American Petroleum Institute (API) guideline #752, "Management of Hazards associated with Process Plant Buildings," been addressed? ▪ Has Human Error been addressed in the previous PHA? ▪ Have there been any near misses recorded since the previous PHA? ▪ Have there been any changes to the environmental protection practices that could require updating of the previous PHA? ▪ Have there been any changes to worker health and industrial hygiene exposure tolerances that could require updating of the previous PHA? ▪ Have there been any changes to mechanical integrity/design standards that could require updating of the previous PHA? ▪ Have there been any changes to methods of plant operation that could require updating of the previous PHA? ▪ Have there been any changes to methods of plant permitting and maintenance that could require updating of the previous PHA?

SUGGESTED READING (Note: URLs current at date of publication)
"PHA Revalidation: A Six-step Approach" by J.Philley and M.Moosemiller, Chemical Process Safety Report, February 1997 www.thompson.com/libraries/environment/chem
"How to effectively revalidate PHAs" by D.K.Crumpler and D.K.Whittle, Hydrocarbon Processing, October 1998, pages 55 to 60 www.hydrocarbonprocessing.com/contents/publications/hp/
"PHA Revalidations –Get More for Your Money" by C.E. Browning and C.N. Garland. CCPS International Conference and Workshop MAKING PROCESS SAFETY PAY: THE BUSINESS CASE, 2001, pages 523 to 534 www.aiche.org/pubcat/seadtl.asp?Act=C&Category=Sect4&Min=50
"A Practical Approach to Reducing the Cost of PHA Revalidations" by K.Sandler, C.Ferdock, K.McEldowney, F.Leverenz,Jr. CCPS International Conference and Workshop MAKING PROCESS SAFETY PAY: THE BUSINESS CASE, 2001, pages 535 to 550 www.aiche.org/pubcat/seadtl.asp?Act=C&Category=Sect4&Min=50
"Making Process Safety Pay Off – Sustaining Performance in the 21st Century – PHA revalidation: can this cost be reduced?", K.Sandler et al., www.battelle.org/environment/irm/pdfs/making%20process%20safety%20pay%20off.pdf
"Revalidating Process Hazard Analysis", AIChE, CCPS, 2000 www.aiche.org/pubcat/seadtl.asp?Act=C&Category=Sect4&Min=70

CHAPTER 12

Management of Change (MOC)

Introduction

When Change occurs, it is important for it to be correctly managed. By Change we mean modifications to one or more of the following:

- Equipment;
- Procedures;
- Raw materials;
- Process conditions;
- Process technology;
- Instrumentation/Control;
- Materials of construction;
- Piping;
- Experimental equipment;
- Computer programs/logic;

other than "Replacement-in-Kind" type changes.

Changes need to be reviewed prior to their implementation. It is not only permanent changes, but also temporary changes, that need to be carefully evaluated, since some of the worst incidents have been caused by temporary modifications.

Proper documentation, reviews and approval processes are particularly important. Both procedures for reviewing and clearance sheets are needed to facilitate Managed Changes.

Clearance sheets need to typically address:

- Description of Change;
- Purpose of Change;
- Technical Basis for Change;
- Safety & Health Considerations;
- Documentation of Changes to Operating Procedures;
- Documentation of Changes to Maintenance Procedures;
- Documentation of Changes to Inspection/Testing Procedures;
- Changes to PFDs, P&IDs;
- Changes to Electrical Area Classifications;
- Changes to Training/Communications;
- Pre-Startup Inspection;
- Duration, if Temporary Change;
- Approvals & Authorizations.

Whether or not a Process Hazards Analysis is required is less well defined, but the following list of proposed criteria can form the basis of requirement. Such a PHA could vary from the use of the Checklist Methodology for relatively simple and minor changes to full-blown HAZOPs for more complex changes.

Changes Justifying PHAs

For a change to justify the need for a PHA, one or more of the following individual items should apply:

Basic Changes

- Are there any Changes to the Process Flowsheets?
- Are there any Changes to the Material/Energy balances?
- Are there any Changes to the compositions of one or more of the streams?
- Are there any Changes to Heat (or other Energy) sources/sinks?
- Could the Change involve any different chemicals that could react with other materials or chemicals (including solvents, additives) already present?
- Could the Change create undesirable byproducts or introduce impurities?
- Could the Change result in additional heat generation or increase in reaction rate, temperature or pressure and, possibly, a runaway reaction?
- If there are Changes to equipment sizing(s) could this impact other systems or compromise safety systems already in place?
- Could the Change result in greater corrosion or erosion?

RMP 40 CFR Part 68 Impacts

- Could the Change result in the exceedance of the Threshold Process Inventories for Regulated Substances?
- Could the Change affect the Worst Case Offsite Consequence Analysis (OCA) Release Scenarios?
- Could the Change affect the Alternative Case OCA Release Scenarios?

Boundary Changes

- Could the Change result in the extension beyond the normal operating envelope of the process (e.g. higher pressures, temperatures)?
- Could the Change result in the introduction of unstable materials (e.g. organic peroxides)?
- Could the Change result in the introduction of more hazardous materials, affecting plant personnel?
- Could the Change result in the introduction of contaminants (e.g. water that can cause freezing problems)?
- Could the Change result in changes to pressure drops/flows/flow regimes in piping and equipment that could present problems?
- Could the Change result in increased loadings on foundations/structural components?

Changes to Safety Systems

- As a result of the Change, are critical safety devices being disabled or bypassed?
- Are there any Changes to Interlocks, Protective Devices or Emergency Shutdown Systems?
- Are there any Changes to Distributed Control Systems (DCS) that could impact logic or sequencing?
- Are there any Changes to Pressure Relief Device(s) set pressure?
- Are there any Changes to Pressure Relief Devices/Flare System Loading?
- Are there any Changes to Occupational Health & Safety requirements?
- Are there any Changes to equipment isolation needed?

DIERS (Design Institute for Emergency Relief Systems) & Upgraded Emergency Relief Systems

- Does the Change introduce concerns not covered by standard pressure relief practices, such as API 520, 521?
- Does the Change introduce runaway reactions and new pressure relief concerns requiring more complex relief modeling?
- Does the Change introduce new multiple phase pressure relief and venting needs?
- Would the Change need unsteady state transient flow systems analysis?
- Would the Change need special analytical methods – often requiring customization?

Training & Procedures Considerations

- Does the Change require additional training of personnel, such as operators, maintenance personnel and others?
- Does the Change require that new operating procedure(s) be written?
- Could the Change introduce new potential for Human Error?

Change Execution Considerations

- Does the Change need reviews by pertinent technical or other specialists?
- Does the Change require fast-track needs that could compromise safety considerations, unless properly checked?
- Does the Change require extensive checks prior to implementation?
- Does the Change need Hot Work Permits for installation?
- Could the Construction phase, e.g. introduction of cranes, tackle, lifting gear or welding operations, compromise safety?
- Could the Change necessitate any special cleaning or maintenance or equipment purging hazards not already considered?
- Could the Change require new tie points that create additional hazards?

Operating & Maintenance Considerations

- Could normal plant operations require additional attention due to the Change?
- Could start-up, standby or shutdown require additional attention due to the Change?
- Could emergency plant shutdown require additional attention due to the Change?
- Would the operating/maintenance personnel need additional protective gear as a result of the Change?
- Would the Change create new and possibly more hazardous working conditions?

Environmental Considerations

- Does the Change introduce new effluents?
- Could the Change result in damage to end-of-pipe treatments (e.g. biox)?

MOCs Implementation

MOCs can be executed in an eight-step process:

Step 1. Initiator Requests Change.

Step 2. The Change is subject to peer/management review.

Step 3. A detailed evaluation of the Change is undertaken.

Step 4. Formal approval (or rejection) takes place.

Step 5. Safe limits are defined.

Step 6. All affected parties are notified.

Step 7. The Change is implemented.

Step 8. Follow checks are completed.

SUGGESTED READING (Note: URLs current at date of publication)
"Management of Change" by I.Sutton, published by KBI, 1997 www.kbintl.com/pubs/swb_moc.html
"Managing Process Changes – A Managers's Guide to Implementing and Improving Management of Change Systems" by M.L.Casada et al., published by Chemical Manufacturers Association, September, 1993 www.spillcontainment.com/industrialresources.htm
"Guidelines for Process Safety Documentation" by AIChE, CCPS, 1995, pages 177 to 189 www.aiche.org/pubcat/seadtl.asp?Act=C&Category=Sect4&Min=30
"Benchmarking MOC Practices in Process Industries" by N.Kerren et al., Process Safety Progress, June 2002, pages 103 to 112 www.aiche.org/safetyprogress/

CHAPTER 13

Estimation of Time Needed for PHAs

Most PHAs (HAZOP, What If/Checklist) start slowly and increase in speed once an adequate database of information has been created.

Average HAZOP Node

- Small: ***Up to 1 hour***
- Medium: ***Up to 2 hours***
- Large: ***Up to 3 hours (or more!)***

What If/Checklist Node:

- Small: ***Up to 1/2 hour***
- Medium: ***Up to 1 hour***
- Large: ***Up to 2 hours or more***

How to estimate the time

In order to estimate how long a PHA will take:

- Obtain all the P&IDs
- Estimate the number of nodes, noting how many are small, medium and large.

Meeting time hours:

HAZOP meeting time is:

(Number of small nodes × 1)

+ (Number of medium Nodes × 2)

+ (Number of large nodes × 3)

What If/Checklist meeting time is:

(Number of small nodes × 0.5)

+ (Number of medium nodes × 1)

+ (Number of large Nodes ×2)

Preparation Time

- Small HAZOP or What If/Checklist: ***1 to 2 days***
- Medium HAZOP or What If/Checklist: ***3 to 4 days***
- Large HAZOP or What If /Checklist: ***5 to 7 days***

Report time preparation

Table 13-1: Suggested Report Time Preparation for Various Sizes of PHA

Size of PHA	Simple	Extensive
Small	2 hours	3 days
Medium	3 hours	5 days
Large	4 hours	7 days +

A performance chart (Figure 13-1) illustrating importance of preparation and experience requirements for a PHA a team is given on the next page.

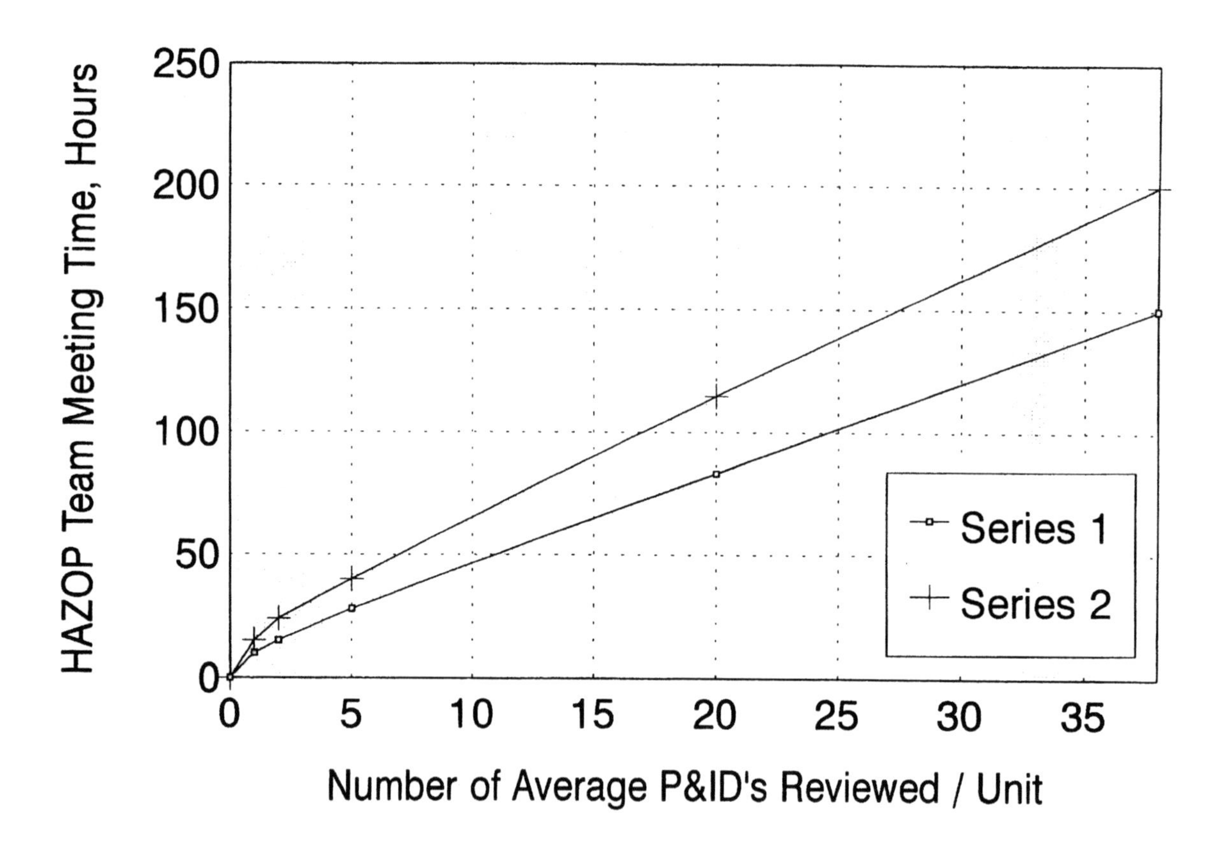

Series 1: Experienced Team
Series 2: Average Team

Figure 13-1

SUGGESTED READING (Note: URLs current at date of publication)
"Mathematical Model for HAZOP study time estimation" by F.I.Khan and S.A.Abbasi, J.Loss Prev. Process Ind., Vol.4 No.4, 1997 www.elsevier.nl/locate/jlp
"Making Process Safety Pay Off – Sustaining Performance in the 21st Century – PHA revalidation: can this cost be reduced?", K.Sandler et al., www.battelle.org/environment/irm/pdfs/making%20process%20safety%20pay%20off.pdf
"Cost Estimation of Industrial Risk in the Bidding Process" by E.Cagno et al.(Website) www.esi2.us.es/prima/Papers/SENETPMR-00.pdf

13

CHAPTER 14

Management of Hazards Associated with Location of Process Plant Buildings

Overview

This chapter provides guidelines in identifying hazards associated with explosion, fire and toxic release that may affect process plant buildings and managing the associated risks. It also introduces the API recommended practice 752 "Management of Hazards Associated with Location of Process Plant Buildings," its relationship with OSHA 29 CFR 1910.119 and the recommended analyses for identifying the above hazards. Detailed checklists are provided at the end of this chapter for verifying if typical routine risk-reduction measures have been implemented.

Major Concerns

The major concerns in a facility are:

- Explosions;
- Fires;
- Toxic releases.

Explosions

Generally, explosions in process industries are of the following types:

1) Physical explosion – resulting from mechanical failure of pressure systems, or unstable pressure and/or temperature conditions in pressure systems;

2) Condensed phase explosion – whereby flammable materials are liquids or solids, such as ammonium nitrate, organic peroxides, sodium chlorate and other types of explosives;

3) VCE – Vapor Cloud Explosion;

4) BLEVE – Boiling Liquid Expanding Vapor Explosion;

5) Reactor explosion due to runaway exothermic chemical reactions;

6) VEEB – Vapor Escape into, and Explosion in, a Building;

7) Dust explosion.

Of these, VCEs are frequently the most serious hazards in the process industries as they cause the largest number of fatalities and most extensive property damage. A VCE occurs when a flammable material experiences the following four stages:

1) The flammable material leaks at a sufficient rate;

2) Some or all of the released material vaporizes to form a vapor cloud of sufficient size over a long enough period and mixes with air in proportions to form a potentially explosive mixture;

3) There is some form of confinement in the area that causes turbulence in the vapor cloud and results in acceleration of the flame front;

4) The flammable material encounters an ignition source (although this is frequently omnipresent).

Typical ignition sources are flames, direct heat, hot work (e.g. welding), incandescent material, hot surfaces, electrostatic/electrical/friction/impact sparks, self-heating and lightning.

Fires

Common types of fires in process plants are as follows:

1) Fires due to leakage or spillage of flammables;

2) Pump fires, flange fires, lagging fires, duct fires, cable tray fires and storage tank fires;

3) Flash fires (vapor cloud fires), pool fires, fireballs, jet flames and engulfing fires;

4) Solid fires – solid combustible materials, dust fires, warehouse fires.

Toxic Releases

Although toxic releases occur less frequently than fires and explosions, the products of combustion, for example carbon monoxide, are frequently toxic. In many instances, such as the burning of hydrogen sulphide, a toxic gas like sulfide dioxide will rise and be more easily dispersed. A large toxic release can produce casualties at very large distances from the point of origin, depending upon the weather conditions and the population density in its path and the density of the vapor being released.

The three modes of exposure are:

1) Inhalation – gases, vapors, fumes, dust, particles;

2) Ingestion – liquids and solids;

3) External contact.

API 752 – Management of Hazards Associated with Location of Process Plant Buildings

OSHA 29 CFR 1910.119 "Process Safety Management of Highly Hazardous Chemicals" requires Process Hazards Analysis (PHA) to address facility siting and identify scenarios that could lead to serious release of toxic or flammable materials or an explosion. In API 752, guidelines are given for identifying the above scenarios in process plant buildings. In addition, the recommended practice provides assistance in managing the associated risk. This recommended practice is not applicable to production facilities surrounded by navigable waters, such as offshore platforms.

Prerequisites in Using API 752

A company must address the following in order to apply the recommended practice:

- Occupancy and emergency role of personnel criteria;
- Evaluation case events;
- Consequence modeling/analysis programs;
- Risk acceptance criteria.

Considerations in Hazards Identification

The following is a list of considerations for identifying the potential hazardous events in the facility:

- Materials of concern;
- Site-specific conditions;
- Occupancy criteria.

Materials of Concern

- ***Flammable materials*** – e.g. LPG or natural gas that if ignited would form a flash, jet or pool fire or fire balls. Some flammable materials can form vapor cloud mixtures in air leading to the formation of VCEs.
- ***Toxic materials*** – e.g. Hydrogen sulfide, Dioxin, Methyl isocyanate or materials that are capable of producing toxic materials, such as on combustion.
- ***Other process materials*** – materials that have the potential for:
    - Runaway reactions, e.g. polymerization of Methyl Methacrylate.
    - Auto-decomposition (chemically or thermally induced), e.g. Tetrafluoroethylene (TFE).
    - Auto-ignition.
- ***Explosive dust.***
- ***Extreme process conditions***:
    - Compressed gases – In case of vessel failures, high-pressure gases can expand explosively leading to the formation of damaging blast waves.
    - Liquids stored well above their atmospheric boiling points (potential for BLEVEs).
    - Hot materials – Rapid combination of hot and cold materials can lead to the explosive vaporization of the colder material.

Site-Specific Conditions

During the process of evaluating the risks to building occupants, it is essential to consider site-specific conditions. Table 14-1 indicates the major site-specific conditions that may affect the severity and likelihood of explosions, fires and toxic releases.

Table 14-1: Site-specific conditions relating to the severity and likelihood of explosions, fires and toxic releases

	Explosions	Fires	Toxic Releases
Plant layout / spacing (distance between process units and buildings)	X	X	X
Degree of congestion and confinement	X	X	X
Building's materials of construction	X	X	
Sources of ignition	X	X	
Drainage	X	X	X
Amount of material released	X	X	X
Release conditions of material	X	X	X
Process conditions	X	X	X
Quantity of process inventories	X	X	X
Physical properties of process material	X	X	X
Effectiveness of emergency response (e.g. isolation, evacuation of inventories)	X	X	X
Weather conditions	X	X	X
Topography	X	X	X
Obstacles			X
Direction/source of the release			X

Occupancy Criteria

The occupancy criteria in plant buildings, established by the company, are used to determine if further evaluation of a potential hazardous event is necessary. If the occupancy level of a building has exceeded the established criteria, further evaluation should be performed. The following are some suggested approaches for defining occupancy criteria:

- ***Definition of occupied and unoccupied*** – qualitative criteria for occupied and unoccupied buildings. For example, occupied buildings could be defined as those that personnel occupy while doing the major part of their work, such as control rooms, laboratories and office buildings. An unoccupied building would then be one that personnel visit infrequently to perform brief tasks or monitor the process.
- ***Occupancy load*** – the total integrated time for the full- or part-time occupants in the building. It is usually expressed as the inhabited time over a specific period, based on an annual average. Some factors to consider in determining the occupancy load are as follows:

    1. The routine presence of additional personnel, such as visitors, contractors and trainees;
    2. Activities performed by personnel on a routine basis such as calibration and maintenance of instruments.

  The occupancy load criteria used by some companies range from 200 to 400 personnel hours per week.
- ***Individual occupancy*** – percentage of an individual's total time spent in a building. The individual occupancy criteria used by some companies range from 25 percent to 75 percent.
- ***Peak occupancy*** – the number of people potentially exposed for a given period. The peak occupancy criteria used by some companies range from 5 to 40 persons.

Analysis Process for an Explosion

API 752 has recommended a three-stage analysis process to identify hazards and manage risk to building occupants from explosion (see Figure 14-1):

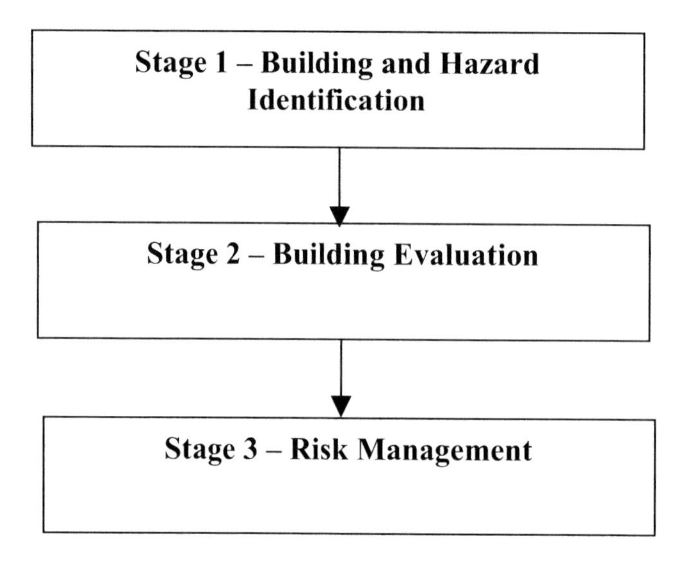

Figure 14-1: Stages for Explosion Risk Analysis

Stage 1 – Building and Hazard Identification

The following procedures can be used to identify buildings that could potentially be impacted by an explosion. The procedures also help to address the siting issues within the PHA requirements of OSHA 29 CFR 1910.119:

1) ***Identify materials that would lead to potential explosion*** – if materials of concern are identified, proceed to the next step.

2) ***Evaluate the site-specific conditions*** – although some facilities may handle materials with VCE potential, site-specific conditions may preclude a VCE, e.g. low degree of confinement, rate of release is sufficiently low. If site conditions cannot preclude explosion concern, proceed to the next step.

3) ***Determine if building exceeds occupancy criteria*** – if the building exceeds the company occupancy criteria, proceed to Stage 2.

Stage 2 – Building Evaluation

Stage 1 provides preliminary identification of buildings that could potentially be impacted by an explosion. Stage 2 provides further screening by evaluating the hazards to buildings based on one or a combination of methods listed below:

- Design or compare to industry and company standards;
- Consequence analysis;
- Screening risk analysis.

Design or compare to industry and company standards – focus on the design of the building and spacing to reduce the effects of explosion – e.g. fire, pressure wave, missiles or toxic release. The following are examples of design and spacing standards:

- o Standards used by Oil Insurance Association;
- o NFPA (National Fire Protection Association) spacing criteria.

Consequence analysis – focus on estimating the magnitude of an explosion, evaluate its effects on a building, and relate the damage sustained by the building to the degree of potential injury or damage to the occupants and/or equipment inside. Depending on the scenarios being studied, evaluation of some or all of the following may be required:

- o Release scenarios;
- o Explosion overpressures;
- o Fire radiation;
- o Site meteorology;
- o Toxic levels, ingress potential.

Screening risk analysis – determine approximate aggregated and individual risk to occupants of a process plant building. This method includes the combined evaluations of event frequency and explosion consequences to determine the risk to a process plant building occupant. The calculated risk is compared with the company's risk acceptance criteria to determine if any additional studies are required.

If the damage of the building and/or occupants is of significant concern and additional evaluation is desired, proceed to Stage 3.

Stage 3 – Risk Management

Stage 3 includes the following elements:

- ***Risk assessment*** – identify specific events or scenarios and determine the associated risk by evaluating the frequency and the consequences of scenarios, consequently defining an overall potential risk to the building occupants;
- ***Risk acceptance criteria*** – compare the risk determined in the above with the risk acceptance criteria established by the company to determine if risk reduction is required;
- ***Risk reduction*** – establish means to mitigate risk to building occupants. There are two broad categories of risk mitigation:
    - **Prevention**: activities that eliminate or reduce process incidents. For example:
        - Reducing the inventory of hazardous material;
        - Engineering controls such as process controls, emergency shutdowns and redundant instrumentation.
    - **Mitigation**: activities that reduce the consequences associated with the occurrence. For example:
        - Relocating personnel or rearranging the room functions within the building;
        - Eliminating windows.

Analysis Process for a Fire

API 752 has recommended the analysis process for identifying hazards and managing risk to building occupants from fire. The process is simpler than that used in the explosion analysis. It can be divided into three stages:

Stage 1 – Building and Hazard Identification

The following procedures can be used to identify buildings that could potentially be impacted by fire. The procedures help to address the siting issue within the PHA requirements of OSHA 29 CFR 1910.119:

- ***Identify flammable materials in the facility;***
- ***Determine if building exceeds occupancy criteria*** – if the building exceeds the company occupancy criteria, proceed to Stage 2.

Stage 2 – Building Evaluation

- ***Design or compare to industry and company standards*** – focus on the design of the building and spacing to reduce the effects of fire.

Stage 3 – Risk Management

- ***Mitigation and Emergency Response*** – determine the appropriate measures for the occupants to take in case of emergency.
- ***Risk reduction*** – establish means to mitigate risk to building occupants. There are two broad categories of risk mitigation:
    - **Prevention**: activities that eliminate or reduce process incidents. For example:
        - reducing the inventory of the hazardous materials;
        - altering the process conditions or materials to reduce the potential for runaway reactions or corrosion.

- **Mitigation**: activities that may reduce the consequences associated with the occurrence. For example:
    - Providing physical separation from the fire event by distance or barriers;
    - Fire and smoke suppression system.

Analysis Process for a Toxic Release

API 752 has recommended an analytical process for identifying hazards and managing risk to building occupants from toxic release. The process is simpler than that used in the explosion analysis. It can be divided into three stages:

Stage 1 – Building and Hazard Identification

The following procedures can be used to identify buildings that could potentially be impacted by toxic releases. The procedures also help to address the siting issue within the PHA requirements of OSHA 29 CFR 1910.119.

- ***Identify toxic materials of concern in the facility;***
- ***Determine if building exceeds occupancy criteria*** – if building exceeds the company occupancy criteria, proceed to Stage 2.

Stage 2 – Building Evaluation

Site Conditions Evaluation – Determine if the site conditions would prevent a toxic release affecting building occupants. Such conditions would include:

- o Weather conditions;
- o Direction and source of the release;
- o Location and size of the entry point in the building.

Stage 3 – Risk Management

- ***Mitigation and Emergency Response*** – determine the appropriate measures for the occupants to take in case of emergency, e.g. the use of personnel protective equipment (PPE);
- ***Risk reduction*** – establish means to mitigate risk to building occupants. There are two broad categories of risk mitigation:
    - o **Prevention**: activities that eliminate or reduce process incidents.

For example:

- Reducing the inventory of hazardous material;
- Providing inspection programs for piping and equipment.

- **Mitigation**: activities that reduce the consequences associated with the occurrence. For example:
  - Ventilation system controls with appropriate detection;
  - Use of elevated intake stack for potential releases of heavier-than-air materials.

API 752 Building Checklist

API 752 provides a sample building checklist for verifying if routine risk-reduction measures are in place.

Table 14-2: API 752 Process Plant Building Checklist
* E = Explosion; F = Fire; T = Toxic.
This checklist may or may not be appropriate for a particular situation.

E.F.T*	Questions
FT	1. Is the building located upwind of the hazard?
EFT	2. Is the building included in an emergency response plan for fire and toxic release? Are the occupants trained on emergency response procedures? Are evacuation instructions posted?
E	3. Are large pieces of office equipment or stacks of materials within the building adequately secured?
E	4. Are the lighting fixtures, ceilings, or wall-mounted equipment well supported? Are process controls mounted on interior walls?
E	5. Is there heavy material stored on the ground floor only?
E	6. Have all the exterior windows been assessed for potential injury to occupants?
EFT	7. Are there doors on the sides of the building opposite from an expected explosion or fire source?
FT	8. Are there exterior and interior fire suppression equipment available to the building?
FT	9. Is there a detection system within the building or in the fresh air intake to detect hydrocarbons, smoke or toxic materials?
FT	10. Is the air intake properly located?
FT	11. Can the ventilation system prevent air ingress or air movement within the building? Are there hydrocarbon or toxic detectors that shut down the air intake? Does the building have a pressurization system?
FT	12. Are there wind socks visible from all sides of the building?
EFT	13. Is there a building or facility alarm or communication system to warn building occupants (of an emergency)?
T	14. Is there sufficient bottled air or fresh supplied air for the occupancy load?
EFT	15. Are all sewers connected to the building properly sealed to prevent ingress of vapors?

Facility Siting Checklists

The following are additional considerations (Tables 14-3 to 14-7) that may be put in place to assist in risk reduction as it applies to siting and layout:

- General siting issue;
- Building protection;
- Spacing;
- Health and safety;
- Location of control rooms/critical buildings.

Table 14-3: General Siting Checklist
Are the following requirements satisfied: ▪ Process needs, e.g. gravity flow where possible, adequate NPSH for pumps? ▪ Ease of plant operation? ▪ Ease of maintenance? ▪ Ease of construction? ▪ Ease of commissioning? ▪ Ease of future expansion (if required)? ▪ Ease of access? ▪ Ease of plant drainage?
If plant contains flammables, are they located outdoors to reduce risks?
Is plant subdivided into areas of high, medium and low risk?
Is plant exposed to hazards from neighboring plants?
Are public or personnel beyond the property line protected against potential hazards?
Does site security prevent access by unauthorized persons while not hindering emergency services (e.g. fire fighters, paramedics)?
Are there below-ground-level locations (pits, ditches, sumps) where toxic or flammable materials can collect?
Can transportation of hazardous materials and impact of spillage be reduced by suitable site location?
Other general concerns (specify)?

Table 14-4: Building Protection Checklist
Is ground or paving sloped so that flammables will not accumulate beneath vessels?
Could drainage system cope with both storm water and fire fighting water?
Are structures that are load bearing fireproofed if they are required to support vessels, equipment or pipework carrying flammable, toxic or hazardous materials?
Are dikes, berms, barricades or containment systems required to protect personnel and equipment against fire or explosion?
Are traffic signs/crash barriers/restrictions required to protect against vehicle or other impacts or injuries to personnel in the vicinity?
In the event of an explosion, would fire fighting water supplies still remain intact (preferably buried)?
Does plant meet requirements of electrical hazardous areas classifications?
Other protection concerns (specify)?

Table 14-5: Spacing Checklist
Are well-established codes being referenced for establishing plant spacing (e.g. Industrial Risk Insurers, NFPA, etc.)?
Does plant layout:
▪ Reduce chances of explosion or fire by minimizing ignition hazards?
▪ Limit the spread of fire or damage caused by flying debris or blast?
▪ Permit access of fire fighting vehicles, equipment and personnel?
▪ Minimize effects of fire/explosion on adjacent facilities?
▪ Separate continuous ignition sources from potential release sources of flammable materials?
▪ Ensure that critical facilities (e.g. fire fighting) are not subject to fire or explosion?
Are high-risk units (e.g. at 1000 psi or more) spaced farther from other units?
Are exothermic reactors located at the periphery of units and away from key facilities, control rooms, etc.?
Are fire heaters located upwind of potential release sources (e.g. pumps with seals that may leak flammables, compressors handling flammables)?
Are cooling towers and utilities located well away from battery limits?
Are control rooms located well outside battery limits and next to an access roadway?

Table 14-5: Spacing (Continued)
Do pumps handling flammables avoid the following locations: ▪ Immediately beneath piper racks or access structures? ▪ Beneath air/fan exchangers? ▪ Beneath drums or exchangers operating at high temperatures?
Is storage area located in a hazardous manner (e.g. uphill of process plant, without adequate diking)?
Have potential risks from pressurized storage (e.g. propane bullets, storage spheres) been assessed?
Is electrical switchgear located at periphery of unit to minimize risk?
Have routing of flare headers through hazardous locations been minimized?
Is equipment adequately spaced to permit maintenance (e.g. pulling of heat exchanger tube bundles, catalyst removal)?
Does layout minimize use of heavy lifting equipment?
Are pipe rack configurations that tend to box units in and make them less accessible avoided?
Other spacing concerns (specify)?

Table 14-6: Health and Safety Checklist
Are there at least two separate means of escape for operating personnel from all locations on the plant?
Are escape routes sign-posted in complicated areas?
Are tripping and bumping hazards eliminated?
Are walkways and accessways wide enough for personnel wearing breathing packs?
Are assembly point stations allocated in emergency situations?
Is head clearance adequate for working areas and walkways?
Are emergency shower and eye bath locations provided?
Is adequate lighting provided?
Other health & safety concerns (specify)?

Table 14-7: Location of Control Rooms/Critical Buildings Checklist
Could control room/critical building be impacted by: ▪ Vapor cloud explosion from facility? ▪ Pool fire, jet fire, fireball, flash fire from facility? ▪ Toxic release from facility?
For control room/critical building subject to potential blast: ▪ Would the building collapse under peak overpressure of, say 10 psi? ▪ Is the building designed for blast protection? ▪ Is the building outside the likely impact range? ▪ Are building windows minimized, blast protected? ▪ Will the building materials used withstand blast forces? ▪ Will the control center remain functional for shutdown purposes? ▪ Could internal components fail?
For control room/critical building impacted by fire: ▪ Are nonflammable construction materials used? ▪ Is ground sloped away from building to prevent ingress of burning liquids? ▪ Are windows minimized, or can they be, to withstand thermal effects?
For control room/critical building subject to potential toxics/asphyxiants including combustion products: ▪ Can fresh air intakes be sealed closed in the event of emergency? ▪ Are fresh air intakes automated to close in event of toxic release? ▪ Are self-contained breathing air packs available to personnel normally within the building? ▪ Other concerns related to control room/critical building (specify)?

SUGGESTED READING (Note: URLs current at date of publication)
"Management of Hazards Associated with Location of Process Plant Buildings, API Recommended Practice 750, May 1995 http://api-ep.api.org/filelibrary/ACF4B.pdf
"Derivation of fatality probability functions for occupants of buildings subject to blast loads" by W.S. Atkins for HSE, UK, Contract Research Report 151/1997 (Website) www.hse.gov.uk/research/crr_pdf/1997/crr97151.pdf
"Occupant response shelter evacuation model" by Electrowatt Engineering (UK) Ltd. for HSE, UK, Contract Research Report 162/1998 (Website) www.hse.gov.uk/research/crr_pdf/1998/crr98162.pdf
"General Recommendations for Spacing" published by IRI, Industrial Risk Insurers www.industrialrisk.com/
"Personnel Safety Review Checklist", University of Florida (Website) http://pie.che.ufl.edu/guides/safety_health/personnel_checklist.html
"Guidelines for Evaluating Process Plant Buildings for External Explosions and Fires" by AIChE, CCPs, 1996 www.aiche.org/pubcat/seadtl.asp?Act=C&Category=Sect4&Min=20
"Guidelines for Evaluating the Characteristics of Vapor Cloud Explosions, Flash Fires and BLEVE's" by AIChE, CCPS, 1994 www.aiche.org/pubcat/seadtl.asp?Act=C&Category=Sect4&Min=20
"Design of Blast Resistant Buildings in Petrochemical Facilities", American Society of Civil Engineers, 1997 www.pubs.asce.org/BOOKbrowsesu.cgi?860
"Spacing in Chemical Plant Design Against Loss by Fire", by R.Robertson, IChemE Symposium Series No. 47 on Process Industry Hazards, 1976, pages 157 to 174 www.icheme.org/framesets/aboutusframeset.htm
"The Siting and Construction of Control Buildings – A Strategic Approach", by V.C.Marshall, IChemE Symposium Series No. 47 on Process Industry Hazards, 1976, pages 187 to 204 www.icheme.org/framesets/aboutusframeset.htm
"Structural Design of Control Buildings", by J.M.Langeveld, IChemE Symposium Series No. 47 on Process Industry Hazards, 1976, pages 205 to 216 www.icheme.org/framesets/aboutusframeset.htm

CHAPTER 15

PHA Protocols and Administrative and Engineering Controls

PHA Protocols

It is very important in a PHA to carry out the analysis using a structured format. For example, with Guide Word HAZOP, once having chosen a specific Deviation, such as High Flow, all the Causes of High Flow should be recorded. Once the list of Causes is completed, the Consequences, Safeguards and required Recommendations should be addressed, starting with the first Cause. The second Cause should be treated in the same manner, and so on for each of the remaining Causes in turn.

When recording Consequences, these should be assessed based upon none of the Safeguards being present or active. The tendency to state how the Safeguards will respond to a Cause in describing the Consequences should be avoided. For example, a Cause for the Deviation of High Pressure may be Blocked Discharge, pertaining to a Vessel being pressurized by a pump or compressor. If there is a Pressure Safety Valve (PSV) on the Vessel, designed to protect against Overpressure, the Consequence should not be recorded as "Overpressure on vessel, causing PSV to open." The Consequence should be more akin to "Overpressure with potential for loss of containment, release of flammables: potential fire/explosion hazard." The Safeguards should indicate protection by the PSV. If considered sufficiently Safeguarded, no Recommendations are considered necessary. If not, one or more Recommendations may be put forward to control the hazard.

Administrative and Engineering Controls

Administrative Controls are normally procedural in nature and considered common practices for many different types of facility. They are heavily reliant upon creation/adoption/updating by plant management, plant operations, plant maintenance and plant safety personnel, who are also responsible for their use, interpretation and execution. They are also subject to the potential for misinterpretation and human error and even more so if they are ignored, not enforced or not taken seriously. Table 15-1 provides a list of Typical Administrative Controls:

Table 15-1: Typical Administration Controls
Accurate labeling of additives
Adequate data logging records
Adequate number of operators for specific task
Adequate number of operators
Adequate time allowed in procedure
Annual review of plant signs
Blinding and isolation procedures
Buddy system in hazardous areas
Car seals logging on critical valves
Checks to ensure raw materials entering plant conform to bill of materials
Color-coded labeling system for process, utilities, etc.
Company organizational procedures
Company policy that prevents experimentation with production plant
Company training courses
Conformance to standard procedures enforced
Design checking procedures
Detailed procedure for handover of operating logs
Documentation and records maintenance
Emergency response planning
Emergency response procedures developed for surrounding area
Enforcement of procedures

Table 15-1: Typical Administration Controls
Independent verification of procedure
Inspection and maintenance program to detect corrosion/erosion and replace affected components
Instrument/maintenance checks
Inventory check
Labeling of lines
Limits on vehicular activity on facility
Lock out/tag out procedures
Logging of locks on critical valves
Maintenance procedure for bolt torquing
Maintenance work permitting
Manual resetting of cutout valves
Enforcement of electrical area classification code requirements
Enforcement of engineering procedures
Enforcement of in-plant construction safety procedures
Enforcement of maintenance procedures
Enforcement of plant security
Enforcement of safety procedures
Operating procedure for operating bypass valves
Operating training and certification
Operational log check
Operator certification
Periodic updating
Permitting, logging and tagging procedures
Permitting/inspection systems enforced
Plant addresses weather forecasts and plans accordingly
Plant perimeter and access reviews for security
Plant security and access procedures
Plant inspection/maintenance procedures for detecting leaks and replacing seal/packing
Plant safety policy overrides production demands

Table 15-1: Typical Administration Controls
Post maintenance inspection/testing procedures
Pre-startup checklist
Pre-operational check procedure
Prevention of vehicle access in area
Previous operating logs being made readily available
Procedures for creating and posting of plant signs to warn of hazards, etc.
Procedural review for loss of utilities
Procedure for checking proposed temporary installations for hazards
Procedure for limiting and/or controlling construction activities
Procedures for testing and inspection
Procedures for vessel entry
Procedure to call for verification by independent operator
Purging and cleaning procedures
Purging medium and procedures and connections
QA/QC and inspection
QA/QC procedures in place
QA/QC program for relief valve repair, inspection and maintenance
Rapid escape routes for personnel
Readily available operating manuals
Record maintained of bypass valve positions
Regular cleaning/flushing program
Regular lubrication schedule
Regular predictive maintenance schedule
Regular testing and inspection procedures
Regular testing of shutdown systems
Regular updating of operating manuals
Regular updating of piping and instrument diagrams
Replacement program for high-wear components, e.g. flexible hoses
Re-qualification procedures for component(s) in new service
Sampling and testing procedures

Table 15-1: Typical Administration Controls
Seal inspection/maintenance procedures
Security requirements for plant
Special construction procedures in operating facilities
Specific guidance in procedures
Start-up procedural checking
Strictly enforced welding safety procedures
Strictly enforced welding safety procedures and use of high-quality welding/hot tap equipment
Supervisory procedure overrules in favor of safe operation
Tagging and procedures for blinding
Task list for shift operators
Testing and inspection procedure for critical instruments
Testing of hazardous environment prior to vessel entry, etc.
Training and qualification of personnel
Training for emergency situations
Use of additional operators
Use of conservative lifting gear and measures to avoid snagging
Use of experienced personnel
Use of portable phones/two-way radios/cellular phones to optimize communications
Verification by independent operator
Well-documented plant logs
Work permitting procedures
Work study of specific task
Working environment checked for asphyxiants, toxics, flammables, etc.

Engineering Controls are provided by Devices, such as Alarms, Trips, Emergency Shutdown Valves, Deluge Systems, Fire Detectors and engineering features such as Adequate Corrosion Allowances, Nitrogen Purging and so forth. These are usually specifically engineered features of the plant in question. Table 15-2 provides a list of Typical Engineering Controls:

TABLE 15-2: Typical Engineering Controls
Additional corrosion allowance
Adequate corrosion allowance
Adequate pipe supports
Adequate tank spacing
Adequate/excess flare system capacity
Adequately sized vents and venting procedures
AFF systems
Air intake located at safe height
Alarm on high mass/volume summation
Alarm on low mass/volume summation
Alternate source of cooling
Anti-Fire Fighting Foam (AFFF)
Area fire detectors and alarms
Area flammable gas detectors
Area toxic gas detectors and alarms
Automated closure of louvers on air-cooled heat exchangers
Automated depressurizing of vessel
Automated emergency shutdown system
Automated high-pressure vent
Automated interlocks
Automated quench cooling
Automated quench system
Automated shut-off of feed on high level
Automated supply valve closure with high temperature

TABLE 15-2: Typical Engineering Controls
Automated/alarmed interlocks
Automatic deluge
Automatic depressurizing valve
Back-up non-return valve
Bird screen on tank vent
Blanket gas on pressure control
Blanks or caps or plugs
Blinds
Burst disc (with downstream vent)
Burst disc downstream monitoring for flow
Car seals on critical valves
Cathodic or other corrosion protection
Check valve
Cold temperatures incorporated in piping specifications
Color-coded lines, connectors
Color coding/labeling of lines
Column grounding to prevent static electrical build-up
Compatible fire fighting media
Comprehensive instruction manuals
Comprehensive operating procedures
Control valves fail in safe position on loss of instrument air
Controlled depressurizing
Controlled filling procedures
Controlled heat input
Corrosion coupons
Dedicated closed drainage system
Deluge system
Deluge system and fire monitors
Depressurizing valve
Double block and bleed valves

TABLE 15-2: Typical Engineering Controls
Double block valves on drains
Double block valves where freeze-up can occur
Drainable water boot
Drained dikes surrounding pump
Electrical area classification code
Emergency alarm and automated shutdown
Emergency depressurizing system
Emergency isolation
Emergency reactor depressurizing system
Emergency response planning
Emergency response procedures developed for surrounding area
Emergency shutdown and de-inventorying capability
Emergency shutdown and/or isolation
Emergency tank vent
Emergency venting of firebox outlet
Environmental monitors to detect leaks
Equipment grounding to prevent static electricity
Equipment supports fireproofed
Equipment thermally insulated
Excess flow valve snaps shut
Existing structure(s) checked for adequacy
Explosion hatches on firebox of furnace
Explosion venting device(s)
External detectors
Feed control of reactants
Feed flow trip
Feed ratio control of reactants
Fire detection and prevention system
Fire monitors
Fire monitors to maintain cooling

TABLE 15-2: Typical Engineering Controls
Flame arrester in line
Flame scanner on pilot gas trips furnace on loss of pilot
Flame scanner trips furnace on loss of main flame
Flammable gas detection and alarm
Flange guards
Flare system for relief/depressurizing loads
Flow indicator/alarm
Foams to limit vaporization of hazardous materials
Freeboard allowance on tanks and vessels
Grounding of equipment to prevent static electrical discharges
Hardware interlocks
Heat tracing/winterization
Heavy duty corrosion allowance
Heavy gauge instrument lines
High bottoms level alarm
High cooling medium temperature alarm/trip
High differential pressure alarm
High differential pressure alarm/trip
High differential pressure indicator/alarm
High feed rate alarm
High feed-stream temperature indicator/alarm
High flow alarm
High flow alarm on feed streams
High flow alarm on quench stream
High flow alarm/trip
High high flow alarm/trip
High high interface level alarm
High high level alarm
High high pressure alarm/trip
High high temperature alarm/trip

TABLE 15-2: Typical Engineering Controls
High interface level alarm
High level alarm
High level alarm on vessel
High level alarm/feed trip
High level alarm/high high level trip
High pilot pressure alarm
High pressure alarm at tank base
High pressure alarm in vapor space
High pressure alarm on column overheads
High pressure alarm on combustion air to furnace
High pressure alarm on firebox of furnace
High pressure alarm/feed trip
High pressure alarm/interlock on heat to column reboiler
High pressure alarm/relief/trip
High pressure alarm/trip
High pressure fuel alarm/furnace shutdown
High pressure indication
High pressure relief/trip
High pressure trip on compressor
High pressure trip on feed
High pressure trip on heating medium to reboiler
High pressure trip on pump
High pressure trip on reboiler
High process temperature alarm/trip
High-quality hot tap gear that minimizes risk
High skin temperature alarm
High speed alarm/trip
High speed trip on compressor
High speed trip on pump
High suction temperature alarm/trip on compressor

TABLE 15-2: Typical Engineering Controls
High temperature alarm
High temperature alarm on feed
High temperature alarm on overheads
High temperature alarm/trip
High temperature initiates quench
High temperature monitor/alarm/trip
High temperature override and shutdown
High upstream pressure indicator/alarm
High viscosity monitor/alarm
Higher differential temperature between hot and cold streams
Impact/vehicle crash barriers
Increased cooling available
Increased cooling medium supply
Increased separation distances
Increased working level in drum supplying pump
Independent automated isolation of flow
Independent flow indicator/alarm
Independent isolation valve actuated by pressure or reverse flow
Independent level alarm
Independent low flow indicator/alarm
Independent manual shutdown procedure in place
Independent monitors/alarms
Independent shutdown interlocks
Independent verification of procedure
Independent visible indication/alarm
Independent visible/audible alarm
Independent visual indication
Indication/alarm of high flow condition
Indication/alarm on low flow condition
Individual safety interlocks

TABLE 15-2: Typical Engineering Controls
Inerts supplied on low pressure trip
In-line analyzers with alarms/trips
Inspection and maintenance procedures
Inspection and maintenance program to detect corrosion and/or erosion and replace affected components
Inspection and radiography
Instrument /maintenance checks
Instrument air receiver with additional capacity
Instrument air receiver(s) supplying extra capacity
Insulation
Insulation to reduce heat influx from external fire
Interlocked emergency shutdown
Inventory reduction of hazardous materials
Isolatable sight/gauge glasses
Isolation valve
Isolation valve actuated on reverse flow condition
Isolation valve in line
Isolation valve (remotely operable)
Isolation valves
Isolation valves on instruments/bridles
Level gauge
Level gauge on interface
Limiting concentration of reactants
Line flushing system
Lines are labelled
Local and area fire detectors and alarm system
Local fire extinguishers
Local isolation valve
Locks on critical valves
Louvers of air cooled exchangers close on fire case

TABLE 15-2: Typical Engineering Controls
Low coolant flow alarm/trip
Low downstream pressure indicator/alarm
Low feed alarm
Low feed stream temperature indicator/alarm
Low flow alarm
Low flow alarm on loss of coolant
Low flow alarm on reflux
Low flow alarm/shutdown
Low flow alarm/trip
Low flow alarm/trip on cooling medium failure
Low flow alarm/trip on desuperheater water supply
Low flow alarm/trip on loss of feed or quench
Low heating medium flow alarm/trip
Low heating medium temperature alarm/trip
Low interface level alarm
Low level alarm
Low level alarm in drum supplying pump
Low level alarm/trip
Low level alarm/trip on upstream vessel
Low low level alarm/trip
Low pressure alarm
Low pressure alarm in vapor space
Low pressure alarm/shutdown
Low pressure alarm/trip
Low pressure alarm/trip on lube oil failure
Low pressure indication/alarm
Low pressure/temperature alarm on desuperheated steam supply
Low process temperature alarm/trip
Low speed alarm on agitator or recirculation pump failure
Low speed alarm/trip

TABLE 15-2: Typical Engineering Controls
Low temperature alarm
Low temperature alarm on bottoms product from column
Low temperature alarm/trip
Low temperature indicator/alarm
Low temperature material of construction
Low upstream pressure indicator/alarm
Main fuel isolation valves
Maintenance and inspection procedures
Maintenance and inspection procedures on critical valves
Maintenance procedure for bolt torquing
Manual bypass around control valve
Material of construction suitable for low temperature
Minimum line size specified
Mixer on linked timer system
Monitor and analyze and trip for unstable components
Monitor conversion of reactants
Monitor/alarm for packing failure
Monitor/alarm high seal pressure
Monitor/analyze and trip with unstable components
Monitoring of feed flows and ratios
Monitors to detect leakage from internal seal/packing
Monitors to detect leakage from seal/packing
Multipoint bed monitoring and alarm for catalyst bed
Multipoint bed temperature alarms for catalyst bed
Naked flame source, e.g. furnace, located upwind of potential release
Nitrogen blanketing of low flash point liquids
Nitrogen purge
Nitrogen purging facilities and connections and procedures
No flow alarm/trip
Non-return valve in line

TABLE 15-2: Typical Engineering Controls
On-motor trip alarm and shutdown
On-line feedstream analyzer and feed trip
On-line sampling or analysis of feed
Open flow path with valves locked/car sealed open
Operating procedures (specific)
Operating procedures for routing critical flows
Operating procedures for routing flows
Operational interlocks to minimize error
Overhead deluge system
Oversized agitator motor
Over-speeding trip on motor
Piping specification adequate for maximum pressures
Piping specification covering full temperature range possible
Plant inspection and maintenance procedures for detecting leaks and replacing internal seal/packing of prime movers
Plant interlocks
Plugs/blanks on vents/drains
Plume plotters for toxic gas releases
Position indicators or pointers on valves
Positive isolation of utility
Post welding stress relief
Pre-shutdown critical review of system isolation
Pre-screening of interlock systems
Pressure control on venting with low temperature override
Pressure differential alarm/trip
Pressure or temperature control on venting
Pressure relief device
Pressure relief valve
Pressure relief valve on compressor discharge
Pressure relief valve on pump discharge (upstream of block valve)

TABLE 15-2: Typical Engineering Controls
Pressure relief valve on suction side of pump (to protect against overpressure from parallel pump)
Pressure relief valve on tower or column
Pre-start-up critical review of instrumentation
Pre-start-up review of critical utilities (including shutdown)
Prevention or pre-separation of contaminants in feed
Procedures for checking catalyst activity and process inventory
Procedures for controlling contaminants in feed
Proposed unit/equipment checked out for adequacy
Protection against vehicle impacts
PSV adequately sized to handle maximum flow
Pulsation dampener in suction line
Pulsation dampeners on suction and discharge
Purge supply of inert coolant
Purging below flammable limits
Purging medium and connectors and procedures
Purging medium and procedures and adequate connections
Quench stream injection
Ratio control/alarm/trip of feed streams
Reactor vessel grounding to prevent static electrical buildup
Reboiler trips on high column pressure
Reduced feed to column
Reduced vapor flow
Redundant instrumentation
Redundant safeguarding to minimize chances of incidents
Redundant temperature alarm/trip system
Relief valve on column
Remote shutdown of pump
Remotely activated snuffing steam
Remotely operable isolation valve on pump suction

TABLE 15-2: Typical Engineering Controls
Remotely operable isolation valve on suction and/or discharge
Remotely operable isolation valves
Removable spool pieces
Reverse rotary lock on impeller of pump
Rod drop monitor/alarm and trip
Rupture disc
Separate alarm/interlock systems
Separation distance limits/reduces hazard
Siphon breaks
Skirts and/or vessel supports fireproofed
Sliding supports/flexible connections where necessary
Sloping ground to prevent accumulation of flammables beneath vessel
Slow closure time on automatic valves to prevent liquid hammer
Snuffing steam (remotely activated)
Space between rupture disc and pressure relief valve vented
Spare cooling medium pump
Spare fire water pump
Spare pump on standby
Spare strainer or filter
Special construction procedures in operating facilities
Special purpose tools inventory maintained/available
Spectacle blinds
Staged depressurizing
Standby agitator
Standby agitator or standby recirculation pump
Standby cooling pump with separate circuit
Standby diesel generator
Standby reactor agitator
Standby recirculation pump
Steam accumulator PSV

TABLE 15-2: Typical Engineering Controls
Stress relief of material under duress
Structural components have adequate support capability
Structure assessed for higher liquid densities in vessels, etc.
Substitution by less-hazardous materials
Suction line sloped back to suction drum (compressor circuits)
Suction line traced and insulated
Supporting structures fireproofed
System designed for full vacuum
Tank cleaning program
Tank dike
Tank insulation
Tank lining
Tank overflow (adequate size and to safe location)
Tank vent (adequately sized)
Temperature control of material entering tank
Temperature control of reaction
Temperature control of reactor bed
Temperature control on equipment
Temperature control on material entering tank
Temperature control on system
Temperature control with hot bypass around cooler
Temperature indication/alarm
Temporary strainer to remove fabrication debris
Thermal pressure relief valve
Thrust bearing monitor/alarm/trip
Tight shut-off isolation
Tight shutoff isolation valve activated by no/reverse flow
Tight shut-off isolation valves
Tracing and insulation
Two-way communication systems between control center and field

TABLE 15-2: Typical Engineering Controls
Uninterrupted power supply (UPS) available
Unique connectors for flexible lines
Unique connectors to prevent hazardous cross connections
Upstream knock-out drum
Upstream restriction orifice
Use of balanced bellows relief valve(s) that are able to use higher backpressure
Use of high-integrity non-return valve
Use of high-quality welding/hot tap equipment
Use of lower pressures, temperatures
Use of non-flammable synthetic lubricants
Use of separate control consoles
Use of tight shut-off valves
Use of training simulator
Use of upgraded fittings at points of high stress and high turbulence
Use of upgraded lighting
Vacuum breaker
Vacuum breaker with inert gas supply
Valve position indicators
Valve travel position indicator check
Valve-by-valve start-up, operating, shutdown review
Valves with directional pointers
Vapor space between rupture disc and relief valve vented
Vapors below flammable limits with adequate spacing between ignition source and release zone
Vehicle crash barriers, alternative routing
Velocities maintained sufficiently high to prevent solids deposition or fouling
Vent between rupture disc and pressure relief valve
Vent leaked gases to flare
Vented pressure control
Vents and drains

TABLE 15-2: Typical Engineering Controls
Vents and drains on column
Vents and drains on reactor
Vents and drains on vessel
Vents and drains plugged/blanked
Vessel grounding to prevent static electrical build-up
Vessel thermally insulated
Vibration analysis performed
Vibration monitor/alarm/trip
Vibration monitoring and high-vibration alarm/trip
Visually accessible level/sight glasses
Warehouse spares for critical equipment
Water monitors and deluge system
Weak roof-to-shell attachment on API 650 tank (overpressure relief)

Administrative and Engineering Controls as Safeguards

Both Administrative and Engineering Controls play a major role in facility Safeguarding. A number of Safeguards may be needed to prevent a specific Cause or mitigate specific Consequences. Each Safeguard may be effective only to some limited degree, and unless there are a number of Safeguards providing adequate back-up, Recommendation(s) for additional Safeguards may be needed.

Not all proposed Safeguards may be valid. Some examples of invalid Safeguarding would include:

- "Operator Awareness;"
- "No problems experienced to date;"
- A local Pressure or Temperature Indicator that is in an inaccessible or hard-to-reach location;
- A vessel Gauge Glass that cannot be read because the contents foul the glass, making it opaque;
- A High Flow Alarm as a part of a Flow Control Loop, which may itself be responsible for the failure;
- Any Safeguard needing a very rapid operator intervention when there is not sufficient time for the operator to respond and react.

Consequences of Failures of Administrative and Engineering Controls

As indicated at the beginning of this chapter, Consequences must be based on the absence of Safeguarding.

Consequences should typically address the potential for the worst possible logical outcome. These may address the potential for loss of containment resulting in the potential for fire, explosion and/or toxic release hazards.

An example of a failure of an Administrative Control would be to have less than the required number of operators. The Consequences of this failure could be running the plant under unsafe conditions, with increased potential for incidents.

If an Administrative Control calls for adequate plant security and this is not practiced, unauthorized personnel may enter the plant, causing damage through vandalism or sabotage. This failure may result in major plant damage and concomitant hazards.

If an Administrative Control sets out a procedure for controlled vessel entry and this is omitted, lives may be lost due to asphyxiation.

The Consequences of failure of a control could also result in additional problems. For example:

- A gauge glass on a pressure vessel containing flammables could fracture, causing a major release and fire;
- A (water) deluge system to mitigate fire could also cause electrical hazards and failure of control systems as well as personnel hazards (electrocution);
- A relief valve, once lifted, could continue to chatter and may not re-seat;
- A check valve designed to prevent reverse flow, of the incorrect type, could slam shut, causing liquid hammer and potential line rupture;

- Fire extinguishers containing carbon dioxide may explode if they are not used and, when full, are subjected to heat or flame;
- Pressure safety valves in a dirty service may plug up and fail to operate when needed (unless backed up by an upstream rupture disk with a small vent between the PSV and the disk);
- Heat tracing, for winterization, may also overheat the line it is designed to protect, resulting in line rupture;
- Valves in liquid lines installed for isolation purposes, but could result in line rupture if the trapped liquid overheats (e.g., lines exposed to sun or other heat sources) and overpressures line;
- A rupture disk may spontaneously fail, although not over-pressurized, causing a potential hazard with loss of containment;
- Flame arrestors, designed to stop a flame front from propagating, may plug a line due to filtration of particulates, causing upstream over-pressuring;
- Flame scanners in furnaces can prompt spurious shutdowns. The re-starting of such furnaces is often considered to be a hazardous process.

In summary, an existing safeguard or a safeguard that is introduced to deal with a specific problem, or potential hazard, may introduce new or unforeseen hazards. The corollary to this is that when a PHA, such as HAZOP is completed and recommendations are made for plant modifications, these modifications themselves need reviewing, or mini PHAs be done, to ensure new hazards are not introduced. The basic concept here is to make sure that the "cure" is not worse than the "disease"!

SUGGESTED READING (Note: URLs current at date of publication)
"Guidelines for Hazard Evaluation Procedures" by AIChE, CCPS, 2nd edition, 1992 plus "Guidelines for Hazard Evaluation Procedures" by AIChE, CCPS, 1st edition, 1985 www.aiche.org/pubcat/seadtl.asp?Act=C&Category=Sect4&Min=20
"What is your corporate perspective on loss prevention?" by R.Scholing and P.Rieff, Hydrocarbon Processing, October 1997, pages 69 to 74 www.hydrocarbonprocessing.com/contents/publications/hp/
"Guidelines for Engineering Design for Process Safety" by AIChE, CCPS, 1993 www.aiche.org/pubcat/seadtl.asp?Act=C&Category=Sect4&Min=20
"Guidelines for Hazard Evaluation Procedures" by AIChE, CCPS, 2nd edition, 1992, Appendix B www.aiche.org/pubcat/seadtl.asp?Act=C&Category=Sect4&Min=20

CHAPTER 16

Human Factors

Introduction

All problems can hypothetically be attributed to human error in some way, shape or form, with the exception of natural disasters, since humans are involved in most activities. The saying "to err is human" may be correct, and the complete elimination of human error is practically impossible, no matter how diligent the attempts may be.

The analyses of near misses and incidents invariably result in some pattern of human error(s) emerging, frequently of great complexity. Human error analysis is very valuable in understanding incidents and near misses, but the ability to predict cases involving human error is virtually impossible.

While machines and automated devices, materials and so forth may exhibit some degree of predictable behavior, human beings are subject to free will and can make mistakes, slip up, err deliberately or fail to diagnose problems correctly. Such behavior can be minimized but rarely, if ever, totally eliminated.

Human Factors in Relation to PHAs

The goal, therefore, of considering Human Factors as part of PHAs is to minimize the potential for failure due to human error by reviewing and possibly amending:

- Equipment and Design Problems;
- Instrumentation and Control Problems;
- Operational Problems;
- Organizational Problems;
- Communications Problems;
- Environmental Problems.

The following tables (Tables 16-1 to 16-6) contain lists of typical concerns and proposed safeguards/recommendations:

Table 16-1: Equipment and Design Problems

Concern	Proposed Safeguard or Recommendation
Design basis not reliable	Review plant design for weak links through trouble-shooting exercise.
Design data not reliable	Check design for consistency.
Design not checked for full range of feedstocks	Check design using process simulator to see if it can handle full range of feedstocks.
Insufficient design margin	Carry out de-bottlenecking study.
Poor access to valves & equipment	Use extension spindles or chain wheels for inaccessible valves. Provide additional structures, where necessary.
Poor control philosophy	Review and amend controls, if necessary.
Poor layout	Provide better access, where possible, and additional active safeguards to reduce potential hazards, e.g. more and better flammable gas detectors/alarms/trips.
Poor plant documentation, e.g. P&IDs out-of-date	Update & improve documentation to reflect modifications, MOCs & revamps.
Poorly displayed controls & instrumentation	Use computerized display systems with alarm prioritization etc.

Table 16-2: Instrumentation and Control Problems	
Concern	**Proposed Safeguard or Recommendation**
Calibration too high/low	Provide regular calibration checks on instruments. Ensure scales are adequate for service.
Disabled instruments not notified	Maintain and update listings of both valid and non-functional instruments.
Field/control room discrepancy	Use two-way communication systems between control center & field.
Incorrect choice of instrument	Use correct choice of property as basis for instrument monitoring.
Lack of setpoint data	Ensure operating manual provides data on setpoint.
Poor calibration	Calibrate instrument against reliable benchmark.
Poor or misleading instrumentation	Review & update mimic representations and identify parameters (flow, pressure, etc.) not adequately monitored.

Table 16-3: Operational Problems	
Concern	**Proposed Safeguard or Recommendation**
Equipment hard to access	Ensure there are adequate stairways & platforms. Install extension spindles & hand wheels on valves.
Heavy production demands	Ensure that plant safety policy overrides production demands.
Opportunities for error	Ensure there are adequate checklists & automation.
Mental tasks are too many	Use adequate number of operators for specific task.
Overly sensitive controls	Limit stops on critical control valves.
Stereotype violations, e.g. left handed	Create procedures to limit stereotype violations.
Too little time to act	Review safeguards and consider introducing automated safeguards to free-up operators.

Table 16-3: Operational Problems	
Concern	**Proposed Safeguard or Recommendation**
Too many interlocks	Review plant interlocks for operability.
Action is too fast/slow	Ensure there is an emergency alarm and automated shutdown.
Action out of sequence	Ensure there is adequate instrumentation for monitoring & feedback purposes. Create operational checklists.
Additional batch added	Perform measurement with interlock to prevent addition of extra batch.
Batch not added	Use batching recorder, weight measurement/recording.
Failure to meet schedule	Use time & motion study to establish schedule requirements.
Fatigue/boredom	Ensure there are adequate breaks, task lists. Use operational interlocks to minimize error.
Incorrect performance of a task	Train operators using plant simulator.
Insufficient knowledge	Provide operator training/certification. Ensure that operating manuals are readily available.
Operator experiments with plant	Ensure that company policy prevents experimentation with production plant.
Operator sick and/or has accident	Use buddy system in hazardous areas.
Many simultaneous alarms	Use time out; make safety critical alarms register first.

Table 16-4: Organizational Problems	
Concern	**Proposed Safeguard or Recommendation**
Conflicting priorities	Update company organizational procedures.
Inadequate instructions	Create well-documented operating instructions.
Poor labeling	Use color-coded labeling system for process, utilities, etc.
Lack of emergency response	Perform emergency response planning.
Lack of guidance/training	Establish company training courses.
Out-of-date procedures	Update procedures regularly.
Lack of preparedness	Use periodic simulation of emergency situations.
Failure to correctly diagnose problems	Ensure there are close ties between operations, maintenance and engineering to provide trouble-shooting analysis.

Table 16-5: Communications Problems	
Concern	**Proposed Safeguard or Recommendation**
Confusion in an emergency situation	Train personnel on plant simulator.
Failure to report problems	Ensure there is frequent supervisory communication.
No communication with operators in field	Provide two-way communication systems.
Shift change problems	Provide well-documented plant logs.
Personal problems reducing performance	Establish human resources group with counseling capability.

Table 16-6: Environmental Problems	
Concern	**Proposed Safeguard or Recommendation**
Excessive noise distracts operators	Provide local hearing protection in the field. Provide sound proofing for control centers, offices, etc.
Excessively hot working environment	Provide sufficient ventilation with fans, etc. in control centers, and provide air conditioning in offices.
Excessively cold working environment	Provide field operators with sufficient warm clothing, gloves, etc. that do not limit mobility. Provide adequate heating in control centers and offices, etc.
Lack of visibility due to fog, mist	Provide TV monitoring for critical equipment, with zoom, night vision capability, etc. back to control center.
Poor housekeeping	Enforce house keeping procedures.
Poor lighting	Review lighting in all areas, including provision for emergency back-up lighting.

SUGGESTED READING (Note: URLs current at date of publication)
"An Engineer's View of Human Error" by T.Kletz, published by Taylor & Francis, 2001 http://harsnet.iqs.url.es/library.htm#books
MIL STD 1472D : Human Engineering Design Criteria, 1989 http://store.mil-standards.com/eproducts/doclist/MIL%20CD%20Power%20User.pdf
"Integrating a Human Factors Approach into Process Safety Management Systems" by D.A.Moore. CCPS International Conference and Workshop MAKING PROCESS SAFETY PAY: THE BUSINESS CASE, 2001, pages 403 to 414 www.aiche.org/pubcat/seadtl.asp?Act=C&Category=Sect4&Min=50
"Don't ignore the Human Side of Plant Safety", by A.Shanley, Chemical Engineering, April 1997, pages 71 to 72 www.che.com/
"Eliminating Potential Process Hazards" by T.Kletz, Chemical Engineering, April1, 1985, pages 67, 68 www.che.com/
"Guidelines for Preventing Human Error in Process Safety" by AIChE, CCPS, 1994 www.aiche.org/pubcat/seadtl.asp?Act=C&Category=Sect4&Min=30
"Designing to Avoid Human Error Consequences", by S.L.N. Chen-Wing and E.C.Davey (Website) http://www.dcs.gla.ac.uk/~johnson/papers/seattle_hessd/sara-p.pdf
"Be a safety champion" by M.S.Terrell, Hydrocarbon Processing, January 2001, pages 95 to 100 www.hydrocarbonprocessing.com/contents/publications/hp/
"Computer Control and Human Error", by T.Kletz, IChemE, 1995 http://harsnet.iqs.url.es/library.htm#books
"Process Safety Analysis, An Introduction" by Bob Skelton, IChemE, 1997 www.icheme.org/framesets/aboutusframeset.htm

CHAPTER 17

Loss of Containment

Loss of Containment refers to the release or escape of material, usually gas or liquid, contained inside plant equipment or piping such that it can enter the immediate environment of the plant and potentially migrate to the surrounding community outside the plant boundaries.

Loss of Containment can vary from small quantities to very large releases.

Examples of small releases would include:

- Fugitive emissions from valve seals/packing;
- Minor emissions from pump seals/packing;
- Minor emissions from compressor seals/packing;
- Minor leaks from screwed fittings, couplings;
- Cracks in welds, equipment;
- Leaking (unblanked) drain, vent valves;
- Flange leaks.

Examples of large releases would include:

- Vessel rupture;
- Heat exchanger tube or shell rupture;
- Reactor disintegration (e.g. due to runaway exotherm);
- Line rupture;
- Vessel BLEVE in a fire situation.

The rate of the release is highly dependent upon:

- The hole or aperture size;
- Whether the hole or aperture is in the vapor or liquid space;
- The upstream pressure of the contained material;
- The normal boiling point and vapor pressure of the contained material (if liquid);
- Material density.

Calculations can be performed that will provide an estimate of the release rate. Usually, as the internal pressure declines there will be a concomitant reduction in the release rate.

Examples of Loss of Containment

Losses from non-pressurized storage tanks and vessels

Losses from a storage tank or vessel where there is a hole in the liquid space will release material dependent upon the location of the hole, the density of the liquid and the hole size. If the tank is contained within a diked area, the dike should contain the full volume of the released material. However, if the dike wall is located too close to the tank or vessel, the potential exists for the liquid arc to fall outside the wall of the dike. This situation needs to be addressed prior to the construction of the dike.

It is also usual where there are multiple tanks within a diked area for the dike to be able to hold the contents of only one tank, namely that with the greatest volumetric capacity.

Losses from pressurized vessels, e.g. anhydrous ammonia vessels, vessels containing liquid propane

The loss of containment for a hole in an anhydrous ammonia vessel is highly dependent upon whether the hole is in the liquid space or the vapor space.

If the hole is in the vapor space, the liquid in the vessel will flash off vapor, i.e. boil, and auto-refrigerate the remaining liquid until the normal atmospheric boiling point of the liquid of about -25°F is reached. Once depressurized, further boil off will be controlled by the external heat influx and release of approximately 1 lb. of vapor for every 587 BTUs of heat influxed. The vapor flashing off, which has a molecular weight of 17 and which is possibly colder than the surroundings, will likely disperse as a Gaussian plume.

If the hole is in the liquid space, there will be a flashing, two-phase flow of anhydrous ammonia, likely as a cold (-25°F) aerosol mist. The mist will behave as a dense gas, with some rain-out of liquid droplets followed by re-evaporation and eventual heat-up so that the ammonia will eventually rise because of its lower-than-air molecular weight. The ammonia is highly hazardous to human health, particularly due to its great affinity for

water. Ammonia vapor, given the right combination of air and ammonia, is also a potential explosion hazard.

Vessels containing liquid propane will behave in a similar fashion, but since the boiling point of propane is around -43°F and the molecular weight is 44, the vapor will hug the ground and not rise because it is denser than air. The propane is very flammable and once it reaches an ignition source, it can catch fire with flash-back to the tank. A major enveloping fire resulting in a fireball, jet fire or even BLEVE is possible, given the right circumstances.

Losses from reactor vessels, e.g. hydrocrackers

Reactor vessels such as hydrocrackers, whether they are the trickle bed or the ebulating bed variety, can experience temperature excursions and runaway reactions that lead to reactor overheating and excessive temperatures at the reactor walls. This can lead to loss of containment and major explosions with accompanying fires and fireballs. Usually, reactors have emergency depressurizing capabilities that provide controlled depressurization to flare, thereby avoiding this problem.

Losses from fired heaters, furnaces

Tubes in fired heaters and furnaces are particularly susceptible to failure, especially if overheated due to flame impingement from burners. If the process system on either side of the fired heaters is under high pressure and contains flammable material, e.g. hydrogen or hydrocarbons, there can be a massive depressurization of process side contents into the firebox leading to extensive damage and extensive fire surrounding the fired heater.

Hydrogen fires

Loss of containment in high temperature systems can pose additional hazards, since hydrogen burns with an invisible flame and poses a major personnel jet fire hazard in the vicinity of leaking flanged joints. In addition, flange joints, especially in pipe racks handling flammables, should be staggered to prevent torching of flanged connections that may be in parallel.

Loss of Containment Calculations

Calculating release rates from vessels is generally complex since, for liquids, the variation in liquid head affects the driving force, for compressed gases and pressurized liquefied gases, pressure and temperature variations affect the driving force and thus the release rate. Furthermore, releases from piping systems can add to the complexity and, if the liquid is flashing, two phase flows must be factored into the calculations.

Calculations to establish release rates of material, with respect to time, are best performed using software which can handle transient conditions and uses small time increments.

Some basic equations are given to assist the reader.

Releases of Non-Boiling Liquids and Releases of Gases Held as Liquids by Refrigeration

For non-boiling, or non-flashing, liquid releases — that is, liquids held at temperatures at or below their respective normal boiling points — following vessel failures, the following discharge situations appertain:

i. Discharge from a small hole or crack on vessel or process equipment
ii. Discharge from attached open piping or hose failures

i. Discharge from a small hole or crack on vessel or process equipment

When liquid escapes from a vessel, the release rate may be computed from the following equation, the volumetric flow rate being given by

$$Q = C_d A_o \sqrt{2g(H_L - H_o) + \frac{2g_c(P_o - P_a)}{\rho_L}}$$

It should be noted that as the head in the vessel drops, so also will the flowrate drop.

ii. Discharge from attached open piping or hose failures

When using the above equation to calculate the release rate from a pipe attached to a vessel a modified overall discharge coefficient (C_d) should be used. The modified C_d is based on the sum of the frictional losses in the attached piping. The overall discharge coefficient is given by

$$C_d = \frac{1}{\sqrt{1 + K_{sum}}}$$ where $K_{sum} = \sum K_i$ and i = individual pipe section and $K_i = \frac{4 f_{Fi} L_i}{D_i}$

Note: Equivalent pipe lengths are calculated based on actual pipe length, fitting losses and entrance/exit effects for the pipe section.

Note: Fanning friction factor (f_F) = Moody friction factor (f_M)/4

Releases of Compressed Gases

The driving force for a gas-type release is the differential pressure between vessel pressure and exit pressure (assumed equal to atmospheric pressure). As the vessel pressure drops the internal temperature also drops . Two scenarios are considered:

i. Discharge from a small hole or crack on vessel or process equipment

ii Discharge from attached open piping or hose failures.

Each scenario requires a different correlation method. Vessel temperature and exit temperature drop are calculated, assuming adiabatic expansion, at each time increment level based on inlet temperature and isentropic expansion factor (calculated by the heat capacity ratio where $C_p/C_v = k$).

$$T_2 = T_1 \cdot \left(\frac{P_2}{P_1} \right)^{\frac{k-1}{k}}$$

i. Discharge from a small hole or crack on vessel or process equipment (adiabatic cooling)

Scenarios for compressed gas storage usually involve the release of gas stored under high pressure through a small hole. A small hole is one where the ratio $Area_{orifice}/Area_{total\ vessel\ surface} < 0.01$. There are two potential conditions: **choked** or **non-choked**. The isentropic expansion factor (k-value) is equal to the ratio of specific heats (C_p/C_v) of the gas. The k-value is calculated using the Meyer equation, and is given by

$$k = \frac{C_p}{C_p - \frac{R}{M}}$$

The critical pressure ratio is defined by

$$\left(\frac{P_2}{P_1} \right)_{crit} = \left(\frac{2}{k+1} \right)^{\frac{k}{k-1}}$$

If the pressure ratio of P_2/P_1 exceeds the critical value the mass flow rate for an instantaneous discharge under non-choked flow conditions is given by

$$G = C_d \rho_2 A_o \sqrt{\frac{2P_1 g_c}{\rho_1} \frac{k}{k-1} \left[1 - \left(\frac{P_2}{P_1} \right)^{\frac{(k-1)}{k}} \right]}$$

If the pressure ratio of P_2/P_1 is below the critical value the exiting mass flow rate is **choked**, or sonic, and is limited to a critical maximum value. The mass flow rate for an instantaneous discharge under choked flow conditions is given by

$$G = C_d A_o \sqrt{P_1 g_c \rho_1 k \left(\frac{2}{k+1} \right)^{\frac{(k+1)}{(k-1)}}}$$

Note: For choked flow conditions, the flow rate is independent of outlet pressure.

ii. Discharge from attached open piping or hose failures (adiabatic cooling)

Consideration of frictional resistance to flow for compressible fluids in pipes often requires the use of graphical techniques to calculate discharge rates.

Mass flow rate for adiabatic gas flow through a pipe section with a constant diameter is given by

$$G = \left[\left(\frac{k}{k+1} \right) \left(\frac{\pi D^2}{4} \right)^2 \cdot \frac{M}{z_1 T_1 R} \cdot P_1^2 g_c^2 \frac{\left[1 - \left(\frac{P_2}{P_1} \right)^{\frac{(k+1)}{k}} \right]}{\left[\frac{f_F L}{2gD} - \frac{1}{gk} ln \left(\frac{P_2}{P_1} \right) \right]} \right]^{\frac{1}{2}}$$

17

Note: Equivalent pipe lengths are calculated based on actual pipe length, fitting losses and entrance/exit effects for the pipe section.

Flow is sonic (maximum critical flow) when P_2 is equal to the critical pressure given by

$$P_{crit} = \frac{4}{\pi}\frac{zG_{crit}}{D^2}\sqrt{\frac{R}{g_c}\cdot\frac{T_1}{kM}}$$

Combining these two equations produces the following equation to calculate the critical outlet pressure for a specific pipe length, upstream pressure and expansion factor:

$$\frac{f_F L}{D} = \frac{2z}{(k+1)}\cdot P_1^{\frac{(k-1)}{k}}\cdot\frac{\left[1-\left(\frac{P_2}{P_1}\right)_{crit}^{\frac{(k+1)}{k}}\right]}{\left(\frac{P_{2_c}^2}{P_1^{\frac{(k+1)}{k}}}\right)} + ln\left(\frac{P_2}{P_1}\right)_{crit}^2$$

Pressurized Liquefied Gas Releases

Pressurized gases stored as liquids are usually in liquid-vapor equilibrium, so vessel pressure is equal to the vapor pressure of the liquid at the vessel temperature. Two-phase flow may occur when a pressurized liquefied gas flows through a pipe and the local pressure in the pipe becomes lower than the saturation pressure of the flowing liquid, due to a decrease of pressure along the pipe as a result of friction.

Typically there may one of the following discharge situations:

i. Discharge from a small hole or crack on vessel or process equipment from the liquid space

ii. Discharge from attached open piping or hose failures from the liquid space (two-phase flashing flow)

iii. Discharge from a small hole or crack on vessel or process equipment from the vapor space (evaporative cooling)

iv. Discharge from attached open piping or hose failures from the vapor space (evaporative cooling)

For discharge cases **iii** and **iv**, evaporative cooling occurs when a liquefied gas is released from a pressurized tank. The vaporizing gas absorbs energy in the form of auto-refrigeration, or auto-cooling, of the vessel contents and partially from the external environment (the latter usually negligible or comparatively low). If we assume that the entire energy absorbed by the exiting gas is removed from the liquid remaining in the vessel we can use the following correlation to calculate the temperature drop of the vessel contents:

$$W \Delta H_v (t - t_o) = (m_o - W(t - t_o)) C_p (T_2 - T_1)$$

The outlet (exit) temperature drop, however, is calculated in the same way as for a compressed gas release, which assumes adiabatic expansion, based on the instantaneous inlet temperature and isentropic expansion factor (as previously described).

i. Discharge from a small hole or crack on vessel or process equipment from liquid space

The standard liquid flow equation to analyze the release of saturated liquids from a hole in the side of a vessel is used, namely:

$$Q = C_d A_o \sqrt{2g(H_L - H_o) + \frac{2g_c(P_o - P_a)}{\rho_L}}$$

As before, it should be noted that as the head in the vessel drops, so also will the flowrate drop.

ii. Discharge from attached open piping or hose failures from the liquid space (two-phase flashing flow)

When a superheated liquid is released through a pipe, the discharged material is usually a two-phase mixture of vapor and liquid. Several methods exist for calculating the critical discharge rate of a two-phase flow from a pipe. For maximum flow modeling, the Homogeneous Equilibrium Model (HEM) based on work by Fauske (1985) may be used. (The homogeneous flow assumption has proven adequate in most engineering design applications).

For two-phase flashing flow systems, this flow regime combined with the assumption of thermodynamic equilibrium between the two phases provides the best prediction of low-quality choked flow data of various fluids (Moody, 1975; Fauske, 1985). For risk analysis and the design of emergency relief systems, this flow model would give conservative values.

The model is based on theoretical and experimental studies conducted on the critical discharge of steam flows from pipes with different reservoir pressures and L/D ratios (Fauske and Associates, 1965). Conclusions from these experiments state that for L/D ratios ≈ 12, thermodynamic equilibrium could be assumed.

Subsequent work by Fletcher (1983), however, suggests that maybe only the absolute pipe length (L), and not the ratio of L/D, determines whether or not fully established flashing flow in thermal equilibrium can occur. According to Fletcher, three inches (75 mm) appears to be the critical pipe length above which thermodynamic equilibrium flow occurs.

Thus, the HEM correlation for liquefied gas releases (from liquid space) can be applied for pipes longer than three inches. For two-phase pipe releases, the critical pressure at the pipe exit, based on the single-phase gas formula may be computed based upon the critical pressure ratio:

$$P_{crit} = P_1\left(\frac{2}{k+1}\right)^{\frac{k}{k-1}}$$

The temperature corresponding to this critical outlet pressure (T_c) based on vapor pressure data may be computed.

The most important factor in two-phase flow calculations is the vapor in liquid quality, which determines to a large extent the mass flow rate and the friction in the pipe. The largest possible discharge rate happens for a pure liquid phase flow. For a two-phase discharge, the mass flow rate may be substantially smaller due to the increased specific volume of the fluid. Assuming thermodynamic equilibrium, the vapor mass fraction that would flash off from the liquid is given by

$$m_v = 1 - exp\left(-\frac{C_p}{\Delta H_v}\left(T_1 - T_{crit}\right)\right)$$

Assuming homogeneous mixing and no slip between the phases, the mixture density is calculated at the critical choke conditions:

$$\rho'_c = \left(\frac{m_v}{\rho'_g} + \frac{1 - m_v}{\rho'_l} \right)^{-1}$$

Finally, a slightly modified standard liquid discharge formula based on a critical outlet pressure and the calculated mixture density may be used to calculate the critical flow rate:

$$Q = \frac{C_d A_o}{\rho'_c} \sqrt{2 g_c \rho'_c \left(P_o - P_c\right)}$$

iii. Discharge from a small hole or crack on vessel or process equipment from the vapor space (evaporative cooling)

The standard compressed gas flow equations to calculate critical flow data for the release of liquefied gases from the vapor space through a hole or crack, is shown above under ***"Discharge from a small hole or crack on vessel or process equipment (adiabatic cooling)"***.

iv. Discharge from attached open piping or hose failures from the vapor space (evaporative cooling)

The equations for adiabatic gas flow through pipe to calculate the critical flow data for the release of superheated liquids from the vapor space from piping attached to a vessel may be used, as shown above under ***"Discharge from attached open piping or hose failures (adiabatic cooling)".***

Nomenclature (see note 5)

Symbol	Description	SI (metric) Units	FPS (English) Units
A_o	Orifice area	m^2	ft^2
C_d	Discharge coefficient – see note 1 below for values	Dimensionless	dimensionless
$C_{p_{vap}}$	Vapor heat capacity at constant pressure	kJ/kg.K	BTU/lb.R
$C_{v_{vap}}$	Vapor heat capacity at constant volume	kJ/kg.K	BTU/lb.R
$C_{p_{liq}}$	Liquid heat capacity at constant pressure	kJ/kg.K	BTU/lb.R
D	Internal pipe diameter	m	ft
f_F	Fanning friction factor = 1/4 × Moody friction factor, f_M	Dimensionless	dimensionless
f_M	Moody friction factor	Dimensionless	dimensionless
G	Mass flow rate	kg/s	lb/s
g	Gravitational constant	9.81 m/s^2	32.18 ft/s^2
g_c	Gravitational conversion factor	1	32.18 ft/s^2
H_o	Height of orifice (hole) above bottom of container	m	ft
H_L	Height of liquid above bottom of container	m	ft
ΔH_v	Latent heat of vaporization of liquid	kJ/kg	BTU/lb
k	Isentropic expansion factor = (C_p/C_v)	Dimensionless	dimensionless
K	Velocity head losses	Dimensionless	dimensionless
L	Equivalent pipe length – see note 2 below	m	ft
m_o	Mass of liquid in vessel	kg	lb
m_v	Vapor mass fraction or quality	Dimensionless	dimensionless
M	Molecular weight	g/mol	lb/mol

17

Symbol	Description	SI (metric) Units	FPS (English) Units
P	Pressure (absolute)	Pa	lb/ft^2 absolute
P_c	Critical pressure (absolute) – see note 3 below	Pa	lb/ft^2 absolute
P_{crit}	Pseudo critical outlet pressure (absolute) – see note 3 below	Pa	lb/ft^2 absolute
P_a	Ambient pressure (absolute)	Pa	lb/ft^2 absolute
P_o	Upstream pressure (absolute)	Pa	lb/ft^2 absolute
$P_{1,2}$	Inlet and outlet pipe section pressures (absolute), respectively	Pa	lb/ft^2 absolute
Q	Volumetric flow rate	m^3/s	ft^3/s
R	Ideal gas constant – see note 4	8.314 kJ/kmol.K	1545 ft.lb /R.lb mol and 1.986 BTU/R.lb mol
t-t_o	Incremental time difference	s	s
T	Temperature (absolute)	K	R
T_{crit}	Pseudo critical outlet temperature (absolute) – see note 3 below	K	R
$T_{1,2}$	Inlet and outlet temperatures (absolute)	K	R
W	Mass flow rate	kg/s	lb/s
z	Compressibility factor for gas	dimensionless	dimensionless
ρ_L	Liquid density	kg/m^3	lb/ft^3
ρ'_c	Mixture density at T_{crit} and P_{crit}	kg/m^3	lb/ft^3
ρ'_g	Vapor density at T_{crit} and P_{crit}	kg/m^3	lb/ft^3

Notes:

1. For sharp edged orifices and where the Reynolds number exceeds 30,000, the coefficient of discharge approximates to 0.61. For a well rounded nozzle the coefficient of discharge approaches unity. For short sections of line attached to a vessel with a length to diameter ratio not less than 3, the coefficient of discharge is approximately 0.81. For discharge rates from attached open piping or hose failures refer to page 17-7. Where uncertain, and for conservative results, a value of 1.0 may be used.

2. Equivalent pipe lengths are calculated based on actual pipe length, fitting losses, and entrance/exit effects for the pipe section.

3. The term *critical* here refers to conditions caused by flow restrictions arising solely due to choked flow.

4. For English units, where energy units are ft.lb, a value of 1545 ft.lb/R.lb mol should be used, but where energy units are BTU, a value of 1.986 BTU/R.lb mol should be used.

5. In cases where the user is uncertain of the units, dimensional analysis should always be used to confirm the values of variables, constants and final results.

SUGGESTED READING (Note: URLs current at date of publication)
"Discharge Rate Calculation Methods for Use in Plant Safety Assessments" by P.K.Ramskill, UKAEA, Safety and Reliability Directorate, report SRD R 352, February, 1986 www.ukaea.org.uk/contact/mail.htm
"Gas-flow Calculations: Don't Choke" by T.Walters, Chemical Engineering, January 2000, pages 70 to 76 www.che.com/
"Fluid flow calculations" (Website), LMNO Research Engineering, (Website) www.lmnoeng.com/Pipes/
"Fluid flow calculations" (Website), UENGINEER.com, (Website) www.virtualdesignbuild.com/DB/calc/fludcalc.htm#Pressure%20Drop
"Classification of Hazardous Locations" by A.W. Cox, F.P.Lees, M.L.Ang, published by IChemE, 1990 http://harsnet.iqs.url.es/library.htm#books
"Liquid and gas discharge rates through holes in process vessels" by J.L.Woodward and K.S.Mudan, J.Loss Prev. Process Ind., 1991, Vol 4 April, pages 161 to 165 www.elsevier.nl/locate/jlp

CHAPTER 18

Managing and Justifying Recommendations

The Dilemma for Management

When Management receives a list of recommendations as a result of Process Hazards Analyses, such as HAZOPs, the assumption that these will be automatically endorsed and enacted should not be taken for granted.

Although they wish for a safe operation, Management's primary concerns are:

- Facility production output;
- Facility production profitability;
- Satisfying the financial demands of shareholders.

While plant safety is deemed as being very important it has, nonetheless, a price. Excessive safety demands could imply high expenditures and significant loss of profits due to upgrades and plant re-vamps. If the plant production and profitability are compromised, the operation may cease to be economically viable. How much safety is too little safety? How much safety is too much safety? The question of justification of recommendations therefore becomes very important.

How to Proceed with Presenting Specific Recommendations to Management

It cannot be assumed that a specific recommendation will be adopted because it appears on what is, effectively, a “wish list.” Indeed, can the recommendation be understood in terms of what hazards it may prevent or mitigate? What are the relative merits of adopting a specific recommendation? Even safety has a price and could that price be too high?

What is clearly missing is an objective assessment of the merits for specific recommendations. This assessment requires more information, in addition to correctly defined recommendations.

Correct Descriptions of Recommendations

The importance of providing well defined, stand-alone recommendations cannot be understated. The following need to be addressed in the description:

- Exactly **what** is being recommended?
- **Why** is the recommendation being made?
- **Is there enough information** for the recommendation to be stand-alone?
- Does the recommendation indicate that it is **preventing** the cause or **mitigating** the consequences?

The Role of Risk Matrices in Indicating Viability of Recommendations

Most organizations use risk matrices as part of their HAZOPs. The method gives a semi-quantitative representation of risk.

Most risk matrices plot likelihood versus severity based on one or more of the following criteria for severity:

- Mortality and degree of harm;
- Capital losses;
- Production losses;
- Environmental impacts to fauna/flora, damage to waterways/soil.

Mortality and degree of physical harm are considered as paramount and should be reduced to negligible levels. However, all human activities carry some level of risk, whether at home or traveling by car or airplane. No activity is risk free.

Risk matrices that show likelihood versus mortality and degree of harm, capital losses, production losses and environmental impacts may present extensive useful information to managers. However, they may be difficult to interpret and endorse because the merits appear as somewhat subjective. Such matrices inform but provide little direction unless the organization can accept nominal direction, regardless of financial considerations.

Validity of Risk Matrices

A typical risk matrix is as follows:

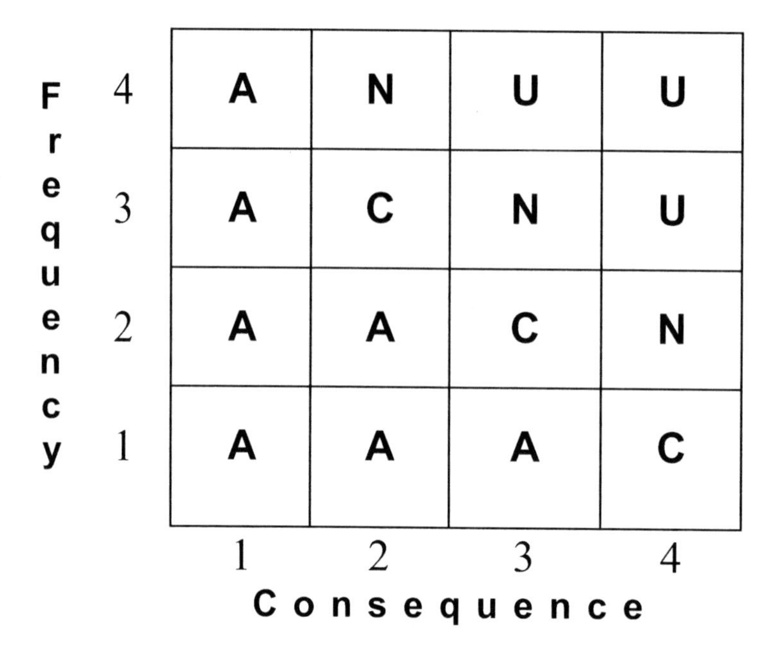

Figure 18-1: Typical Risk Matrix

A:	**Acceptable:**	No risk control measures are needed
C:	**Acceptable with Control:**	Risk control measures are in place
N:	**Not desirable:**	Risk control measures should be introduced within a specified time period
U:	**Unacceptable:**	Risk control measures should be introduced at the earliest opportunity

And where Frequency Categories are:

Category	Description
1	Not more than once in facility lifetime
2	Occurs several times in facility lifetime
3	Occurs on an annual basis
4	Occurs frequently, e.g. monthly

And where Severity Categories are:

Category	Description
1	No health impacts
2	Minor Injury
3	Major Injury
4	Deaths (one or more)

Such a risk matrix has no clear established basis and the merit is purely nominal. We can, however, create a matrix based on generally accepted and publicized mortality values, using individual risk expressed in deaths/annum. For likelihood, these can be expressed as the inverse of multiple annual intervals. For severity, these can be expressed as the fractional probability of death, based upon the following generally accepted levels of individual industrial risk:

Table 18-1: Risk Levels

Levels of Individual Risk in Deaths/Annum	Designated Risk Level
$RR \geq 10^{-2}$	Very High
$10^{-2} > RR \geq 10^{-3}$	High
$10^{-3} > RR \geq 10^{-4}$	Medium
$10^{-4} > RR \geq 10^{-5}$	Medium Low
$10^{-5} > RR \geq 10^{-6}$	Low
$RR < 10^{-6}$	Very Low

Note: RR – Risk Ranking

A 5 x 5 risk matrix with ranged values of likelihood versus fractional probability of death is shown at the top of the next page.

Likelihood					
10 times a year	10^{-4}	5×10^{-4}	10^{-3}	5×10^{-3}	10^{-2}
Once a year	10^{-5}	5×10^{-5}	10^{-4}	5×10^{-4}	10^{-3}
Once every 10 years	10^{-6}	5×10^{-6}	10^{-5}	5×10^{-5}	10^{-4}
Once every 50 years	2×10^{-7}	10^{-6}	2×10^{-6}	10^{-5}	2×10^{-5}
Once every 100 years	10^{-7}	5×10^{-7}	10^{-6}	5×10^{-6}	10^{-5}
Fractional Probability of Death	0.00001	0.00005	0.0001	0.0005	0.001

Figure 18-2: Risk Matrix of Likelihood vs Fractional Probability of Death

Such a matrix is difficult to interpret because the fractional probability of death is hard to assess. If we consider an individual risk level of 10^{-4} deaths per annum, for example, this would imply a risk of death of 0.01% per annum. For 1,000 people exposed to the risk, this would mean one death over a 10-year period (without specifying, of course, exactly when it would occur).

Use of Financial Risk Matrix

The most useful form of risk matrix for managers is a financial risk matrix, which sums up and takes into account all forms of risk under a single parameter, namely that of cost. Cost includes the following:

- Mortality and injury (assigning $ value);
- Environmental cleanup costs, penalties, etc.;
- Capital loss of plant;
- Production loss.

On the surface, mortality could be assigned as "priceless" and no amount of reparation may be considered as adequate. However, based on the concept of acceptable risk criteria, most workers are exposed to some level of risk, however low, and reparation might reflect aggregated earnings over a significant portion of a lifetime, say in the region of \$1 to \$2 million total per individual.

Environmental costs can be evaluated based on location, whether the event is a spill or gaseous release, and whether it affects flora, fauna, soil and/or waterways.

Capital costs of damage to the plant may be calculated from demolition/rebuild estimates.

Lost production during shutdown, loss of market share, etc. can be estimated.

The following is a financial risk matrix of likelihood, ranging from once in a millennium to 10 times per annum, versus severity, ranging from \$10 to \$10 million overall loss. The risk ranking is therefore as follows:

Table 18-2: Potential Loss Levels/Annum

$ Level of Loss Per Annum	Designation	Risk Index Power
$100 millions	Ultra High Risk	8
$10 millions	High Risk	7
$1 millions	Medium High Risk	6
$100,000	Medium Risk	5
$10,000	Medium Low Risk	4
$1,000	Low Risk	3
$100	Very Low Risk	2
$10	Ultra Low Risk	1
$1	Nil Risk	0

Thus, if severity is defined as total $ lost, we can use this formula:

Total $ Loss = **$ Mortality Costs + $ Capital Costs + $ Production Losses + $ Environmental Costs**

Likelihood	$1,000	$10,000	$100,000	$1MM	$10MM
10 times a year	4	5	6	7	8
Once a year	3	4	5	6	7
Once every 10 years	2	3	4	5	6
Once every 100 years	1	2	3	4	5
Once every 1000 years	0	1	2	3	4

Severity in $ Total Loss

Figure 18-3: Financial Matrix of Likelihood vs $ Total Loss

Note: (MM denotes millions)

Justification of New Risk Measures

In essence, new risk measures can be justified financially provided that they save more money than they cost to implement. We will assume for the case in point that a one-year payback is reasonable, although a two-, three-, four- or even five-year payback could also be considered, depending on the profitability of the facility.

Assigning dollar values to risk mitigation costs allows us to create a Cost Factor Index:

Table 18-3: Risk Mitigation Cost

Risk Mitigation Cost	Cost Factor (As Index)
$100 millions	8
$10 millions	7
$1 millions	6
$100,000	5
$10,000	4
$1,000	3
$100	2
$10	1

Thus, by dividing the risk per annum, expressed in total $/annum, by the mitigation cost, we can define a justification score:

$$\textbf{Justification Score in \$/\$} = \frac{\textbf{Total risk in \$/annum}}{\textbf{Mitigation cost}} = \frac{10^{\text{Risk Index Power}}}{10^{\text{Cost Factor Index}}}$$

This type of justification scoring can serve two purposes:

(a) determining whether, and to what extent, dollars spent mitigating risk are justified;

(b) ranking recommendations based on economic viability.

For example, if the addition of a software alarm as a high-level switch costs only $100 to implement, but could save $1 million per annum, it may be far more viable than another recommendation that costs $10,000 and also saves $1 million. In other words, dollars spent on higher justification give a better $/$ return on investment, where the investment is in reduced facility risk.

The benefits of risk justification scores and ranking can be summed up as follows:

(a) Providing dispassionate, objective economic evaluations to management;

(b) Providing a method of effective ranking for recommendations to be implemented in order to "get the best bank for the buck."

(c) Providing Risk justification scoring is a measure of the return on investment (ROT), based upon risk.

The following example demonstrates the use of justification scoring for ranking recommendations.

Table 18-4: Example of Recommendations Report with Justification Scoring

Recommendation	Place(s) Used	Drawings	Max RR	Cost Factor	Justification Score $/$	Order of magnitude justification based on $ risk/annum
16. To guard against maintenance hazards and positive isolation of absorber, add spectacle blind upstream of CSO valve on line 10"-K20-3412.	1.9.12.1	X-32-1274	6	3	1000	$1,000 cost spent in mitigation could potentially save $1MM per annum
12. Caustic could enter flare system resulting in personnel hazard as well as possible caustic embrittlement problems. Recommend LAH on LC-570 for caustic overfilling of vapor scrubber.	1.9.1.1	X-32-1275	5	2	1000	$100 cost spent in mitigation could potentially save $100,000 per annum
15. Review potential for overpressuring of line sections of 10"-P20-4127" with trapped liquid butane exposed to sunlight. Provide thermal relief in these cases.	1.10.1.1	X-32-1275	6	3	1000	$1,000 cost spent in mitigation could potentially save $1MM per annum
2. To reduce loss of conversion due to loss of feed cooling prior to reactors, add high temperature (soft) alarm on TI-839.	1.1.6.1	X-32-1359	5	2	1000	$100 cost spent in mitigation could potentially save $100,000 per annum
8. Incorrect alignment on setting up could lead to maloperation, possible reverse/misdirected flow and loss of performance. Provide a matrix that shows which manual valves (which need to be labeled on both P&IDs and in the field) should be opened/closed, and the sequence for Summer/Winter operations around the debutaniser overheads. Matrix should be part of operating procedures.	1.7.1.1	X-32-1528	5	3	100	$1,000 cost spent in mitigation could potentially save $100,000 per annum
14. Existing absorber may be undersized resulting in potential hazardous release of acid spray. Recommend review of sizing basis for existing absorber and whether suitable for new service.	1.9.1.2	X-32-1549	5	3	100	$1,000 cost spent in mitigation could potentially save $100,000 per annum
20. Ensure that dissolved propane in water is used in NPSH calculation for P-2575/A sizing. 6'-0" could be too low for min. height of V-379 above grade.	1.11.10.1, 2	X-32-1649	4	2	100	$100 cost spent in mitigation could potentially save $10,000 per annum

Recommendation	Place(s) Used	Drawings	Max RR	Cost Factor	Justification Score $/$	Order of magnitude justification based on $ risk/annum
[Guard against cavitation and pump damage].						
4. In the event of compressors C-457 & 327 being in parallel, there could be the potential for pressure equalization and damage to compressors if one of the compressors trips. This could also result in reverse flow. Provide emergency interlocks and instrumentation to prevent this from potential.	1.3.2.1	X-32-1274	6	4	100	$10,000 cost spent in mitigation could potentially save $1MM per annum
5. Recommend review of load sharing and effect on compressors to prevent overloading for flow from depropaniser bottoms. Determine how best to vary (e.g. variable frequency drive on new compressor).	1.4.1.1	X-32-1389	7	5	100	$100,000 cost spent in mitigation could potentially save $10MM per annum
9. Recommend extensive review at detail design stage of integrating both units' emergency shutdown systems. Also ensure that units are protected by adequate isolation through ESD valves (e.g. MOVs) so that one unit does not endanger the other.	1.7.1.2	X-32-1399	7	5	100	$100,000 cost spent in mitigation could potentially save $10MM per annum
1. To prevent loss of conversion in reactors, add high flow alarms on FC-398/396/397.	1.1.1.1	X-32-1399	5	3	100	$1,000 cost spent in mitigation could potentially save $100,000 per annum
3. With new configuration, the load on the existing compressor may be too high and result in shutdowns. Operating procedures need to allow for reduced feed rate to prevent potential overloading.	1.3.1.1	X-32-1265	5	3	100	$1,000 cost spent in mitigation could potentially save $100,000 per annum
11. Increased load on flare system due to tying two systems together needs to be re-assessed to check (a) sizing of flare header, (b) sizing of flare KO drum and (c) sizing of flare itself.	1.8.1.2; 1.8.2.2; 1.9.4.1	X-32-1266	5	3	100	$1,000 cost spent in mitigation could potentially save $100,000 per annum

Recommendation	Place(s) Used	Drawings	Max RR	Cost Factor	Justification Score $/$	Order of magnitude justification based on $ risk/annum
6. Provide depressuring valve on existing depropanizer to flare in addition to PSV to guard against BLEVE situation with fire case. [Note: all vessels handling light ends should be reviewed for their BLEVE potential and possible need to add additional depressuring valve(s).]	1.5.1.1	X-32-1346	6	4	100	$10,000 cost spent in mitigation could potentially save $1MM per annum
7. Make bottoms isolation valve on existing depropanizer a remotely actuated motor (or pneumatic) operated isolation valve so that, in the event of fire, it may be closed.	1.5.2.1	X-32-1439	6	4	100	$10,000 cost spent in mitigation could potentially save $1MM per annum
10. Consider need to provide depressuring valve on existing deisobutanizer to flare in addition to PSV to guard against BLEVE situation with fire case. [Note: all vessels handling light ends should be reviewed for their BLEVE potential and possible need to add additional depressuring valve(s). Also ensure that deisobutanizer column support skirt is fire-proofed.]	1.7.2.1	X-32-1368	6	4	100	$10,000 cost spent in mitigation could potentially save $1MM per annum
18. If LV-347 or controller fails CV open, the performance could be severely affected with subsequent damage to de-propanizer bottoms pumps. Consider LLL trip on pumps.	1.11.1.1; 1.11.4.1; 1.11.10.2	X-32-1525	4	3	10	$1,000 cost spent in mitigation could potentially save $10,000 per annum
19. For sizing PSV 5890 ensure that 2 phase flow and flashing condition are examined for fire case. This may be a candidate for DIERS technology.	1.11.6.1; 1.11.8.1	X-32-1525	4	3	10	$1,000 cost spent in mitigation could potentially save $10,000 per annum
17. Some concern over potential misdirection of butane to pentane storage causing overpressure and contamination of pentane storage. Therefore recommend spool piece in 4"P-45-4763 in addition to double block and bleed as more positive means of line isolation. [This eliminates need for additional check valve in line].	1.10.2.1	X-32-1255	4	3	10	$1,000 cost spent in mitigation could potentially save $10,000 per annum

Recommendation	Place(s) Used	Drawings	Max RR	Cost Factor	Justification Score $/$	Order of magnitude justification based on $ risk/annum
21. With calculation for P-4890/A ensure motor sized to handle run-out condition – could occur if they ever see pure propane.	1.11.10.2	X-32-1525	4	3	10	$1,000 cost spent in mitigation could potentially save $10,000 per annum
13. Undersizing of caustic make-up pump could lead to poor control and under-dosing. From an operability standpoint, recommend make-up caustic pumps be rotary design with range of around 8 to 15 usgpm	1.9.2.1	X-32-1479	4	4	1	$10,000 cost spent in mitigation could potentially save $10,000 per annum
22. Provide additional trim cooler to ensure maximum production during Summer peak temperatures.	1.1.3.1	X-32-1376	4	5	0.1	$100,000 cost spent in mitigation could potentially save $10,000 per annum

SUGGESTED READING (Note: URLs current at date of publication)
"Using Quantitative PSM Techniques to Improve Safety and Save Dollars" by M.Boult, M. Moosemiller, S. Rout. CCPS International Conference and Workshop MAKING PROCESS SAFETY PAY: THE BUSINESS CASE, 2001, pages 269 to 286 www.aiche.org/pubcat/seadtl.asp?Act=C&Category=Sect4&Min=50
"Quantified risk assessment: Its input to decision making" published by UK Health & Safety Executive, 1980, (Website) www.hsebooks.co.uk/homepage.html www.hse.gov.uk/dst/ilgra/minrpt1.htm#CONTENTS
"Approximate risk assessment prioritizes remedial decisions" by E.P.Bergman, Hydrocarbon Processing, August 1993, pages 111 to 116 www.hydrocarbonprocessing.com/contents/publications/hp/
"Guidelines for Hazard Evaluation Procedures" by AIChE, CCPS, 2nd edition, 1992, pages 208, 209 www.aiche.org/pubcat/seadtl.asp?Act=C&Category=Sect4&Min=20
"Understanding quantitative risk assessment" by R.K.Goyal, Hydrocarbon Processing, December 1994, pages 106,107 (specifically) www.hydrocarbonprocessing.com/contents/publications/hp/

CHAPTER 19

PHA Team Leadership

Objectives of PHA

Primary Objective of PHA

The main objectives are to identify mechanisms and routes (may be process, mechanical, electronic, human failures, etc.) by which hazardous events or incidents may be initiated.

Question: Why examine multiple areas?

Answer:

- If a plant cannot be operated easily, it is unsafe.
- If a plant cannot be maintained, it is unsafe.
- If a plant cannot be controlled, it is unsafe.
- If a plant is unreliable, it is unsafe.
- If a plant is hard to start up, it is unsafe.
- If a plant is hard to shut down, it is unsafe.

Hence, a plant that is unreliable, poorly maintained or poorly controlled is, by default, unsafe. Therefore, the PHA should not address side issues and avoid redesigning the plant but needs to maintain focus.

The key issues that need to be addressed in a PHA are summarized as follows:

- Process safety;
- Operability issues (includes instrumentation & control);
- Reliability issues;
- Maintenance issues;
- Environmental release issues.

Secondary Objective of PHA

The secondary objective is to identify ways in which the plant can fail, resulting in loss of production.

Opposition to PHAs

Performing PHAs may be regarded as a somewhat hostile activity because:

- Process design engineers feel their design capabilities are being questioned;
- On new designs, project managers believe that it will increase costs and delay schedules;
- On existing facilities, plant managers believe it may lead to plant modifications, plant shutdowns and loss of production.

However, once a Process Safety Management (PSM) program is in place most parties endorse it because it:

- Improves quality assurance;
- Assists in personnel training, especially operators;
- Improves on-stream plant performance;
- Improves safety record.

FEW PEOPLE WILL OPENLY OPPOSE PSM PROGRAMS, BUT MANY WILL ATTEMPT TO MINIMIZE COOPERATION OR POSTPONE THESE ACTIVITIES, WHICH ARE REGARDED AS A "NECESSARY EVIL."

BUT, attitudes are beginning to change in a number of cases…

Driving Forces Behind PSM

The major forces that drive Process Safety Management are as follows:

- Legislation;
- Insurance industry;
- Urban communities who may feel threatened;
- Enlightened individuals within organizations;
- Bodies, e.g. CMA, who impose standards on their members;
- Some pressure from unions, especially in the U.S.A.

Role of PHA Leader (Facilitator)

- Maintain an objective view of the facility without bias as to the merits, or lack of them, of the design/facility.
- Be informed with respect to:
  - Type of design/facility being reviewed;
  - Process of planning for, executing and documenting PHAs.
- Stay aware of compliance requirements, if any (e.g. OSHA 1910.119).
- Ensure ***quality*** of PHA is maintained:
  - Thoroughness;
  - Dedication to key issues;
  - Accurate documentation;
  - Full team participation.
- Educate team members about PHA methodology.

PHA Team

Team Content

Assemble a team consisting of the ***right*** people for the PHA. As most PHAs either recommend or analyze the need for changes, appropriate engineering personnel are a necessity.

Consider having:

- Process designer engineer;
- Project engineer;
- Mechanical specialist;
- Instrumentation engineer;
- Operations personnel;
- Maintenance personnel.

Suggested Numbers (Optimal)

PHAs can be performed with as few as 3 people or as many as 15 people:

- Typical HAZOP or What If/Checklist: **6 people**
- "Mini" HAZOP or What If/Checklist: **3 people**
- Failure Mode & Effect Analysis (FMEA): **3 people**
- Preliminary Hazards Analysis (PHA): **3 to 6 people**

Choice of PHA & Factors in Determining Choice

There are a number of factors in determining which method should be applied. Not all methods are universally applicable.

Example: A plant is about to be commissioned. What type of hazard analysis should be performed?

METHOD	ADVANTAGES	DISADVANTAGES
HAZOP	THOROUGH	TOO LENGTHY, NOT PRACTICAL
WHAT IF/ CHECKLIST	GOOD	TOO LENGTHY
FAULT TREE	THOROUGH	VERY LENGTHY
SPOT CHECKS	QUICK	NOT THOROUGH
CHECKLIST	QUICK, GOOD	NONE
SAFETY AUDIT	QUICK, GOOD	NONE

Factors Involved

The correct method is affected by a number of factors:

- What stage are you at?
- What type of equipment is involved?
- New or existing plant?
- Type of process?
- Proven/unproven design?
- Hazardous or not?
- What is being analyzed?
- Component intensive?

Stages Can Be Defined As

- Conceptual;
- Basic design;
- Detail design;
- Construction issue;
- As built.

Conceptual stage – At the conceptual stage, screening-type tools are most applicable, such as:

- Preliminary Hazards Analysis;
- What If analysis;
- Kepner Tregoe (if choice is required between systems);
- Screening tool (e.g. Dow/Mond Indices).

Basic design – With basic design, more structured tools are needed, including:

- What If/Checklist;
- HAZOP;
- FMEA (useful if equipment components are known).

Detail design – With detail design, use the same tools as for basic design.

Existing plant – With an existing plant, use the same tools as for basic design, as well as safety audits and checklists.

Types of Equipment/Units Include

- New equipment;
- Grass roots design;
- Existing equipment;
- Revamped (modified) equipment.

Other Factors Affecting Choice of PHA Tool

- Is process batch or continuous?
- Is process established or new?
- Are process materials highly hazardous?
- Do you need to analyze operating manual/procedures?
- Is equipment mechanically/electronically component intensive (e.g. aero engine)?

Thoroughness of Analysis

Question: How do you know if your analysis is thorough/complete?

Answer: Thoroughness depends on:

- Validity of method chosen;
- Experience of team (including insight);
- Stage at which analysis is performed;
- Type of process/equipment;
- Degree of complexity of process.

The greatest obstacles to thoroughness are:

- Unsuitable methods (e.g. HAZOP on a compressor rather than FMEA);
- Indifference, lack of enthusiasm of team;
- Lack of documentation;
- Poor documentation, lack of adequate recording tools;
- Tendency to rush analysis (especially at end of day).

Major Grass Roots Design

The ***best*** approach is to use a ***combination matrix*** of different methodologies:

- **CONCEPTUAL DESIGN** – use Preliminary Hazards Analysis;
- **BASIC DESIGN** – use Checklist;

- **DETAIL DESIGN** – use What If/Checklist;
- **CONSTRUCTION ISSUE** – use Guide Word HAZOP.

If one method overlooks something, another method is likely to recognize the shortfall.

Existing Facilities

- **Safety audit** – a good approach where ***clear/obvious*** violations can be identified.
- **HAZOP** – Guide Word HAZOP is a very good tool and widely used for existing facilities. It can be used with/without an experienced team but should have an experienced facilitator.
- **What If/Checklist** – also good and widely used for existing facilities ***provided*** that team is experienced.
- **FMEA** – widely used where equipment has many components, and most widely used in automobile manufacturing and defense industries.

You don't have to use one method exclusively for a unit.

Types of PHA for an Existing Process Unit

Example: Hydrotreater

For the Main Process:

Use Guide Word HAZOP or Knowledge Based HAZOP (if team is experienced).

For Recycle/Feed Compressors:

Use FMEA (following HAZOP, where compressors are regarded as process equipment items). In this case, detailed equipment drawings, detailed specifications, etc. would be required.

For off site, e.g., Cooling Water, Storage, Steam, etc.:

Use What If/Checklist or Checklist.

What is Best to Use and When?

Preliminary Hazards Analysis

Best used at the conceptual phase for identifying major hazards.

What If/Checklist

Can be used at most stages because it is very versatile. It is also good with mechanically intensive systems, e.g. conveyors, mechanical handling, etc.

HAZOP

Best used with a detail design or an existing plant.

FMEA

Best used on prime movers (e.g. pumps, compressors) where multiple (moving) component failures can occur.

Analysis of Operating Manuals

Hard to analyze unless manual is broken down into succinct stages:

- Can use HAZOP, e.g. No charging (of vessel), More charging, Part of charging, etc.
- Can use What If/Checklist to consider alternatives such as "What happens if vessel isn't charged at the right time?"
- Can use Checklist if the process is well known, similar to another and it's largely a repeat exercise.

Overview

- Determine how thorough your PHA needs to be, so you don't use a sledge hammer to kill a fly. For instance, Checklist may be better, on occasions, than detailed HAZOP.
- How many PHAs do you intend to run? If only one, use a more rigorous method, such as Guide Word HAZOP.
- The PHA methodology may be sufficient, but if your team cannot support it adequately, use a more friendly method, e.g. What If/Checklist in place of HAZOP.
- Use screening tools, e.g. Kepner Tregoe, where conceptual choices are needed and no "clear" route is obvious. Use Dow/Mond indices for general risk ranking.

Manage the Time Spent on PHAs

- When you start, you will likely be ***slow*** until you have built up an adequate database;
- Avoid excessive repetition;
- Consider "High Productivity" HAZOP. Be realistic about time needed: estimate, but ***do not guess.***

Length of a PHA Session

- HAZOPs should ***not*** exceed 5 to 6 hours since the team will become very exhausted and will be ineffective;
- What If/Checklist could last 6 to 7 hours;
- FMEA could last 6 to 7 hours.

Preparation Before PHA Sessions

Collect Information

The success of a good PHA session largely depends on how well you have prepared for it. Check that you have:

- PFDs, P&IDs;
- Available layout drawings;
- Appropriate equipment specifications & data sheets.

Size the Nodes

When nodes are too small, it can cause a great deal of repetition and lead to much frustration.

Start by using small nodes and expand until you feel comfortable with larger-sized nodes. You can start with single nodes, such as a Line, a Pump and a Heat Exchanger, and ***later*** you may wish to combine them as a compound node, such as Line + Pump + Heat Exchanger.

The optimum size for a node is determined by its common function (also see Chapter 7 on Choosing & Sizing of Nodes for HAZOP). For example, a feed system could be a single node.

PHA LEADERSHIP: RESPONSIBILITY

Why a Responsible Attitude is Required

- Need for "Ownership" of the PHA.
- For guidance of:
  - Team members (as a whole);
  - Individual team members.
- Provide advice to those in Management who are requesting PHA.
- Concurrently meet:
  1. Standards;
  2. Schedule;*
  3. Budgetary requirements.*

  **Take exception where these are inadequate to meet standards or quality is compromised.*
- To provide ***focal point***, i.e. leadership.
- Avoidance/elimination of "laissez faire" attitude, i.e. prevent it from being run on an "as it comes" basis.

Results of "Laissez Faire" Approach

May minimize team effort but can result in:

- Lack of coverage;
- Making your organization liable through negligence and lack of conformance;
- Poor documentation;
- Non-auditable reports;
- Loss of safety on facility if key issues are not identified;
- Greater frequency of incidents;
- Lack of adequate safeguarding;
- Loss of credibility of your organization due to failure to exercise due diligence.

Key Points for Exercising Responsibility

(1) Initial Setup of PHA

- Advise management ***how long*** PHA will take to execute, bearing in mind:
    - Preparation hours needed;
    - PHA session hours;
    - PHA reporting hours.
- Advise management ***who*** should be present for PHA sessions, e.g.:
    - Facilitator and Scribe;
    - Process Designer Engineer;
    - Project Engineer;
    - Mechanical specialist;
    - Instrumentation Engineer;
    - Operations personnel;
    - Maintenance personnel.

  Set up PHA in conjunction with a process/system/mechanical specialist who is most responsible/knowledgeable about unit.

- Emphasize that too many people, such as 10+, may make PHA too cumbersome and that too few may result in inadequate coverage.
- Advise management which type of PHA methodology will give best coverage – either Checklist, What If/Checklist, HAZOP, FMEA or Preliminary Hazards Analysis.
- ***Provide rationale*** to management for choice, i.e. ***not*** "we will use FMEA because I think that's best" but "We should use FMEA for the following reasons..."
- If you review the process and then feel that your initial estimates were incorrect, advise management of your new evaluation. Don't wait until it's too late and find that the PHA team cannot stay later to complete the analysis.

- Determine optional configuration for analysis:

  1. How many units to be "HAZOPed"?
  2. Which units are highest priorities?
  3. Should unit be split up into main process, off sites, etc.?
  4. How many nodes/subsystems?
  5. Can some nodes be compounded (e.g. pump & line & heat exchanger)?
  6. With HAZOP, what deviations should be used?
  7. With What If/Checklist, is there enough information to prepare a Checklist or do you need additional assistance?

(2) Educating Team and Explaining Purpose of PHA

- Does team know how to perform the type of PHA methodology chosen? For example, HAZOP has the longest learning curve.
- Emphasize importance of orderly approach to specifying:
    - Consequences;
    - Safeguards;
    - Recommendations.

  The tendency always exists for team members to jump to Recommendations without looking at Causes, Consequences and Safeguards.
- Make team aware that they are responsible, both as individuals and as a whole.
- Make a list of major hazards relating to the facility beforehand and draw team's attention to it.
- Counteract negative comments by individual team members at the start, such as: "Why are we here? Plant is safe and I could be doing something else now."

  After your presentation, request show of hands to indicate those who consider the PHA to be unnecessary. Anyone raising a hand should be given serious consideration for discharge from session, otherwise they may prove to be uncommitted.
- Emphasize the benefits of PHA:
    - Assists in training operators and others in plant features/operations;
    - Makes design safer (if Recommendations are incorporated);
    - Makes design more reliable and gives better start-up (if new) and better on-stream time;

- Better for plant, the personnel and the surrounding community.

(3) Conducting PHAs

- Encourage full team participation. There are many instances where PHA sessions are dominated by a few people and not everyone contributes. Encourage team members who have views to express themselves. The Team Leader should ask individuals for their opinions, especially when they have a special area of expertise.
- Do you have the right people?
- Will some people destructively interact to destroy the PHA review?
- Are all the areas (process, mechanical, electrical, instrumentation) covered by full-time/part-time participation?
- Can everyone see the right drawings, are they adequately marked up beforehand, and are extra drawings needed for reference purposes?
- Are you getting balanced participation? If one person is talking all of the time, are you losing team participation?
- Is there a tendency to address side issues? Bring team back on track, but don't ban side issues off hand because they can identify new problem areas.
- Don't be prey to a "wish list" philosophy. The PHA is not there to revamp a plant design but to give an objective assessment. Too many "wish list" items can discredit more important issues with large numbers of Recommendations.
- Make Recommendations that are "stand alone" as far as possible. They must be brief, to the point, self evident, well referenced and accountable by a specified person by a specified time.

- Choose from a group of styles for conducting PHAs:

Authoritative – Treats the PHA session like a military operation.

Advantages – Keeps to schedule. Fixed focus.

Disadvantages – Loss of ideas that could assist progress. Confrontational style.

Open Approach – Treats the PHA as though it's an open debate, with no limitations.

Advantages – Many issues discussed. Extensive coverage.

Disadvantages – Schedule not met. Poor performance as a result of loss of focus. Hard to document accurately.

Recommended Approach – Use a blend of the open and authoritative approaches, so you emphasize the advantages of each approach while downplaying the disadvantages.

(4) Recording PHAs

- Use consistent language.
- Make sure equipment, lines, drawings, etc. are all correctly identified and traceable.
- Be grammatically correct, and avoid cryptic statements. Could someone else, in say five years time, understand the PHA?
- Ensure team participation in recording PHA by using monitors, liquid crystal display, etc.
- Be accurate. Check with the originator of a comment, item or contribution to ensure that the point is adequately recorded.
- Do not accept situations that clearly expose your organization to risk. Even if your team says it's okay, take exception and explain the potential consequences and liabilities involved with taking unnecessary risks.

(5) Documentation of Proceedings

- Issue preliminary listing of report ***as soon as possible*** after the PHA session.
- Invite criticism of and comments about the preliminary report by a specific date, which is usually two days to one week later.
- The final report should include the following items:
  1. A brief description of the process of how the PHA was conducted, where and by whom.
  2. All major components of the report:
     a. Outline;
     b. Worksheets (detail);
     c. Recommendations Report;
     d. Risk Matrix used;
     e. Attendance Register;
     f. List of Team Members.
- Include drawings and computer printouts.
- Reference standards of compliance.
- Provide executive summary.

For ***compliance*** purposes, e.g., OSHA 1910.119 and PHA Compliance, do the following:

- Complete PHA, e.g. HAZOP, FMEA, etc., making documentation fully auditable.
- Address issues not covered by PHA methodology and ensure that compliance is met, e.g., incidents with potential to cause hazards, operating history, siting, etc.
- Have one or more third parties who are well versed in PHA review your documentation before submission.

(6) Non-Responsibilities of Facilitator

A Facilitator is:

- ***Not responsible*** for the specific design/plant being analyzed (more like an arm's length relationship);
- ***Not responsible*** for the follow-up of Recommendations or action items created during the PHA session ***other than*** for clarification purposes;
- ***Not responsible*** for the results of the PHA – this is a shared PHA team responsibility.

A responsible Facilitator should not undertake a specific PHA if he or she feels that the standards/methodologies are being compromised. If this is the situation, the Facilitator should explain to management the position and limitations. If management understands these concerns, the PHA can be either modified to what it should be or executed on a limited basis.

(7) Combining Facilitator and Scribe Roles

This point should remain an open issue.

Advantages of combined roles: Gives better documentation, initially, provided that the Facilitator can type sufficiently fast.

Disadvantages of combined roles: Does not allow Facilitator to exercise the leadership role so extensively, and there is less involvement with process issues that need discussion.

Although there are ***some*** incentives for combining both Facilitator and scribing roles, they should ***not*** be judged on an economic basis. A good scribe may be a lot cheaper than a Facilitator.

You may be losing economic performance by combining roles.

IF YOU DO NOT FEEL CONFIDENT IN COMBINING FACILITATOR AND SCRIBE ROLES, INSIST THAT A SCRIBE BE SEPARATELY APPOINTED TO ASSIST.

Analyze Your Performance

Once you have completed the PHA, ask yourself:

- Did we have the right team players and number of people?
- Did we focus our attention on the important issues?
- Did we manage the time correctly?
- Were we adequately prepared?
- Was the type of PHA method used adequate?
- How does the final report read and what did we learn from the PHA?

STEPS FOR PERFORMING PHA

Table 19-1: PHA Steps

STEP	ACTIVITY
1	Obtain work package that includes: ▪ PFDs; ▪ P&IDs; ▪ Material and Energy Balances; ▪ Specification Sheets; ▪ Plot Plans, etc.
2	Determine optimum choice of PHA: ▪ HAZOP; ▪ What If/Checklist; ▪ Checklist; ▪ Preliminary Hazards Analysis; ▪ FMEA; ▪ Safety Audit.
3	Select team members in addition to Facilitator/Scribe, typically: ▪ Process; ▪ Mechanical; ▪ Instrument; ▪ Operations; ▪ Maintenance; ▪ Project; ▪ (Other).

STEP	ACTIVITY
4	Estimate time required for PHA: ▪ Preparation time; ▪ Team Sessions; ▪ Reporting and Final Documentation.
5	Organize timing of PHA so that team members are present and arrange meeting room: ▪ Will key personnel be present? ▪ How many daily sessions are needed so that team is not excessively fatigued? ▪ How long will each session last? ▪ Is meeting room location away from main plant to minimize interruption? ▪ Does meeting room have adequate space to hang up drawings? ▪ Have you arranged for adequate computer and graphic display systems as well as a printer? ▪ What about meals at midday and coffee breaks? ▪ If you run over in time can you extend the room booking?
6	Facilitator and Process Engineer get together and prepare PHA Outline: ▪ Divide plant up into nodes or systems, etc.; ▪ Assign deviations, prepare Checklists, etc.; ▪ Identify and mark up full scale P&IDs; ▪ Prepare Outline document with full lists of Deviations, Checklists, etc., as applicable.

STEP	ACTIVITY
7	Process Engineer to provide: ▪ Reduced sets of PFDs + Heat & Material Balance; ▪ Reduced sets of P&IDs.
8	Begin team sessions: ▪ Create Attendance Sheet, which is passed around (Name, Title, Company, Location); ▪ Introduce team members and their responsibilities; ▪ Facilitator explains PHA methodology; ▪ Process engineer explains plant design; ▪ Process engineer explains first node, first system as applicable.
9	Facilitator progresses PHA by encouraging participation and controlling proceedings. Limit non-related discussions and side issues. Typical questions include the following: ▪ "Have you considered such and such…?" ▪ "Isn't there is a real concern over…?" ▪ "I don't understand such and such, can you explain…?" ▪ "But what about…?" ▪ "Isn't such and such a real hazard…?" ▪ "What are the Causes…?" ▪ "Are these the full Consequences…?" ▪ "Have we identified all the valid Safeguards…?" ▪ "Can this Recommendation be understood by the Responsible individual…?" ▪ "Are there any more Recommendations needed here…?"

STEP	ACTIVITY
10	Session needs: ▪ Limit length of sessions; ▪ Maintain focus; ▪ Avoid redesigning during sessions; ▪ Avoid "end of day" type rush, where concerns can be overlooked; ▪ Identify "orphan"/interface areas that can be overlooked; ▪ Document information accurately and consistently; ▪ Ensure that documentation is self explanatory; ▪ Reference all Actions and Recommendations correctly; ▪ Prioritize Actions/Recommendations as well as identification of Responsible person(s) for enactment.
11	At end of sessions: ▪ Check over all printouts of PHA for any significant errors; ▪ Prepare and issue Preliminary Report; ▪ Distribute Preliminary Report to Attendees and Responsible persons identified.
12	Prepare final report, including: ▪ Executive Summary; ▪ Plant Description; ▪ PHA Methodology/Procedures used; ▪ Conclusions/Recommendations; ▪ Reduced drawings; ▪ Computer printouts of sessions, etc.; ▪ Copy of disk with files.

MAIN GOAL OF THE PHA: RECOMMENDATIONS & REMEDIAL ACTIONS

Specifying Consequent Remedial Actions

We identify Safeguards, in the first place, as a check as to whether the hypothetical problem area/hazard, etc., has been accounted for in mitigative terms.

Safeguards are of four types:

1. Safeguarding against the Cause or Failure occurring in the first place (can be regarded as a 1st-level Safeguard);
2. Providing remedial action in the event that the Cause or Failure is not prevented (can be regarded as a 2nd-level Safeguard);
3. Mitigation of the consequences in the event that an incident occurs (can be regarded as a 3rd-level Safeguard);
4. Post-incident response (can be regarded as a 4th-level Safeguard).

Examples of Safeguards

1st-Level Safeguard

a. Tripping of a level switch that closes a control valve on low level in a vessel;

b. Tripping of an electric motor on overload;

c. Alarming a high temperature in reactor followed by emergency shutdown.

2nd-Level Safeguard

a. Pressure relief valve opening in the event of overpressure;

b. Temperature/pressure/level monitor;

c. Introduction of a quench stream to cool an overheated reactor;

d. Manual override of a control valve.

3rd-Level Safeguard

a. Fire detection/protection monitors on release of flammables;

b. Flammable gas detectors that alarm flammable gas release;

c. Increased equipment spacing (to reduce fire/explosion impacts).

4th-Level Safeguards

a. Emergency Response Plan in place;

b. Training of Employees in emergency situations;

c. Plant-wide intercom/warning systems.

How Effective are the Safeguards?

- What is likelihood of event occurring?
- What is potential severity of incident?
- How much time is there for someone to react? Don't assume people can immediately understand/react to complex situations that could occur due to a ***number*** of causes.

More Safeguards don't necessarily guarantee protection – effectiveness counts.

Preferred Approach is to have Safeguards at All 4 Levels

Example: A hydrocracker in a refinery.

1st Level: High temperature alarm/shutdown in event of runaway reaction.
2nd Level: Manual depressuring to flare system (blowdown) to reduce hydrogen content.
3rd Level: Fire monitors and deluge systems.
4th Level: Emergency Response Plan in refinery in event of a major incident.

Specifying Remedial Actions

These actions must increase/improve on existing Safeguards. With an ***existing*** plant, you can rarely introduce new passive features (e.g. increased plant spacing) and may have to increase active features (e.g. alarms, trips and shutdowns).

How to Specify a Remedial Action

- Make sure it is concise, to the point and self-explanatory.
- Reference pertinent documents (e.g. P&IDs, etc.).
- Avoid vague and indecisive wording (e.g. "Consider the possibility of studying...").
- Record your ***best opinion*** of what needs to be done – someone else may reject it ***after*** the PHA session anyway if they don't like it!
- Avoid "wish list" type suggestions. Too many recommendations/proposed actions can reduce credibility.
- Be practical and realistic.

What Needs to be Specified

- The Recommendation/Action item itself.
- ***Who*** is to be responsible for implementation.
- ***When*** it is to be implemented by (target date).
- ***Status*** of item. Is it something you definitely want to do? Should it be studied? Do you want to put it on hold? If it is to be incorporated, indicate that it is INCOMPLETE.
- How important is it in terms of Risk or Severity (or schedule) priority?
- Are there any comments you need to include?
- When it is finally resolved, RECORD RESOLUTION.

AUDITING OF PHAs

What needs to be audited?

Typical Issues:

- Correct choice of PHA methodology?
- Thoroughness?
- Are safety issues fully addressed?
- Are consequences made clear?
- Are toxic, flammable and explosive hazards identified?
- Are safeguards fully addressed?
- Do *effective* safeguards exist for the more serious issues?
- Are recommendations understandable and well referenced?
- Was any portion of the proceedings clearly rushed or inadequately covered?
- Could any of the issues be liability type problems?
- Is the documentation accurate and representative?
- Does the PHA meet the legislated requirements, e.g. those of OSHA 1910.119, if applicable?

Who Audits PHA?

Third party, non-involved, who is experienced in PHAs and who understands the subject matter.

SUGGESTED READING (Note: URLs current at date of publication)
"ARCO Chemical's Hazop Experience" by J.C.Sweeney, Process Safety Progress, Vol. 12, No.2, April 1993, pages 83 to 91 http://www.aiche.org/safetyprogress/
"Guidelines for Process Safety Documentation" by AIChE, CCPS, 1995, pages 73 to 105 www.aiche.org/pubcat/seadtl.asp?Act=C&Category=Sect4&Min=30
"Lessons from HAZOP experiences" by D.W.Jones, Hydrocarbon Processing, April, 1992, pages 77 to 80 www.hydrocarbonprocessing.com/contents/publications/hp/
"Utilization and Results of Hazard and Operability Studies in a Petroleum Refinery" by A.S.Pully, Process Safety Progress, Vol. 12, No.2, April 1993, pages 106 to 110 www.aiche.org/safetyprogress/
"Guidelines for Hazard Evaluation Procedures" by AIChE, CCPS, 2nd edition, 1992, pages 24 to 50 www.aiche.org/pubcat/seadtl.asp?Act=C&Category=Sect4&Min=20
"Managing the PHA Team" by A.M.Dowell III, Process Safety Progress, January 1994, pages 30 to 34 www.aiche.org/safetyprogress/
"HAZOP: Guide to best practice" by F.Crawley, M.Preston, B.Tyler, IChemE, 2000 www.icheme.org/framesets/aboutusframeset.htm
"Hazard and Operability Studies", by M.Lihou, (Website) www.lihoutech.com/hzp1frm.htm
"Process Hazards Analysis" by I.Sutton, published by SW/Sutton & Associates, 2002 http://www.swbooks.com/books/book_prha.shtml
"Some Features of and Activities in Hazard and Operability (Hazop) Studies", by J.R.Roach and F.P.Lees, The Chemical Engineer,October, 1981, pages 456 to 462 www.icheme.org/framesets/aboutusframeset.htm
"Management of Change – The Systematic Use of Hazards Evaluation Procedures and Audits", by N.Sankaran, Process Safety Progress, July 1993, pages 181 to 192 www.aiche.org/safetyprogress/

Chapter 20

Safety Integrity Levels (SILs)

Standards

There are three standards pertinent to the concept of safety integrity levels. They are:

- ANSI/ISA S84.01 – 1996 (herein referred to as 'S84.01'): *Application of Safety Instrumented Systems for the Process Industries*
- IEC 61508 – 2000 (herein referred to as '61508'): *Functional safety of electrical / electronic / programmable electronic safety-related systems*
- IEC 61511 – 2003 (herein referred to as '61511'): *Functional safety - Safety Instrumented Systems For The Process Industry Sector*

Addressing each of these in turn:

- IEC 61508 was developed by the International Electrotechnical Commission (IEC) and is ***performance based*** rather than ***prescriptive***. It has seven parts, as follows:
    - 61508-1: General requirements
    - 61508-2: Requirements for electrical/electronic/programmable electronic safety-related systems
    - 61508-3: Software requirements
    - 61508-4: Definitions and abbreviations
    - 61508-5: Examples of methods for the determination of safety integrity levels

- 61508-6: Guidelines on the application of IEC 61508-2 and IEC 61508-3
- 61508-7: Overview of techniques and measures

61508 was developed in parallel with the ANSI/ISA-84.01-1996 by the Instrumentation, Systems, and Automation Society (ISA), and later adopted by the American National Standards Institute (ANSI).

- IEC 61511 contains the following three Parts:
  - 61511-1: Framework, definitions, system, hardware and software requirements
  - 61511-2: Guidelines for the application of IEC 61511-1
  - 61511-3: Guidance for the determination of the required safety integrity levels

The IEC standards 61508 and 61511 require that SIL be assigned to the safety instrumented ***functions*** (SIF) of the safety instrumented ***systems*** (SIS) for processes, that have insufficient mitigation from the potential hazards. According to the IEC standards, a SIF is a "safety function with a specified SIL which is necessary to achieve functional safety and which can be either a safety instrumented protection function or a safety instrumented control function." A SIS is an "instrumented system that is used to implement one or more SIFs. It is composed of any combination of sensors, logic solvers, and final elements." SIS is devoted to responding to an emergency situation. SIS consists of instrumentation for emergency shutdown and thus brings the process to a safe state in the event of an upset. Instrumented emergency shutdown systems including flammable gas, toxic gas and fire protection systems are SIS.

Examples include;

- High high level of liquid (LPG) in a knockout drum, which initiates shutdown of emergency shutdown (ESD) inlet feed valve. This protects against liquid

carry-over from entering a compressor suction line, which could result in compressor damage/disintegration and subsequent personnel hazards.

- Another example could be closure of a vessel bottom outlet ESD valve to protect against a loss of containment situation on downstream piping/ equipment, which could also lead to loss of containment/fire hazards.

Levels of SIL

There are four levels of SIL. SIL 1 is the lowest and SIL 4 is the highest level of safety integrity. The assignment of SIL addresses the need to provide safeguards or mitigation matching the potential hazards of the processes including the failure of the instrumented systems. ***SIL is a measure of reliability of the respective SIS.***

Table 20-1. SIL Correlations with Availability and (PFD)

Safety Integrity Level	IEC 61508 / 61511	ISA/ANSI S84.01	Availability Required	Probability to Fail on Demand (PFD)	1/PFD
4	Yes	No	> 99.99 %	10^{-5} to 10^{-4}	100,000 to 10,000
3	Yes	Yes	99.90 - 99.99 %	10^{-4} to 10^{-3}	10,000 to 1,000
2	Yes	Yes	99.00 - 99.90 %	10^{-3} to 10^{-2}	1,000 to 100
1	Yes	Yes	90.00 - 99.00 %	10^{-2} to 10^{-1}	100 to 10

The terms 'SIL' and 'availability' represent the integrity of the SIS when a process demand occurs. Consider that a particular SIF is assigned a value of SIL 1, as an example. Assigning SIL 1 to a particular SIF means that the level of risk is considered to be sufficiently low and that the SIF with a 10% chance of failure (90% availability) is acceptable. The availability of 90% would mean that there would be one statistical failure of that SIF out of every 10 demands for that function. If this risk is not acceptable, the SIL may need to be raised to a level 2 or level 3. In other words it might be more prudent to have a SIL corresponding to one failure in 100, 1,000, 10,000, or more demands, if it can be justified.

Safety Life Cycle

The safety life cycle (SLC) (see Figure 20-1) can be used for any SIS design to mitigate potential hazards during design, installation, commissioning, operation, maintenance, testing and modification phases.

The general sequence of steps in a typical SIL study as per the SLC are:

- Determine whether 61511 or S84.01 is to be used.
- Identify the SIFs using previous PHA studies (PrHA, HAZOP, Hazard Analyses, etc.) for 61511, or the need for SIS if S84.01 is to be used.
- Assign target SILs to the SIFs using one of the many methods (Risk Graph, Consequence based, Risk Matrix, Layered Risk Matrix or Layer of Protection Analysis, LOPA - Note that LOPA is only recommended in 61511, but not by S84.01. See Chapter 21 "Layer of Protection Analysis" for details of the methodology), as per 61511 (S84.01 does not include LOPA as does 61511).

Verify the performance of the SIS with reference to the established target SILs. (SIS is only one of the protective layers. It is important to make a comprehensive assessment of the other layers of protection, as per 61511, that are relevant to the SIFs for SIL estimation).

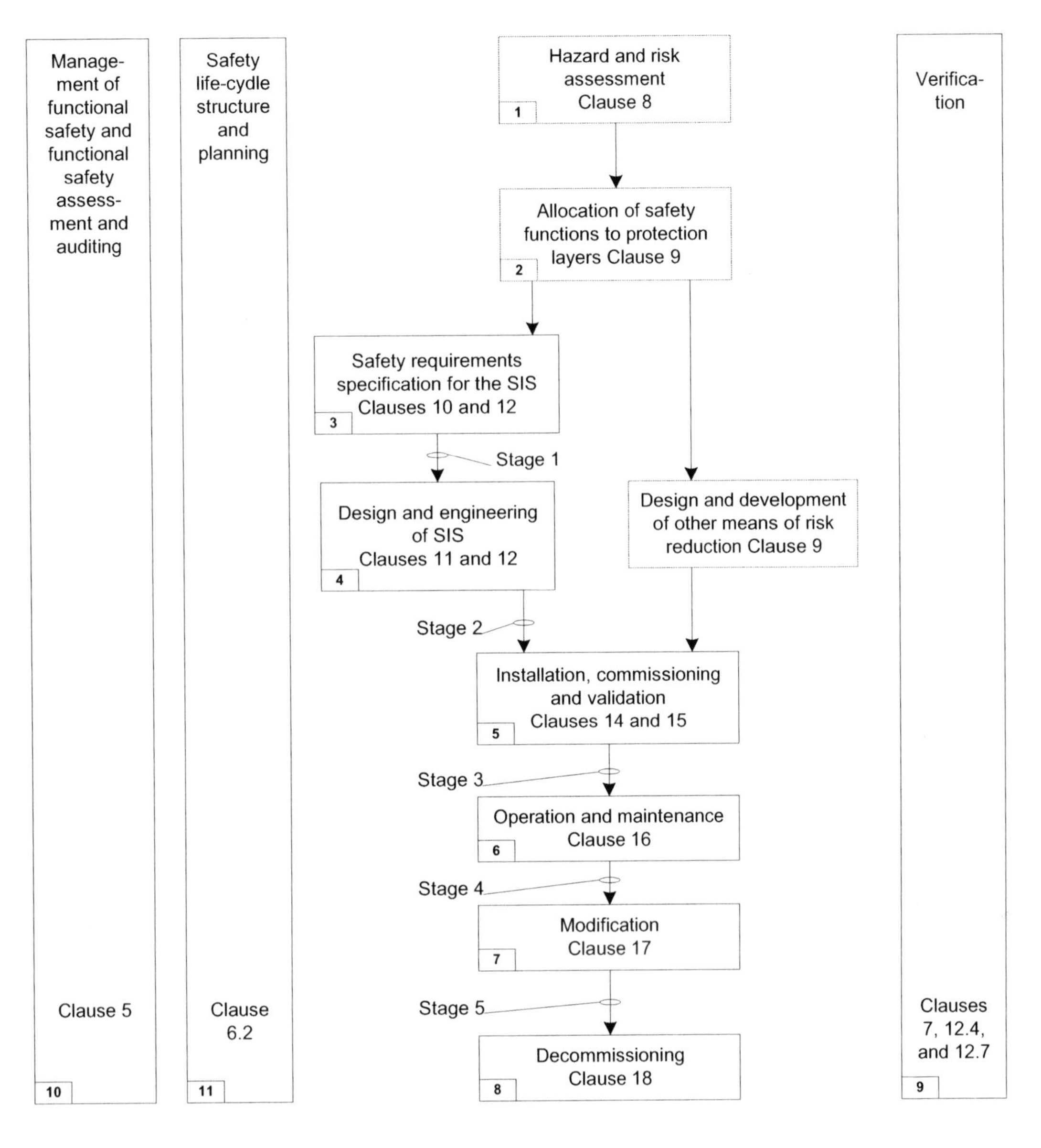

Key:

Typical direction of information flow

No detailed requirements given in this standard

Requirements given in this standard

NOTE 1 Stages 1 through 5 inclusive are defined in 5.2.6.1.3

NOTE 2 All references are to Part 1 unless otherwise noted

SIS safety life-cycle phases and functional safety assessment stages (IEC 61511-1, 2003, p. 33)

Figure 20-1 Safety Life Cycle

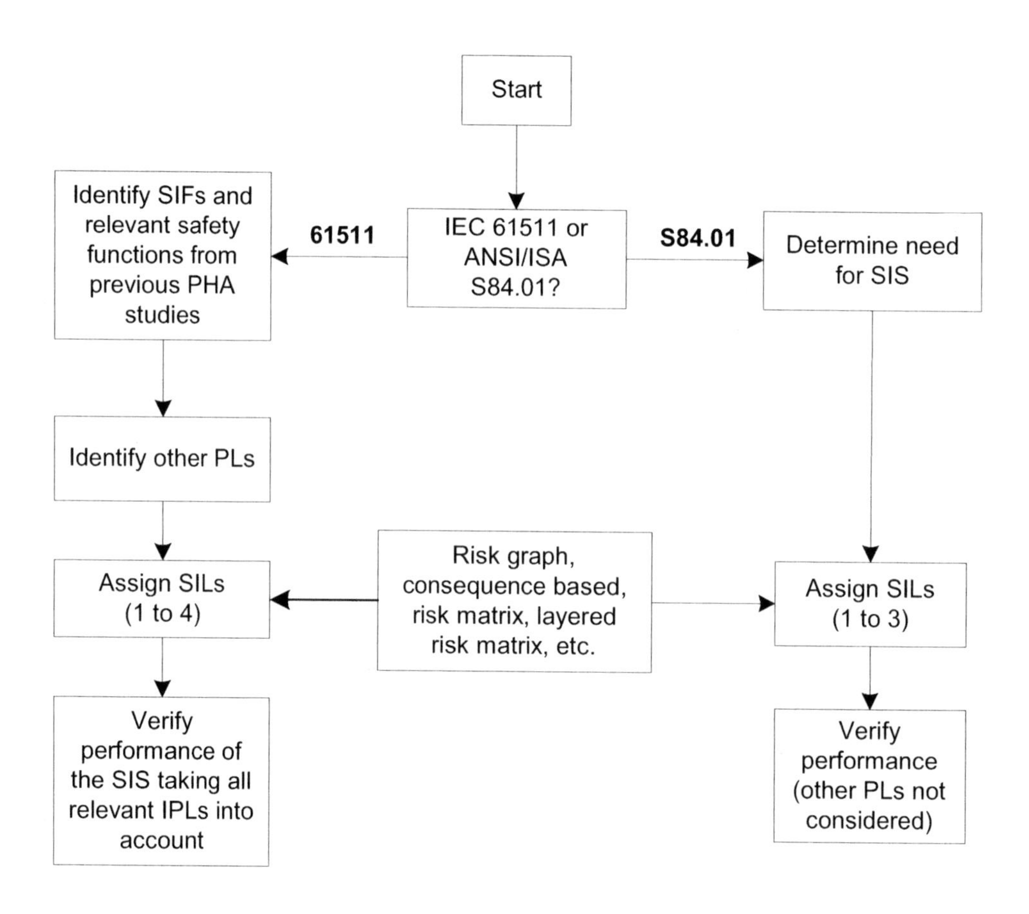

Figure 20-2 General Sequence of Steps for Assigning SIL

As per 61511, SIL estimation also takes into account the other layers of protection (PL) in the process. SILs are calculated for the SIF, which may include one or more protection layers and may be dependent or independent of one another (clearly, greater protection is afforded by totally independent as opposed to dependent protection layers identified for a particular SIF). Setting and meeting SIL targets can be viewed in two basic ways. If the user decides to use only ANSI/ISA 84.01 and ignore other layers of protection, then SIL targets can only be met by upgrading SIS components, e.g. upgrading emergency shutdown systems (ESD). However this can be a very costly business and thus the wisdom of sticking with ANSI/ISA 84.01 and ignoring the other possible protection layers (offered by IEC 61511) is questionable. See "Typical risk reduction methods found in process plants" in figure below:

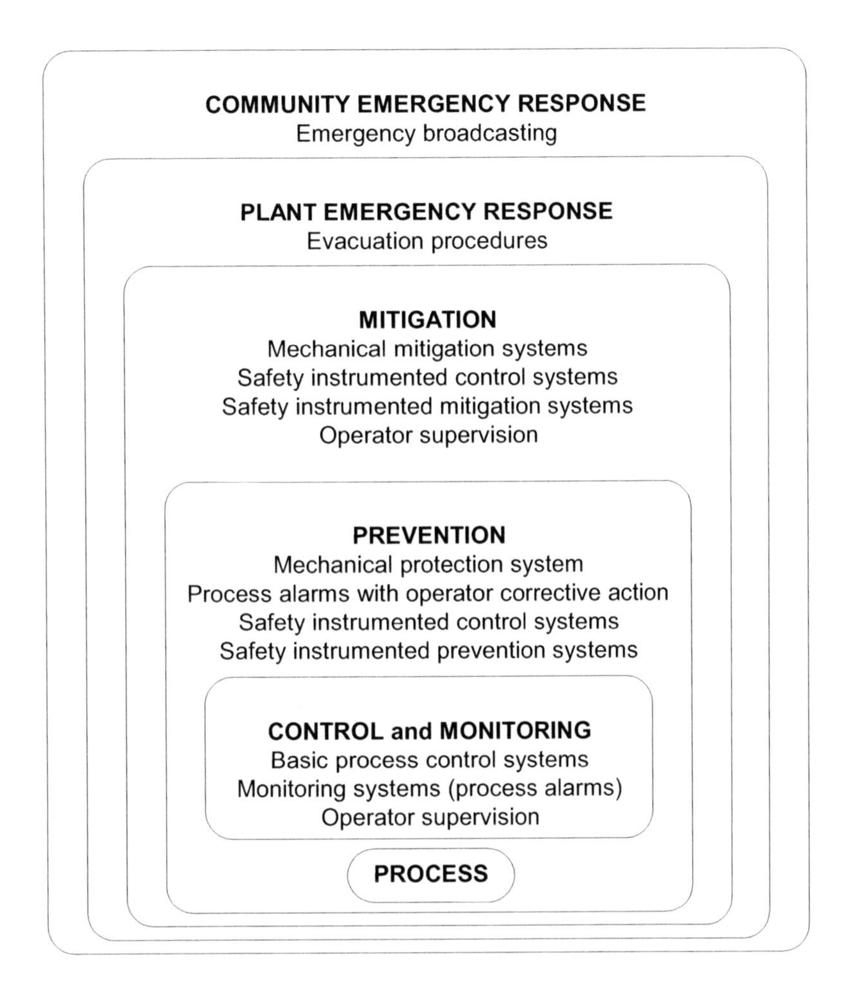

Figure 20-3 Typical Risk Reduction Methods Found in Process Plants

SIL Assignment Methodologies

Various methodologies are available for assignment of SILs. As in the case with PHA studies, this must involve people with the relevant expertise. The Risk Graph, Consequence-based (as recommended by S84.01 only), Modified HAZOP (as recommended by S84.01 only), the Risk Matrix, and the Layered Risk Matrix discussed below are the most common methods used to determine the target SIL. The Layer of Protection Analysis (LOPA) methodology could also be used to assign SILs (see Chapter 21., Layer of Protection Analysis). SILs assigned to SIFs in this manner represent the *target* (for existing or new systems) for the level of performance required to provide a certain level of reliability.

Consequence Based Method (S84.01)

This is the simplest of all SIL assignment methods in that it requires only relating the consequences directly to the SIL values, as shown in a typical SIL and consequence correlations table below.

Table 20-2 SILs Related to Consequences

Consequence	Prescribed SIL Values
Catastrophic community impact	SIL 4
Employee and community impact	SIL 3
Major property and production	SIL 2
Minor property and production	SIL 1

This method is not truly risk based as it only considers consequences. The disadvantage of this technique is that it does not take into account likelihood, is ultra-conservative, and limits the user, possibly prohibitively.

Modified HAZOP (S84.01)

At the design stage of the project, a modified HAZOP technique (or HAZOP Risk Matrix method, which is different from Layered Risk Matrix method) is another simple method to assign SIL values to SIS designs. It needs to be emphasized that if very conservative SIL values are assigned throughout the study, excessive and unnecessary costs can be incurred. This is most likely here because the simplicity of this technique allows this to happen.

The following table shows modified HAZOP-type entries, whereby the SIL values are assigned based on risk ranking.

Table 20-3 SIL Estimation Using Modified HAZOP Method

Deviation	Causes	Consequences	Safeguards	HAZOP Risk Matrix			Recommendations	Required SIL
				S	L	RR		
High Temperature in Reactor R-123	Runaway reaction	Over-temperature and possible reactor rupture leading to explosion & multiple fatalities	(1) Automated depressurizing system (2) Pressure relief valves	3	2	6	Safeguards are adequate	SIL 3
High Level in Storage Tank T-546	Overfilling by operator	Non-hazardous material spill inside dike	(1) Tank overflow (2) Level gauge on tank (3) High level alarm on tank	1	2	2	Safeguards are adequate	SIL 1
High Pressure in Intermediate Vessel V-793	Gas blow through on control valve FV-203 failure	Overpressure of vessel, loss of containment, employee injury	Pressure relief valve on Intermediate Vessel	2	2	4	Install low low trip on control valve FV-203 to prevent gas blow through from upstream vessel if level is lost	SIL 2

Risk Graph Method

61511 recognizes the value of considering multiple protection layers. Typically, this can be reflected by the application of say the Risk Graph technique combined with the different protection layers to modify the actual SIL requirements. These other layers may offer sufficient overall protection. A SIL in the risk graph is determined based on four factors as shown in the following tables and figure:

Table 20-4 Descriptions of Process Industry Risk Graph Parameters (IEC 61511-3, 2003, Annex D, p. 34)

Parameter		Description
Consequence	C	Number of fatalities and/or serious injuries likely to result from the occurrence of the hazardous event. Determined by calculating the numbers in the exposed area when the area is occupied taking into account the vulnerability to the hazardous event.
Occupancy	F	Probability that the exposed area is occupied at the time of the hazardous event. Determined by calculating the fraction of time the area is occupied at the time of the hazardous event. This should take into account the possibility of an increased likelihood of persons being in the exposed area in order to investigate abnormal situations, which may exist during the build-up to the hazardous event (consider also if this changes the C parameter).
Probability of avoiding the hazard	P	The probability that exposed persons are able to avoid the hazardous situation, which exists if the safety instrumented function fails on demand. This depends on there being independent methods of alerting the exposed persons to the hazard prior to the hazard occurring and there being methods of escape.
Demand rate	W	The number of times per year that the hazardous event would occur in the absence of the safety instrumented function under consideration. This can be determined by considering all failures, which can lead to the hazardous event and estimating the overall rate of occurrence. Other protection layers should be included in the consideration.

Table 20-5 Example Calibration of General Purpose Risk Graph (IEC 61511-3, 2003, Annex D, p. 37-38)

Risk parameter		Classification	Comments
Consequence (C) Number of fatalities This can be calculated by determining the numbers of people present when the area exposed to the hazard is occupied and multiplying by the vulnerability to the identified hazard. The vulnerability is determined by the nature of the hazard being protected against. The following factors can be used: V = 0.01 Small release of flammable or toxic material V = 0.1 Large release of flammable or toxic material V = 0.5 As above but also a high probability of catching fire or highly toxic material V = 1 Rupture or explosion	C_A C_B C_C C_D	Minor injury Range 0.01 to 0.1 Range > 0.1 to 1.0 Range > 1.0	1. The classification system has been developed to deal with injury and death to people. 2. For the interpretation of C_A, C_B, C_C and C_D, the consequences of the accident and normal healing should be taken into account.
Occupancy (F) This is calculated by determining the proportional length of time the area exposed to the hazard is occupied during a normal working period. NOTE 1 If the time in the hazardous area is different depending on the shift being operated then the maximum should be selected. NOTE 2 It is only appropriate to use F_A where it can be shown that the demand rate is random and not related to when occupancy could be higher than normal. The latter is usually the case with demands which occur at equipment start-up or during the investigation of abnormalities.	F_A F_B	Rare to more frequent exposure in the hazardous zone. Occupancy less than 0.1 Frequent to permanent exposure in the hazardous zone.	3. See comment 1 above.
Probability of avoiding the hazardous event (P) if the protection system fails to operate.	P_A P_B	Adopted if all conditions in column 4 are satisfied. Adopted if all the conditions are not satisfied.	4. PA should only be selected if all the following are true: ▪ Facilities are provided to alert the operator that the SIS has failed; ▪ Independent facilities are provided to shut down such that the hazard can be avoided or which enable all persons to escape to a safe area; ▪ The time between the operator being alerted and a hazardous event occurring exceeds 1 hour or is definitely sufficient for the necessary actions.

Risk parameter		Classification	Comments
Demand rate (W) The number of times per year that the hazardous event would occur in absence of SIF under consideration. To determine the demand rate it is necessary to consider all sources of failure that can lead to one hazardous event. In determining the demand rate, limited credit can be allowed for control system performance and intervention. The performance which can be claimed if the control system is not to be designed and maintained according to IEC 61511, is limited to below the performance ranges associated with SIL 1.	W_1 W_2 W_3	Demand rate < 0.1D per year 0.1D < Demand rate < D per year D < Demand rate < 10D per year For demand rates higher than 10D per year, higher integrity shall be needed.	5. The purpose of the W factor is to estimate the frequency of the hazard taking place without the addition of the SIS. If the demand rate is very high, the SIL has to be determined by another method or the risk graph recalibrated. It should be noted that risk graph methods may not be the best approach in the case of applications operating in continuous mode, see 3.2.43.2 of IEC 61511-1. 6. D is a calibration factor, the value of which should be determined so that the risk graph results in a level of residual risk which is tolerable taking into consideration other risks to exposed persons and corporate criteria.
NOTE This is an example to illustrate the application of the principles for the design of risk graphs. Risk graphs for particular applications and particular hazards will need to be agreed with those involved, taking into account tolerable risk, see D.1 to D.6.			

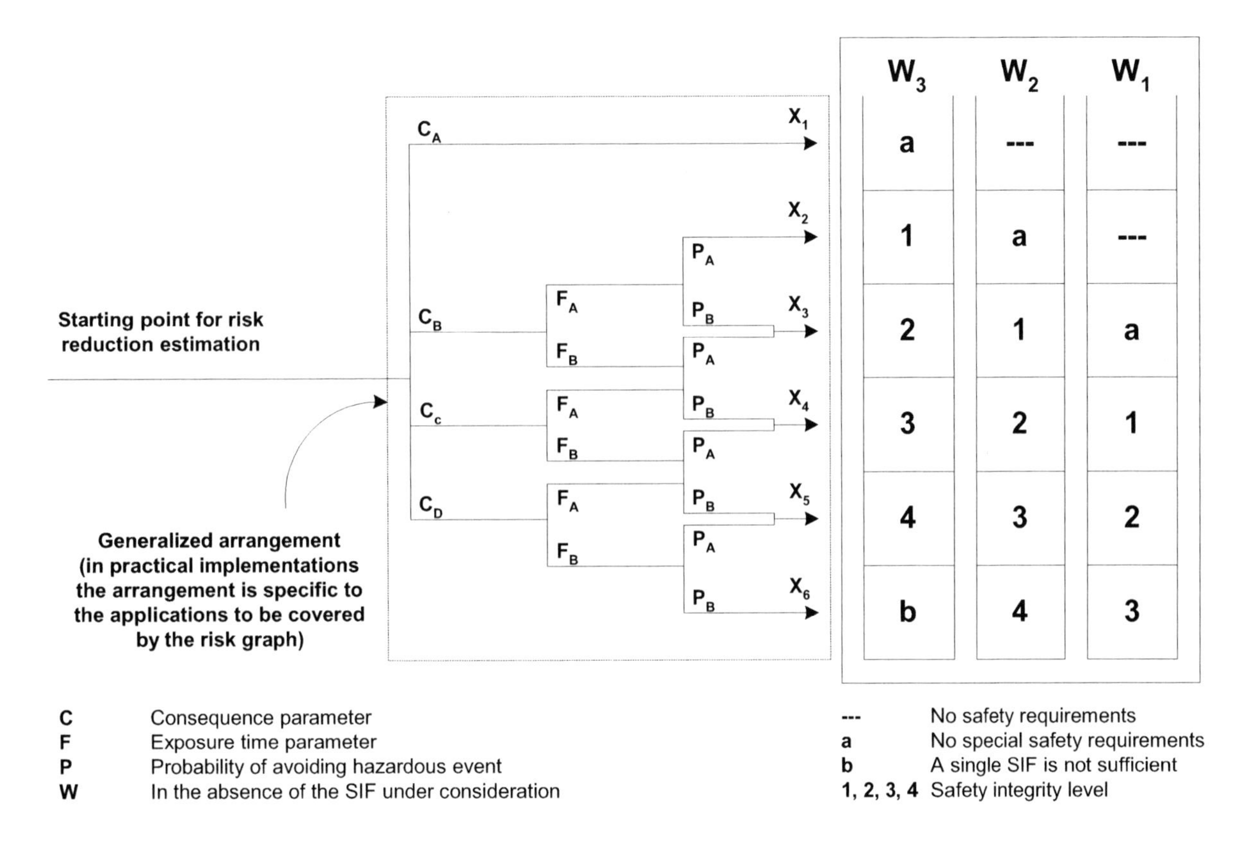

Figure 20-4 Risk Graph: General Scheme (IEC 61511-3, Annex D, p. 37)

If the consequence based route (alone) is chosen as opposed to the risk based methods, it makes mitigation options very limited as it discounts both frequency and probability of unwanted occurrences as contributing factors. It is therefore preferable to consider using the Risk Graph method, which is shown in Figure 20-4, above. This illustrates how the four parameters (C, F, P, and W) generate the target SIL values in the table, as follows. As per 61511, assume that no SIS exist, even though non-SIS may be in place for the process.

Table 20-6 SIL Estimation Using Risk Graph Method

Critical Hazardous Scenario	Consequence	Cause	Existing Safeguards	Target SIL					Required Actions (SIL)
				C	F	P	W	Target SIL	
1. > HHL in KO-101 with entrainment going to compressor	1. Damage to compressor	1. Failure of level control system loop 102	BPCS	C_3	F_1	P_1	W_1	SIL 2	1. Improve reliability of LT 102 such that only a SIL 2 level is required.
2. Loss of LPG containment on pumps	LPG released causing flammable gas release and fire potential	1. Seal failure on P-101A/B on single seal	Maintenance	C_3	F_2	P_1	W_1	SIL 3	2. Double seals instead of single.

Critical Hazardous Scenario	Consequence	Cause	Existing Safeguards	Target SIL					Required Actions (SIL)
				C	F	P	W	Target SIL	
3. Fire beneath KO-101	Potential BLEVE situation	1. Low level in KO-101 and flame impingement on unwetted portion of vessel	Operator training	C_4	F_1	P_1	W_2	SIL 4	3. Heat resistant insulation along the sides of the vessel and the bottom. Stainless steel cladding, spray skirt with concrete, crown the area (sloping), concrete ground instead of pebbles.

Safety Layer Matrix Method

An example of the Safety Layer Matrix (Layered Risk Matrix) is given below. The target SIL is assigned on the basis of the risk ranking value and the number of PLs for that scenario. A difference of the risk ranking and the PLs is correlated with SIL values. This approach consists of matrices for each of the various consequence categories such as Personnel, Operations, and Ecological factors, that are integrated with the HAZOP study and ***incorporates PLs***. The highest of the three SIL values is selected.

According to 61511, the required SIL values are matched with a combination of the frequency and severity of impact of the hazardous events. See the tables and figure below.

Table 20-7 Frequency of Hazardous Event Likelihood - without considering PLs (IEC 61511-3, 2003, Annex C, p. 30)

Type of Events	Likelihood
	Qualitative Ranking
Events such as multiple failures of diverse instruments or valves, multiple human errors in a stress free environment, or spontaneous failures of process vessels.	Low
Events such as dual instrument, valve failures, or major releases in loading/unloading areas.	Medium
Events such as process leaks, single instrument, valve failures or human errors that result in small releases of hazardous materials.	High
* The system should be in accordance with this standard when a claim that a control function fails less frequently than 10^{-1} per year is made.	

Table 20-8 Criteria for Rating the Severity of Impact of Hazardous Events (IEC 61511-3, 2003, Annex C, p. 30)

Severity Rating	Impact
Extensive	Large-scale damage of equipment. Shutdown of a process for a long time. Catastrophic consequence to personnel and the environment.
Serious	Damage to equipment. Short shutdown of the process. Serious injury to personnel and the environment.
Minor	Minor damage to equipment. No shutdown of the process. Temporary injury to personnel and damage to the environment.

Number of PLs	SIL Required								
3							c)	1	1
2	c)	c)	1	c	1	2	1	2	b) 3
1	c)	1	2	1	2	b) 3	b) 3	b) 3	a) 3
Hazardous Event Likelihood	Low	Med	High	Low	Med	High	Low	Med	High
	Minor			Serious			Extensive		
	Hazardous Event Severity Rating								

a) One level 3 SIF *does* not provide sufficient risk reduction at this risk level. Additional modifications are required in order to reduce risk (see d).
b) One level 3 SIF *may* not provide sufficient risk reduction at this risk level. Additional modifications are required (see d).
c) SIS independent protection layer is probably not needed.
d) This approach is not considered suitable for SIL 4.

Figure 20-5 Safety Layer Risk Matrix (IEC 61511-3, 2003, Annex C, p. 31)

Table 20-9 SIL Estimation Using Layer Risk Matrix

Possible Causes	Consequences	Hazardous Scenario	Existing Systems & Procedures (Safeguards)	Layered Risk Matrix					Max Required SIL	Recommendations
				Consequence Category	S	L	PL	SIL		
1. Exchanger EX-103 tube or tube sheet rupture	1. Overpressuring of stripper.	Explosion due to overpressure of the vessel	1. PSV-105 relieving to flare.	Personnel	E	M	2	2	SIL 2	9. Check PSV-105 sizing to handle (a) fire case, (b) tube rupture on reboiler, (c) total loss of reflux to stripper, (d) loss of cooling to condenser EX-102, (e) instrument or controller failure, (f) instrument air failure, (g) power failure etc.
			2. PV-106 opens to flare.	Operations	S	M	2	1		
				Ecological	S	L	2	c)		

New and Existing Systems

The first step for assignment of target SILs is to use the (updated) PHAs or conduct new PHAs to screen for the potential hazards. HAZOP is the most commonly used method. If the risk is unacceptable then it is preferable to reduce it to an acceptable level using non-SIS and SIS elements. However, SISs are considered ***only after*** all the non-SIS protection layers have been considered. HAZOPs identify the potential hazards, using risk matrices in terms of the likelihood and the severity of the hazards. Required SILs are assigned to SIFs identified in the PHA studies.

As introduced in the 61511, the intent of safety functions is to achieve or maintain a safe state for the specific hazardous event in a process. Only those safety functions that are assigned to the SIS are called SIF. According to 61511, the BPCS, relief systems, and other layers of protection may be defined as safety functions for SIL analysis. A SIS may contain one or many SIFs and each is assigned a SIL. As well, a SIF may be achieved by more than one SIS as may be accomplished using components (or systems) deemed to be redundant. Safety functions may be performed by a non-SIS technology such as the basic process control system (BPCS), safety valves, operator intervention, and alarms (these alarms being independent of BPCS). However, there are limits to how much the SIL

component of the BPCS can be taken into account. The BPCS is not credited for a SIF with a greater than SIL 1, as per 61511.

For an existing facility, where SIL values have not been assessed, the exercise is more complex. Although, the "desired SILs" may be identified, the actual in situ SIL values can only checked using reliability modeling, such as fault tree analysis (FTA) or reliability block diagrams supported by applicable failure rate data.

It may not be mandated for an existing facility to assess SIL values as per the standards, however, in the event of plant modifications or for the introduction of new units or grassroots facilities SIL values almost certainly need to be assessed as per the standards. In addition, if there is an incident (accident or near miss), which could be attributed to lack of reliability of SIS, then the standards for assessing SILs are recommended.

SIL Verification

Compliance with ANSI/ISA S84.01-1996 and IEC 61511, requires verification of the performance of SIS. Typically, it is practicable to study only the critical safety functions for a SIL study as there are usually too many safety functions and only those that are deemed important can be considered depending on the allocated resources. The established SILs (from previous steps) are now used as measures for verification purposes when complying with 61511. SIL verifications may require full quantitative assessments (using fault tree analysis - FTA, failure rates, reliability block diagrams, etc.) to check if the performance of the SIS exemplified by the overall ESD system indeed meets the established target SIL values based on unit wide overall scenarios (e.g., fire, toxic release etc.)

A simple example of one shutdown sequence consisting of detectors, logic solver, and final elements is given below. Logic solvers are considered very highly reliable, thus may not be a part of the failure rate calculation per se.

Example:

Consider a shutdown loop consisting of 3 pressure transmitters (connected so that 2 out of 3 must be functional), connected to a high-pressure switch, which in turn is connected to a shutdown valve.

Overall failure rate, $\lambda_{overall}$ = [Failure rate of transmitters] + [Failure rate of pressure switch] + [Failure rate of shutdown valve]

The PFD is calculated using the following equations:

$PFD = 1 - Availability$

Where:

$Availability = 1/(1 + \lambda_{overall} \times downtime)$

RRF = Risk Reduction Factor = 1/PFD (to be used in the SIL Correlations table)

For Transmitters:

Individual failure rate = 0.97 faults per year = 1.1×10^{-4} faults per hour

Assume downtime is 4 hours for repair, the equation for calculating the failure rate of a component with 2 out of 3 voting system is given below (Smith and Simpson, 2001):

[Failure rate of transmitters] = $6 \times (\lambda_{transmitter})^2 \times$ downtime

Hence,

[Failure rate of transmitters] $= 6 \times (1.1 \times 10^{-4})^2 \times 4$

$= \underline{2.9 \times 10^{-7}}$ faults per hour

For Pressure Switch:

Individual failure rate = 0.14 faults per year = $\underline{1.6 \times 10^{-5}}$ faults per hour

For Shutdown Valve:

Individual failure rate = 0.5 faults per year (inc. solenoid) = $\underline{5.7 \times 10^{-5}}$ faults per hour

Thus, the overall failure rate calculated as follows:

$\lambda_{overall}$ $= 2.9 \times 10^{-7} + 1.6 \times 10^{-5} + 5.7 \times 10^{-5}$

$= \underline{7.33 \times 10^{-5} \text{ faults per hour}}$

Availability $= 1/(1 + \lambda_{overall} \times \text{downtime})$

$= 1 / (1 + (7.33 \times 10^{-5}) \times 4)$

$= \underline{0.9997}$

PFD $= 1 - \text{Availability}$

$= \underline{0.0003}$

1/PFD $= \underline{3333}$

This corresponds to a SIL 3 level (from the correlations table).

The above example is a simple illustration of the principle of SIL verification, which only considers revealed failures, failures that can be immediately detected and repaired. In practice, failure rate data used in SIL verification are affected by the type, size and functionality of components being reviewed together with the corresponding failure modes. The failure modes describe the loss of required system function(s) that result from failures. The failure modes can be broken down into four types (Dowell and Green, 1998):

- Hidden dangerous;
- Hidden safe;
- Revealed dangerous; and
- Revealed safe.

The dangerous failure modes result in loss of protection, but the revealed dangerous failures can be immediately detected and repaired. The hidden dangerous failures can only be revealed by a demand or a proof test. The two revealed modes usually result in a false shutdown. A spurious trip is a trip of the ESD system that occurs without a demand. Dowell and Green (1998) provide detail discussions on the concept of hidden and revealed dangerous failures.

For revealed failures, the downtime used to calculate the PFD (as illustrated in the example) consists of the active mean time to repair plus any logistic delays. For unrevealed failures, the downtime is related to the proof test interval plus the active mean time to repair plus any logistic delays.

Important Aspects of SIL Application

- There is danger of placing complete reliance on any one PL to cover hazards. For example, the notion that pressure relief systems alone can protect against all loss of containment situations. If for example, toxic or flammable gas releases can occur without overpressure, e.g., through flange gaskets or seals leaking, then other forms of protection are almost certainly required.

- Full compliance with 61511 is an extremely onerous responsibility requiring considerable deployment of resources. It would be highly undesirable to undertake this exercise with too limited resources. Full planning as would occur for a major project would involve qualified personnel with adequate expertise.

- The earlier standard, S84.01, offers fewer options than the current (as of date) 61511 as (a) it does not recognize SIL 4 and (b) it does not permit/address the contributions made by PLs.

SUGGESTED READING (Note: URLs active at date of publication)
ISA, Technical Articles on www.isa.org. The following URL is active for this link at the time of issuing this manual. http://www.isa.org/Content/NavigationMenu/Members_and_Leaders/Leader_Resources/Section_Leader_Resources/Resources/Technical_Articles.htm
The Comprehensive information site for Instrumentation, Control, Fire & Gas Engineers at http://www.iceweb.com.au. See SIS under http://www.iceweb.com.au/home.html and refer to articles at http://www.iceweb.com.au/sis/sis_index.html
"Improving Safety in Process Control" by C.M. Fialkowski, Control Engineering, September 1, 1998 www.manufacturing.net/ctl/index.asp?layout=article&articleId=CA185727&text=sil
"Partial-Stroke Testing of Safety Block Valves" by A. Summers and B. Zachary, Control Engineering, November 1, 2000 www.manufacturing.net/ctl/index.asp?layout=article&articleId=CA190350&text=sil
"The Complete Safety System", W.L. Mostia, Control for the Process Industries, December 4, 2000 www.controlmag.com/web_first/ct.nsf/ArticleID/RDAT-4RPN79?OpenDocument&Highlight=0,The,Complete,Safety,System
"Ins and Outs of Partial Stroke Testing" by W.L. Mostia, Control for the Process Industries, September 5, 2001 www.controlmag.com/web_first/ct.nsf/ArticleID/PSTR-4YQTAL?OpenDocument&Highlight=0,The,Complete,Safety,System

CHAPTER 21

Layer of Protection Analysis

Introduction

What is LOPA?

Layer of Protection Analysis (LOPA) introduces the concept that protection against an untoward or serious consequence, such as fire, may not simply be at a single level, or layer, but rather that there are likely to be multiple levels or layers of protection. Consider, by way of example, a fire situation. The Emergency Shutdown System (ESD) will constitute one layer, the Pressure Relief and Flare System will constitute another layer, the Fire Protection System involving deluge will be another layer, Emergency Response another layer and so forth. The analysis of the layers is referred to as LOPA. Figure 21-1 illustrates some common layers of protection for a process.

LOPA is a semi-quantitative risk analysis methodology. It is used to evaluate the risk of a selected hazardous scenario by establishing an order of magnitude approximation of risk. It is semi-quantitative as it requires numerical inputs such as event frequency and probability of failure, which are selected with the intent to provide conservative risk estimation.

The estimated risk is then compared with risk tolerance criteria (as established by the company) to decide if the existing layers of protection are adequate, and if additional risk reduction is needed. Without risk tolerance criteria, there is a tendency to keep adding risk mitigation measures in the belief that this would offer greater safety. More risk mitigation measures may well offer greater safety but, at some stage, may add significantly greater cost without adding significantly greater mitigation. Also mitigation measures may be added that are unnecessary and may add to the complexity of the facility that can result in potential new unidentified hazard scenarios and possibly, additional spurious shutdowns. LOPA helps to focus the limited resources on the most critical risk mitigation (and prevention) measures.

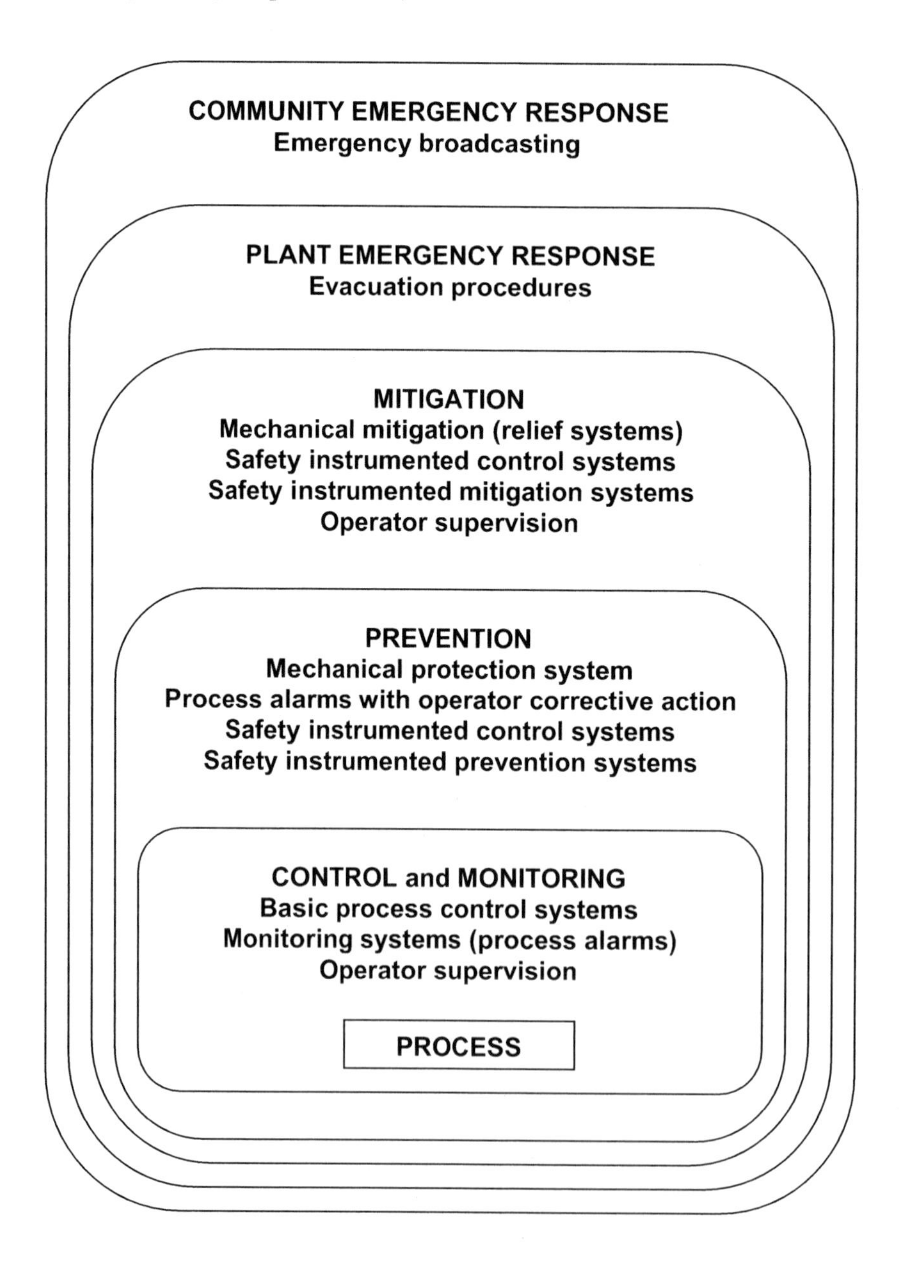

Figure 21-1: Common Layers of Protection in Process Plants (IEC 61511, 2003)

LOPA and Process Life Cycle

LOPA can be applicable throughout the process life cycle. Figure 21-2 illustrates the main phases in the process life cycle.

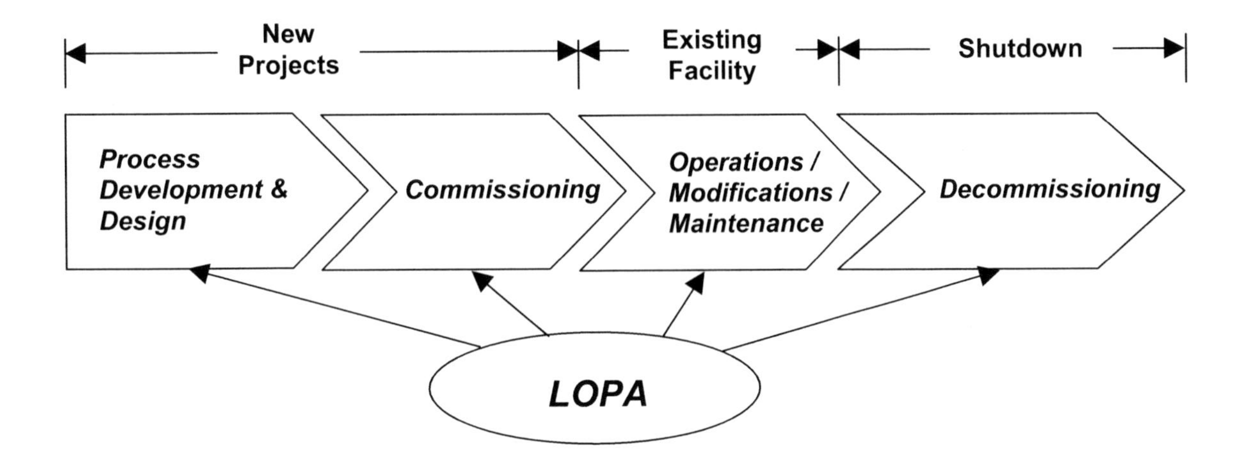

Figure 21-2: Illustration of how LOPA fits into Process Life Cycle

Some applications of LOPA at various phases are given below:

Process Development & Design

- *Overpressure Protection System* - LOPA can determine the existing IPLs and their failure probabilities to help define the controlling case for the relief system design basis for sizing pressure relief devices as when using ASME Code 2211 "Overpressure Protection by System Design" or API 520 "Sizing, Selection and Installation of Pressure Relieving Devices for Refineries".
- *Establishing Target Safety Integrity Levels (SIL)* - LOPA is recognized by IEC 61508 and IEC 61511 as one of the recommended methods for establishing target Safety Integrity Level (SIL) for a Safety Instrumented Function (SIF).
- *Evaluate Process Design Options* - LOPA can be used to examine basic design alternatives and select designs that have lower initiating event frequencies, or lesser consequences. It helps to design inherently safe processes by objectively and quickly comparing alternative designs.

- *Safety Cost Planning* – The LOPA method, integrated with a cost-benefit method, assists with the decision as to which safeguards to select. This helps to realize the financial benefits of reducing risk and to prioritize allocation of resources and comparison of different projects on a common playing field.
- *Emergency Isolation Systems* - LOPA is used to evaluate the need for isolation systems in processes where loss of containment situations e.g., leaks in piping systems, can occur.

Commissioning / Operations / Maintenance / Modifications

- *Evaluate Human Factors During Start-up* – LOPA can be used to examine human failure related scenarios during start-up of processes.
- *Bypassing Safety Systems* – LOPA helps to determine whether a critical *Independent Layer of Protection* (IPL) safety system can be temporarily bypassed or taken out of service for a short duration and what additional layers of protection would be required, if at all.
- *Management of Change* - LOPA identifies the safety issues involved in the modification of processes, procedures, equipment, instrumentation, etc., and whether the modification meets corporate risk tolerance criteria.
- *Mechanical integrity programs* - Safety critical equipment maintains the process within tolerable risk criteria as specified by an organization. LOPA can significantly decrease the need for superfluous safety critical equipment components where an over-conservative approach to safety could result in unreasonably high amounts of such equipment. This can have a drastic impact on costs in new plants and revamps.
- *Safety Training and Operating Manuals* - LOPA can identify operator actions and responses that are critical to the safety of the process. This helps to better define the training and testing needed during the life of the process and improves the clarity of the operating manuals.

How does LOPA work?

LOPA is a scenario-driven methodology. Hence, it is based on pre-identified scenarios from studies such as qualitative Process Hazard Analyses (PHAs), e.g. HAZOP, What-if study, Management of Change evaluation, or design review. LOPA is then applied to one scenario at a time.

A *scenario* is defined by a single *cause-consequence* pair. If a consequence has several causes, each cause-consequence pair is analyzed as a separate scenario. Similarly, if a cause can result in different consequences, additional scenarios should be developed. The cause-consequence pairs are screened further usually on the basis of consequence *severity*. Different severity categorization methods ranging from indirect reference to human harm to quantitative estimation of human harm can be used. A further criterion could be the financial costs incurred as a result of an incident (see Chapter 18 "Managing and Justifying Recommendations"). Further discussion on this topic is presented later in this chapter.

The following is an outline of LOPA procedural steps:

1. Identify and define scenarios
2. Select an incident scenario
3. Identify the initiating event of the scenario and determine the initiating event frequency (events per year)
4. Identify the IPLs and estimate the probability of failure on demand (PFD) of each IPL
5. Estimate the risk of the scenario by the combination of the consequence, the initiating event, and IPL data (PFD).

Scenario Development

Figure 21-3 shows the components in a scenario. The items in solid lines are needed to make up a scenario; the optional items are represented in dotted lines.

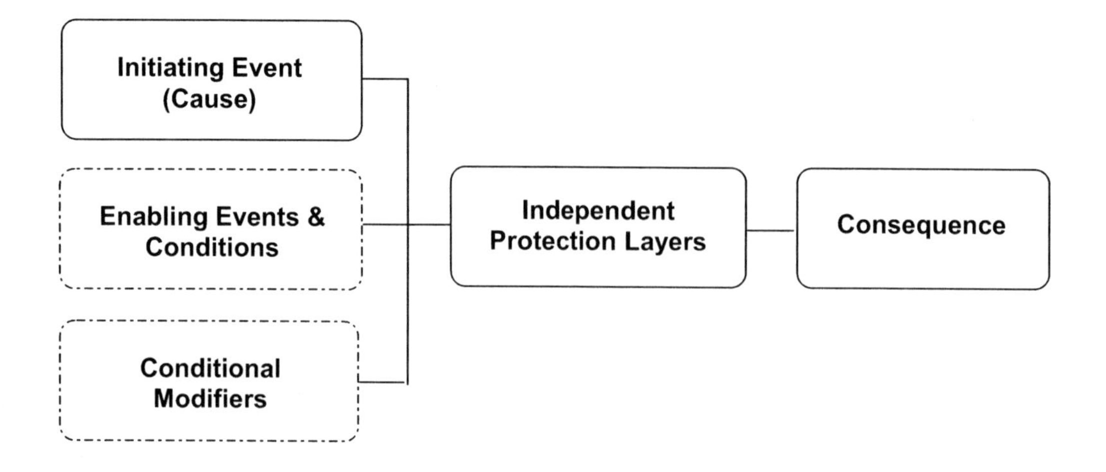

Figure 21-3: Components in a LOPA scenario

The *initiating event* is the single *cause* of the scenario leading to the specified *consequence*.

In some cases, if the initiating event alone cannot result in the specified consequence, it may require other conditions or events to take place. These are the *enabling events and conditions*.

If the categorization of consequence severity is referring to fatalities, or harm to business or the environment, the *conditional modifiers* can be used to refine the outcome of the scenario. Typical modifiers might include:

- Probability of ignition
- Probability of fatal injury
- Probability of personnel being in the affected area
- Probability of personnel escaping from the incident

- Probability of personnel being rescued

An *Independent Protection Layer* (*IPL*) is a safeguard capable of preventing a scenario from proceeding to its undesired consequence. It is independent of the initiating event or the action of any other layers of protection associated with the scenario.

In order to illustrate the concept of LOPA, let us consider the two-phase hydrocarbons separator shown in Figure 21-4.

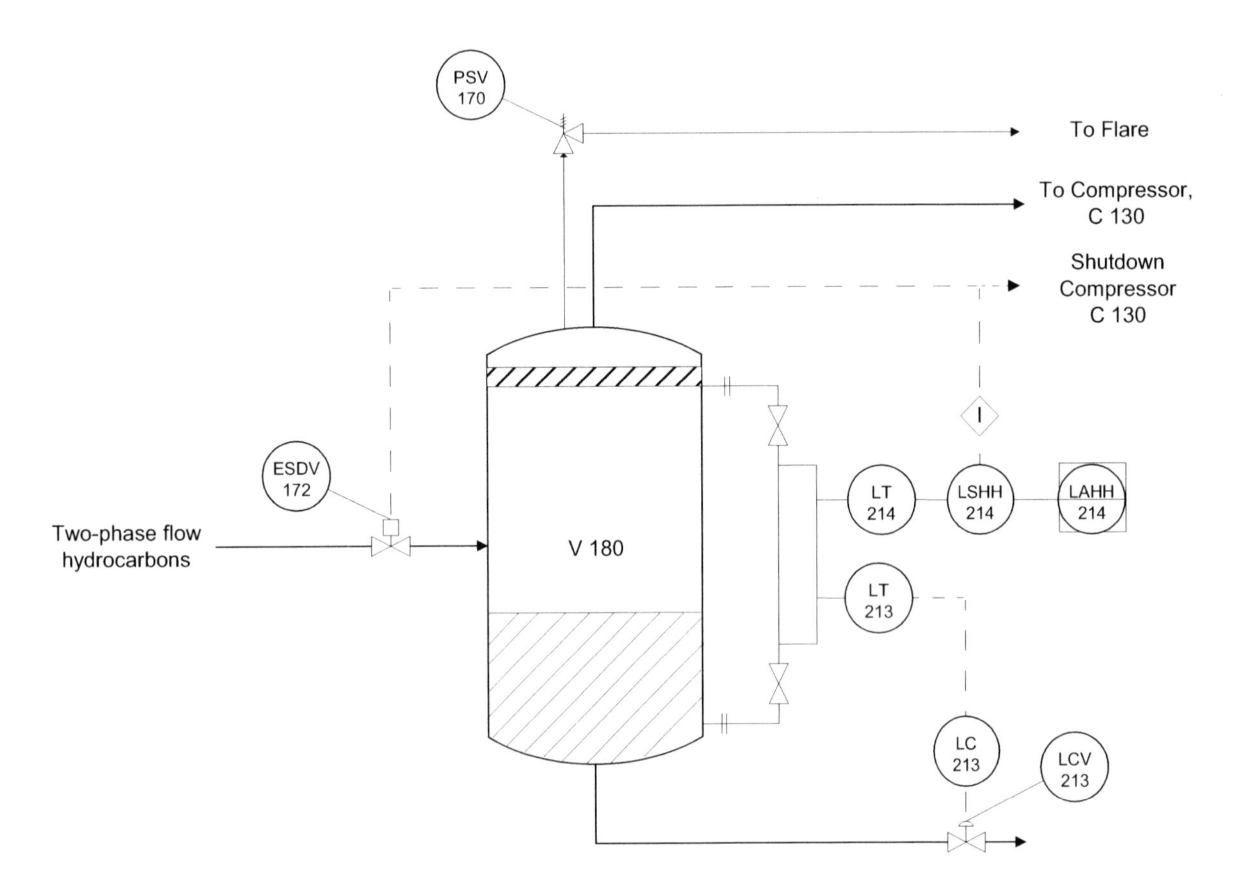

Figure 21-4: Two-phase separator and controls

The two-phase separator V 180 is under level control (Level control LC 213). In case of high high liquid level, the level switch LSHH 214 would close emergency shutdown valve ESDV 172 and shutdown compressor C 130 downstream of V 180. This is to prevent carrying liquid over to the compressor leading to compressor damage. During the HAZOP study, the following hazardous scenario is identified:

Node:	Two-phase separator V 180
Deviation:	High Level
Cause:	Level control loop 213 failure
Consequence:	Potential for liquid carry-over to the compressor, C 130 leading to compressor damage, possible disintegration and potential for fire and personnel injury
Safeguards:	Level switch LSHH 214 interlocks to alarm LAHH 214 and closes ESDV 172 and shuts down compressor C 130 downstream of V 180

Assuming it is selected for further analysis, it would look like this in LOPA:

Initiating Event:	Level control loop 213 failure
Enabling Events:	LCV 213 trends to closure thus leading to accumulation of liquid in the vessel
Conditional Modifiers:	In the event of loss of containment due to compressor destruction or severe damage, the following need to be evaluated as conditional modifiers: • Probability of personnel in the area • Probability of ignition • Probability of injury
IPLs:	Safety Instrumented System (SIS): Level switch LSHH 214 interlocks to alarm LAHH 214 and closes ESDV 172 and shuts down compressor C 130 downstream of V 180
Consequence:	Damage of compressor leading to personnel injury

In other words, the scenario goes like this: The level controller LC 213 fails ***AND*** this leads to failure of LCV 213 in such a way that it won't allow sufficient flow out of the separator ***AND*** SIS (Level switch LSHH 214 interlocks to alarm LAHH 214 and closes

ESDV 172 and shuts down compressor C 130 downstream of V 180) fails to act correctly ***RESULTING IN*** carry-over of liquid to the compressor ***LEADING TO*** potential injury / fatalities.

Once the scenario is built, the major questions are:

- What is the likelihood of this undesired event ?
- What is the risk associated with this scenario?
- Are there sufficient risk mitigation measures?

In order to answer the above questions, numerical values need to be assigned to the scenario components. Figure 21-5 shows what numerical values are required for the scenarios components. In order to evaluate the adequacy of risk mitigation measures, the *risk tolerance criteria* need to be established. The criteria are usually based on benchmark values from industry data, company history and/or statistical data.

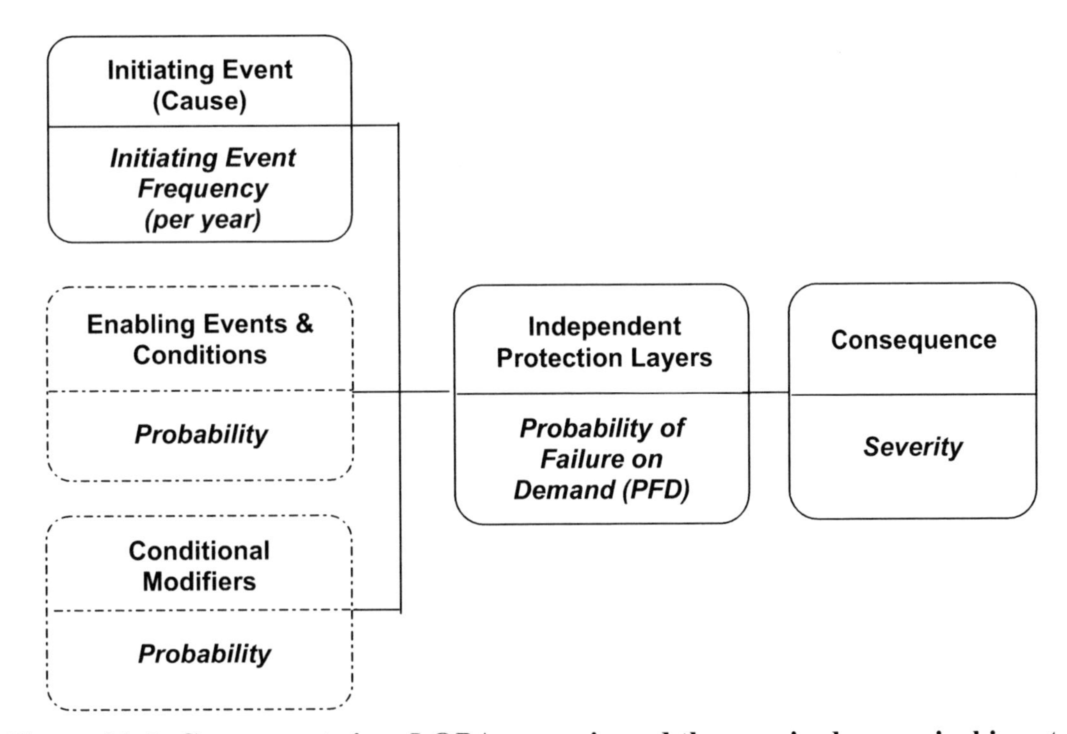

Figure 21-5: Components in a LOPA scenario and the required numerical inputs

For scenarios in which the initiating event frequency is less than twice the test frequency for an IPL i.e. "low demand mode", the frequency (likelihood) for the undesired consequence is calculated by the following equation (CCPS, 2001).

$$f_i^{\mathrm{C}} = f_i^{\mathrm{I}} \times \mathrm{PFD}_{i1} \times \mathrm{PFD}_{i2} \times \cdots\cdots \times \mathrm{PFD}_{ij}$$

Where f_i^{C} = Frequency for consequence C for initiating event I (per year)

f_i^{I} = Initiating event frequency for initiating event I (per year)

PFD_{ij} = Probability of failure on demand of the jth IPL that protects against consequence C for initiating event I

For "high demand mode" scenarios, i.e. the challenge frequency to an IPL is higher than twice the test frequency for the IPL, for example, the IPL is tested once a year and there are more than 2 demands per year, the following equation should be used to calculate the frequency for undesired consequence (CCPS, 2001):

$$f_i^{\mathrm{C}} = 2 \times (\mathrm{IPL}_{i1} \text{ test frequency, per year}) \times \mathrm{PFD}_{i1} \times \mathrm{PFD}_{i2} \times \cdots\cdots \times \mathrm{PFD}_{ij}$$

Hence, in the first equation, the terms for the initiating event frequency, f_i^{I} and the first IPL PFD, PFD_{i1}, are replaced by $2 \times (\text{IPL test frequency, per year}) \times \mathrm{PFD}_{i1}$. This approach provides more realistic frequency results (For further explanation, see CCPS, 2001).

If there are enabling events and conditions and/or conditional modifiers, the above equations are modified to the following:

For *Low Demand Mode*:

$$f_i^{\mathrm{C}} = f_i^{\mathrm{I}} \times \mathrm{PFD}_{i1} \times \mathrm{PFD}_{i2} \times \cdots\cdots \times \mathrm{PFD}_{ij} \times P_{Enabling\ event} \times P_{Condition\ modifier}$$

Where $P_{Enabling\ event}$ = Probability of the enabling event to take place

$P_{Condition\ modifer}$ = Probability of the outcome of modifying factors

For *High Demand Mode*:

$$f_i^C = 2 \times (\text{IPL}_{i1} \text{ test frequency, per year}) \times \text{PFD}_{i1} \times \text{PFD}_{i2} \times \cdots \times \text{PFD}_{ij} \times P_{Enabling\ event} \times P_{Condition\ modifier}$$

The Probability of Failure on Demand (PFD) is estimated for each IPL, typically using available data or look-up tables. Each IPL reduces the frequency of the consequence.

The frequency of each identified initiating event for the scenario, i.e. cause, of the scenario is estimated, usually from failure rate data or from a look-up table.

The selection of appropriate data and sources will be addressed in the next sections. For the purpose of illustration, assuming the following severity categories for consequence are used and severity ranking of 4 is selected. The selected values for other scenario components are given in Table 21-2.

Table 21-1: Qualitative Categorization of Severity ***N.B: The following table of values illustrates the methodology***

Severity	Description	Simplified Injury/Fatality Categorization
1	Low Consequence	Same as Category 2
2	Low Consequence	Minor injury or no injury, no lost time
3	Medium Consequence	Single injury, not severe, possible lost time
4	High Consequence	One or more severe injuries
5	Very High Consequence	Fatality or permanently disabling injury

Table 21-2: Numerical Values used in Two-Phase Separator Scenario ***N.B: The following table of values illustrates the methodology***

Scenario Component	Description	Value
Consequence (Severity)	Damage of compressor leading to personnel injury	Cat.4
Initiating event frequency (per year)	Level control loop 213 failure	1×10^{-1}
Enabling event or condition	LCV 213 trends to closure thus leading to accumulation of liquid in the vessel	0.5
Conditional modifiers (Probability)	Probability of ignition	0.7
	Probability of personnel in the area	0.5
	Probability of injury	0.8
IPLs	SIF (Level switch LSHH 214 with alarm LAHH 214 interlock to close ESDV 172 and shutdown compressor PM 130 downstream of PV 180)	1×10^{-2}

f_i^C for the above scenario is calculated:

$$f_i^C = 1\times10^{-1}\times1\times10^{-2}\times0.5\times0.7\times0.5\times0.8$$

$$f_i^C = \underline{\underline{1.4\times10^{-4}\,\text{per year}}}$$

The risk matrix method is used to assign risk tolerance criteria in this example.

Table 21-3: Risk Matrix Used in Two-Phase Example ***N.B: The following table of values illustrates the methodology***

Conseq. Cat. / Freq. (per yr)	Category 1	Category 2	Category 3	Category 4	Category 5
1 to 10^{-1}	Optional (evaluate alternatives)	Optional (evaluate alternatives)	Not Desirable – Risk control measures to be introduced within a specified time period	Unacceptable	Unacceptable
10^{-1} to 10^{-2}	Acceptable with control	Optional (evaluate alternatives)	Optional (evaluate alternatives)	Not Desirable – Risk control measures to be introduced within a specified time period	Unacceptable
10^{-2} to 10^{-3}	Acceptable – No actions are needed	Acceptable – No actions are needed	Optional (evaluate alternatives)	Not Desirable – Risk control measures to be introduced within a specified time period	Not Desirable – Risk control measures to be introduced within a specified time period
10^{-2} to 10^{-3}	Acceptable – No actions are needed	Acceptable – No actions are needed	Optional (evaluate alternatives)	Optional (evaluate alternatives)	Not Desirable – Risk control measures to be introduced within a specified time period
10^{-3} to 10^{-4}	Acceptable – No actions are needed	Acceptable – No actions are needed	Acceptable – No actions are needed	**Optional (evaluate alternatives)**	Optional (evaluate alternatives)
10^{-4} to 10^{-5}	Acceptable – No actions are needed	Acceptable – No actions are needed	Acceptable – No actions are needed	Acceptable – No actions are needed	Optional (evaluate alternatives)

Based on the risk matrix, it is categorized as "Optional to evaluate alternatives" for the current settings in this example. Other alternatives can also be considered in this case, such as:

- Improving reliability of level control loop 213
- Improving reliability of SIS
- Possible additional IPLs

Consequences and Severity Estimation

There are various methods for evaluating consequences:

- Category Approach without direct reference to human harm
- Qualitative estimates with human harm
- Qualitative estimates with human harm with adjustments for post-release probabilities
- Quantitative estimates with human harm
- Overall cost resulting from potential incident (e.g., capital losses, production losses etc.)

Category Approach Without Direct Reference to Human Harm

This approach has the following characteristics:

- Focuses on preventing the release itself rather than mitigating the consequences.
- Does not use human injury / fatality as end points for risk tolerance criteria.
- Typically uses matrices to differentiate consequences into various categories.

The following is an example of consequence categorization.

Table 21-4: Consequence Categorization Sample (CCPS, 2001). ***N.B: The following table of values illustrates the methodology***

Release Characteristic	1- to 10- lb release	10- to 100- lb release	100- to 1000- lb release	1000- to 10,000- lb release	10,000- to 100,000- lb release	> 100,000 lb release
Extremely toxic above boiling point	Cat. 3	Cat. 4	Cat. 5	Cat. 5	Cat. 5	Cat. 5
Extremely toxic below boiling point or highly toxic above boiling point	Cat. 2	Cat. 3	Cat. 4	Cat. 5	Cat. 5	Cat. 5
Highly toxic below boiling point or flammable above boiling point	Cat. 2	Cat. 2	Cat. 3	Cat. 4	Cat. 5	Cat. 5
Flammable below boiling point	Cat. 1	Cat. 2	Cat. 2	Cat. 3	Cat. 4	Cat. 5
Combustible liquid	Cat. 1	Cat. 1	Cat. 1	Cat. 2	Cat. 2	Cat. 3

Each consequence is assigned a numerical category from 1 to 5. Category 5 is the most severe. The above consequence categorization can be used in conjunction with a risk matrix like the one used in the two-phase separator example (Table 21-3).

Qualitative Estimates with Human Harm

This approach has the following characteristics:

- Focuses on the final impact to humans. The severity is established based on qualitative judgment.
- The resulting risk can be compared directly to a fatality risk tolerance criterion.

The following is an example of consequence categorization.

Table 21-5: Qualitative Categorization – Combined Loss Categories (CCPS, 2001)
N.B: The following table of values illustrates the methodology

	Low Consequence
Personnel	Minor or no injury; no lost time
Community	No injury, hazard, or annoyance to public
Environment	Recordable event with no agency notification or permit violation
Facility	Minimal equipment damage at an estimated cost of less than $100,000 and with no loss of production.
	Medium Consequence
Personnel	Single injury, not severe; possible lost time
Community	Odor or noise complaint from the public
Environment	Release that results in agency notification or permit violation
Facility	Some equipment damage at an estimated cost greater than $100,000 and with minimal loss of production.
	High Consequence
Personnel	One or more severe injuries
Community	One or more minor injuries
Environment	Significant release with serious offsite impact
Facility	Major damage to process area(s) at an estimated cost greater than $1,000,000 or some loss of production
	Very High Consequence
Personnel	Fatality or permanently disabling injury
Community	One or more sever injuries
Environment	Significant release with serious offsite impact and more likely than not to cause immediate or long-term health effects.
Facility	Major or total destruction of process area(s) at an estimated cost greater than $10,000,000 or a significant loss of production

Qualitative Estimates with Human Harm with Adjustments for Post-release Probabilities

This approach is similar to the previous method with additional considerations such as:

- Probability that the event will result in a flammable or toxic cloud
- Probability whether an individual will be present in the area
- Probability of injury / fatality

Quantitative Estimates with Human Harm

This approach requires detailed analyses and mathematical modeling to determine the effects of a release people and equipment. (Chapter 22 "Quantitative Risk Assessment" gives more details on quantitative modeling).

Overall Cost of Potential Incident

An incident can also be equated to financial impacts, such as capital losses, lost production etc. When these are totaled, the overall sum can be considered as a financial measure of risk. (See Chapter 18 "Managing and Justifying Recommendations" for further details).

Initiating Events and Frequency Estimation

The following table provides a list of typical initiating events that can pre-empt an incident. They do not necessarily result in severe or catastrophic impacts, although they can do so.

Types of Initiating Event

Type of event	Examples
Mechanical failures	o Corrosion o Vibration o Erosion o Flow surge or hydraulic hammer o Seal/gasket/flange failure o Relief device stuck open o Puncture o Fracture o Fabrication defects o Brittle fracture
Control systems failures	o Sensors failure o Logic solver failure o Final elements failure o Field wiring failure o Communication interface failure o Software failures or crashes
Utility failures	o Power failure o Loss of instrument air o Loss of plant nitrogen o Loss of cooling water o Loss of steam
Natural external events	o Earthquakes o Tornadoes o Hurricanes o Floods o High winds o Lightning
Human external events	o Major accidents in adjacent facilities o Incidents in adjacent processes o Incidents within the process o Mechanical impact by motor vehicles
Human failures	o Operational error o Maintenance error o Critical response error o Programming error

Examples of Inappropriate Initiating Events

Not all events can be categorized as being the direct or indirect cause for an incident. Some events may be suspect but cannot be confirmed. However, if there is a clear indication that the initiating event and the final incident are quite definitely related, then it is appropriate to use them in the analysis. Typical examples of inappropriate initiating events might be:

- *Inadequate operator training / certification* – Possible underlying cause of an initiating event.
- *Inadequate testing and inspection* – Possible underlying cause of an initiating event.
- *Unavailability of protective devices such as safety valves or overspeed trips* – Requires initiation of other events before protective devices are challenged.
- *Unclear or imprecise operating procedures* – Possible underlying cause of an initiating event.

Verification of Initiating Event

Before assigning initiating event frequencies to the cause of a scenario, it is critical to ensure the cause-consequence relationship is valid. The following are typical criteria that need to be met.

- Need to verify that the cause-consequence relationship for each scenario is unique.
- Try to reduce cause into discrete failure events, e.g. "Loss of cooling" can be due to a number of possible failures such as:
    - Coolant pump failure
    - Failure of cooling fans on air cooled exchangers
    - Power failure

  - Control loop failure, causing coolant failure or bypassing of coolant around exchangers.

Enabling Events/Conditions

Enabling events or conditions are operations or conditions that do not directly cause the scenario, but which must be present or active as scenario components. They should be used when the mechanism between the initiating event and the consequences needs to clarified.

Initiating Event Frequency Estimation

It is important to obtain or derived meaningful estimates of event frequencies. Usually these are obtained from one or more different sources. More importantly their order of magnitude, when different sources are compared, should be the same or similar. Typically failure rate data may be obtained from the following sources:

- *Industry data* – For component failures:
  - Guidelines for Process Equipment Reliability Data, CCPS (1986)
  - Guide to the Collection and Presentation of Electrical, Electronic, and Sensing Component Reliability Data for Nuclear-Power Generating Stations. IEEE (1984)
  - OREDA (Offshore Reliability Data)
- *Industry data* – Human Error Rates:
  - Inherently Safer Chemical Processes: A life Cycle Approach, CCPS (1996)
  - Handbook of human Reliability Analysis with Emphasis on Nuclear Power Plant Applications, Swain, A.D., and H.E. Guttman, (1983)
- *Company experience* – This includes historical data for the process and the experience of plant personnel/logged failure rate data.

- *Vendor data* – Typically optimistic as the data are developed in clean, well-maintained (factory) settings.

The following table lists typical initiating event frequencies.

Table 21-6: Typical Frequency Values (CCPS, 2001)

Initiating Event	Frequency Range (per year)
Pressure vessel residual failure	10^{-5} to 10^{-7}
Cooling water failure	1 to 10^{-2}
Pump seal failure	10^{-1} to 10^{-2}
Atmospheric tank failure	10^{-3} to 10^{-5}
Gasket / packing blowout	10^{-2} to 10^{-6}

Other Considerations

For operations that are not continuously operated, e.g. loading, unloading, startup/shutdown, batch processes, and maintenance, the failure frequencies must be adjusted to reflect the exposure time (or "dwell time"). For example, in a batch reactor operation, the cooling system needs to be switched on for 2 hours when an exothermic reaction takes place in the reactor. Assuming 2 batches are prepared per day, the facility operates 5 days a week and the frequency of cooling system failure is 1×10^{-2} per year, the actual frequency of cooling system failure throughout the year needs to be adjusted to reflect the actual exposure time for the potential failure:

$$f^I = 1 \times 10^{-2} \text{ (cooling system failure rate)} \times \frac{2 \times 2 \times 5 \times 52}{24 \times 365}$$

$$f^I = \underline{\underline{1.19 \times 10^{-3} \text{ per year}}}$$

Independent Protection Layers

All IPLs are safeguards, but not all safeguards are necessarily IPLs. An IPL has two main characteristics:

- The effectiveness of the IPL in preventing the scenario.
- The independence of the IPL from the initiating event and other IPLs.

3D's, 4E's and "Big I" rules

Dowell (2002) provided the following guidelines in evaluating IPLs:

- The "Three Ds" help determine if a safeguard is an IPL. They are – *Detect, Decide and Deflect.*
    - Can the IPL *detect* a condition in the scenario?
    - Can the IPL *decide* to take action or not?
    - Can the IPL *deflect* the undesired event by preventing it?
- The "Four Es (Enoughs)" help evaluate the effectiveness of an IPL. They are – *Big Enough, Fast Enough, Strong Enough and Smart Enough.*
    - Is the IPL *big enough* to handle the undesired event and prevent the undesired consequence (i.e. Is the IPL adequately sized? e.g. relief valve orifice, dike volume, pump capacity etc).
    - Is the IPL *fast enough* in detection, decision and deflection? (i.e. Does the IPL have enough time to detect the condition, process the information, make the decision, take the required effective action?)
    - Is the IPL *strong enough* to withstand the undesired event? (e.g. strength of flare sub-header to withstand relief valve forces on initial opening, sufficient strength of piping to withstand overpressures for short durations, ability for process buildings to withstand forces generated by explosions)

- Is the IPL *smart enough* to prevent the undesired consequence from happening? (An initiating action for a safeguard may be such that for it to be effective, the timing / sequence must also be compatible with other system requirements. For example, if an emergency shutdown valve, on the upstream side of a pump were to close prior to pump shutdown, severe pump cavitation could result.)

- The "Big I" – The IPL must be *independent* of the initiating event and all other IPLs. This is the main assumption in LOPA. It is important to look out for common cause failures. Common cause failure is the failure of more than one component, item, or system due to the same cause or initiating event. If common cause failure exists in a scenario, all of the safeguards affected by the common cause failure should only be considered as a single IPL.

Characteristics of Various Layers of Protection

Typical layers of protection are:

- Process Design
- Basic Process Control System (BPCS)
- Critical Alarms and Human Intervention
- Safety Instrumented System (SIS)
- Physical Protection
- Post-release Protection
- Plant Emergency Response
- Community Emergency Response

Process Design

There are usually two ways of crediting inherently safer process design in LOPA:

- Eliminate some scenarios by the inherently safer process design e.g. greater spacing, reduced inventories etc
- Treat some inherently safer process design features as IPLs but assign nonzero PFDs to them. This approach allows comparison of risk associated with various process / equipment designs based on different engineering standards / practices.

In order to ensure consistency between LOPA studies, either approach must be applied consistently.

Basic Process Control Systems, BPCS

The BPCS continuously monitors, controls and maintains the process within safe operating limits. A BPCS loop usually includes the following components (Figure 21-6):

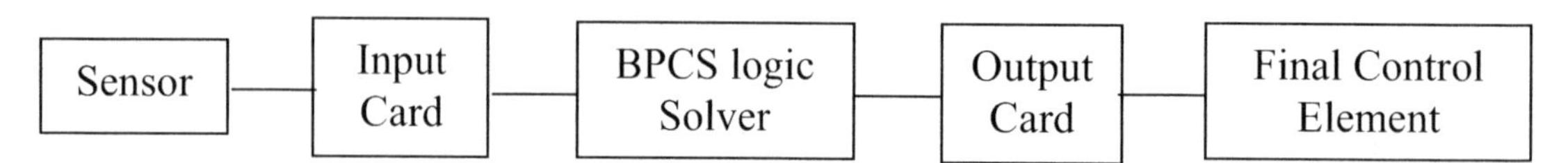

Figure 21-6: Simplified components of a BPCS loop (CCPS, 2001)

There are three different types of safety functions provided by BPCS that can be IPLs:

- *Continuous control actions* – These keep the process within the normal operating limits. For example a level controller, which maintains the liquid level in a tank, prevents overflow of the tank etc.
- *Alarm actions* – Logic solver or alarm trip units, which identify process deviations from normal operating limits and alert the operator, typically as alarm messages, to perform corrective action(s).
- *Return process to stable state* – Logic solver or control relays, which would take automatic action(s) to return the process to a stable state (e.g. A distillation

unit could be put on total recycle if unacceptable deviations in performance occurred).

The following factors should be considered in determining how much credit should be assigned to a BPCS as an IPL:

- *Adequacy of security and access procedures* – Many BPCS installations are deliberately made accessible to personnel who can change set-points, bypass alarms and interlocks. This makes BPCS susceptible to human error and this can degrade the anticipated performance of BPCS if security and control are not adequate.
- *Level of redundancy* – BPCS usually has little redundancy. However, for some sophisticated designs such as hydrocrackers and also offshore oil and gas separation (governed by API 14C), the level of redundancy of BPCS components is higher than that found in normal process control. The use of redundancy will decrease the overall PFD of the BPCS loop.
- *Historic failure rate* – In order to calculate PFD of a BPCS loop, it is essential to review failure rate data of logic solvers, input/output cards, sensors, final control elements, human response etc.
- *Effective test rate* – The reliability of a BPCS also depends on the test frequency and the effectiveness of testing.
- *Other factors* – Other factors to be considered include design, manufacture, installation and maintenance.

Note: IEC 61511 does not allow taking credit for a BPCS PFD ≤ 0.1

Critical Alarms and Human Intervention

These systems are usually activated by BPCS. Consider a furnace where the fuel gas flow control loop is not pressure compensated. The BPCS would generate an alarm on high fuel gas pressure. The operator would then take the appropriate action to control the gas pressure or shutdown the furnace. The IPL here would be the BPCS loop *and* the operator action.

The following factors should be considered in determining how much credit should be assigned to human action as an IPL:

- Detection – How will the condition be detected ? (e.g. alarm)
- Decision – How will the decision to act be made ?
- Action – What action is required to prevent the consequence?

Safety Instrumented System (SIS)

A SIS is a combination of sensors, logic solver, and final elements. It is also called a safety interlock. A SIS is functionally independent from the BPCS. The reliability of a SIS is defined in terms of its PFD and SIL. For further details, refer to Chapter 20 "Safety Integrity Levels" .

Physical Protection

Physical protection usually refers to relief valves and rupture discs. The following factors should be considered in determining how much credit should be assigned to physical protection as an IPL:

- Sizing (that includes controlling cases e.g., fire, power failure etc.)
- Design
- Installation (e.g. piping arrangement)
- Quality of inspection and maintenance
- Cleanness of process fluid (e.g. corrosive services)

Post-release Protection

Typically these refer to dikes and blast walls. These are passive IPLs usually with high reliability. The same considerations listed for physical protection should be considered in determining how much credit should be assigned to physical protection as an IPL.

Plant Emergency Response and Community Emergency Response

They are not normally considered as IPLs as they are activated after the initial release.

What may be perceived or designated as an IPL may not in fact be an IPL at all. However, there are factors which can greatly affect IPLs and PFDs and some of these are listed below:

Table 21-7: Factors relating to IPLs

Factors	Comments
Training and certification	These factors may be considered in assessing the PFD for operator action, but are not – of themselves – IPLs.
Procedures	These factors may be considered in assessing the PFD for operator action, but are not, of themselves – IPLs.
Normal testing and inspection	These activities are assumed to be in place for all hazard evaluations and form the basis for judgment in determining PFDs. Normal testing and inspection affects the PFD of certain IPLs. Lengthening the testing and inspection intervals may increase the PFD of an IPL.
Maintenance	This activity is assumed to be in place for all hazard evaluations and forms the basis for judgment to determine PFDs. Maintenance affects the PFD of certain IPLs.
Communication	It is a basic assumption that adequate communications exist in a facility. Poor communications affects the PFD of certain IPLs.
Signs	Signs by themselves are not IPLs. Signs may be unclear, obscured, ignored, etc. Signs may affect the PFD of certain IPLs.
Fire Protection	The effectiveness of fire protection as an IPL is limited to post-release scenarios and also is highly instrumental in reducing the consequences and domino effects through fire spreading. However, if a company can demonstrate that it meets the requirement of an IPL for a given scenario it may be used (e.g., if an activating system such as plastic piping or

Factors	Comments
	frangible switches are used). Fireproof insulation can be used as an IPL for some scenarios provided that it meets the requirements of API and corporate standards.
Requirement that Information is Available and Understood	This is a basic requirement and does not constitute an IPL.

Probability of Failure on Demand (PFD)

The causes of an IPL failing to perform could be due to:

- A component of an IPL being in a failed or unsafe state when the initiating event occurs (typically this could be a reflection of poor maintenance practices).
- A component failing during the performance of its task (typically due to inadequate design or lack of maintenance or factory defects)
- Human intervention failing to be effective, etc

The following table provides typical values of PFDs for various types of IPLs used in LOPA.

Table 21-8: PFD values (CCPS, 2001)

IPL	Comments	PFD
BPCS	Can be credited as an IPL if not associated with the initiating event being considered (See IEC 61508 and IEF 61511 for additional discussion)	1×10^{-1} to 1×10^{-2} (> 1×10^{-1} allowed by IEC)
Safety Instrumented function	See Chapter 20	See Chapter 20
Dike	Will reduce the frequency of large consequence (widespread spill) of a tank overfill / rupture / spill / etc.	1×10^{-2} to 1×10^{-3}
Blast-wall / Bunker	Will reduce the frequency of large consequences of an explosion by confining blast and protecting equipment / buildings / etc.	1×10^{-2} to 1×10^{-3}
Human action with 10 minutes response time	Simple well-documented action with clear and reliable indications that the action is required	1.0 to 1×10^{-1}
Human response to BPCS indication or alarm with 40 minutes response time	Simple well-documented action with clear and reliable indications that the action is required (The PFD is limited by IEC 61511; IEC 2001)	1×10^{-1} (> 1×10^{-1} allowed by IEC)
Human action with 40 minutes response time	Simple well-documented action with clear and reliable indications that the action is required.	1×10^{-1} to 1×10^{-2}

Applications of LOPA

Implementing LOPA

- *When to conduct LOPA?* - Can be conducted during or immediately following a PHA such as HAZOP or What-If.
- *Who can conduct LOPA?* - Can be applied in a team setting, usually smaller than a PHA team including the analyst, who is familiar with the LOPA methodology, and a process engineer or production specialist. The study can then be reviewed independently by one or more persons with equivalent or greater expertise.
- *Criteria for selecting scenarios used in LOPA* – Typically based on a number of factors:
    - Where there is sufficiently high severity of consequence and likelihood of a scenario generated by a PHA or equivalent.
    - Need to reduce risk to acceptable levels of criteria.
    - Uncertainty of the frequency of the final consequences for critical cases.
    - Uncertainty of the consequences for critical cases.
    - Complexity of the scenarios.
- *Establish risk tolerance criteria* – Typical methods for establishing risk tolerance criteria included:
    - Matrix Methods
    - Numerical Criteria Method
    - Number of IPL Credits

Further details are discussed in the following section.

Making Risk Decisions

After the scenarios in LOPA are established and the existing risk has been calculated, decision making takes place to determine:

- Whether the existing risk is tolerable?
- Whether the existing risk mitigation is adequate?
- How much risk mitigation is required to reduce the risk to an acceptable level?

To answer the above questions, it is essential to understand the relationship between risk and risk reduction. Figure 21-7 illustrates such a concept.

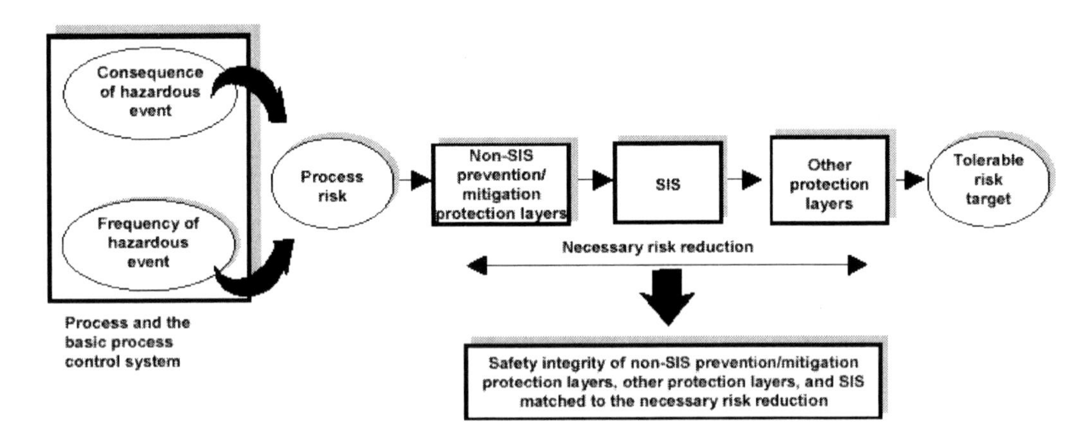

Figure 21-7: Risk and risk reduction concepts (IEC 61511-3, 2003)

The essence of risk and risk reduction concepts is the establishment of the tolerable risk target or criteria. Without the tolerable risk criteria, there may be a tendency to keep adding safeguards believing that safety is continually being improved. This could lead to a number of issues, such as:

- Adding unnecessary IPLs.
- Reducing focus on the IPLs that are critical to achieving tolerable risk.
- Taking credit for IPLs that may not be effective.

When establishing the risk tolerance criteria, the ALARP (As Low As Reasonably Practicable) principle can be applied. The risk associated with industrial activities can be classified into three regions:

- *Unacceptable region* – The activity has such a high risk that it is unacceptable .
- *Broadly acceptable region* – The activity has very low risk that is insignificant. Usually no further measures are required to reduce the risk.
- *Tolerable region* – The level of risk associated with the activity falls between the above two categories and it has been reduced to the lowest practicable level.

ALARP is based on the principle of reducing risk "so far as it is reasonably practicable" or to a level which is "As Low As Reasonably Practicable". When a risk lies between the unacceptable and broadly acceptable regions, the ALARP principle can be applied to achieve a tolerable risk for this specific application. Figure 21-8 illustrates the three regions of risk.

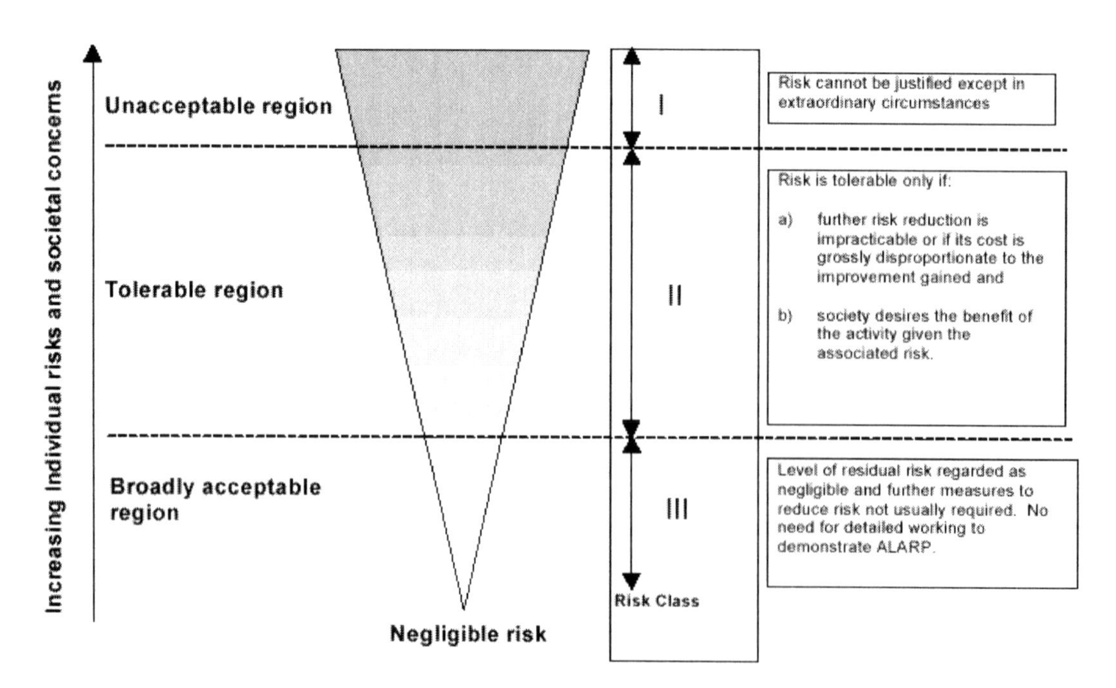

Figure 21-8: Tolerable risk and ALARP (IEC 61511, 2003)

The application of the ALARP principle requires the definition of the three regions, as shown in Figure 21-8, in terms of the likelihood and consequence of an incident. Table 21-9 is an example showing how the three risk classes (I, II, III) in Figure 21-8 are defined based on likelihood and consequence.

Table 21-9: Example of risk classification of incidents (IEC 61511-3, 2003)

	Risk Class			
Probability	**Catastrophic Consequence**	**Critical Consequence**	**Marginal Consequence**	**Negligible Consequence**
Likely	I	I	I	II
Probable	I	I	II	II
Possible	I	II	II	II
Remote	II	II	II	III
Improbable	II	III	III	III
Incredible (i.e. non-credible)	II	III	III	III
Note 1 – See Table 21-10 for interpretation or risk classes I to III Note 2 – The actual population of this table with risk classes I, II and III will be applicable dependent and also depends upon what the actual probabilities are, for example “likely”, “probable”, etc. Therefore, this table should be seen as an example of how such a table could be populated, rather than as a specification for future use.				

Table 21-10: Interpretation of risk classes (IEC 61511-3, 2003)

Risk class	**Interpretation**
Class I	Intolerable risk
Class II	Undesirable risk, and tolerable only if risk reduction is impracticable or if the costs are grossly disproportionate to the improvement gained.
Class III	Negligible risk

Typical approaches for comparing the existing risk with a predetermined risk tolerance criteria are:

- Matrix Method
- Numerical Criteria Method
- Number of IPL Credits

For all of the above approaches, cost-benefit analysis may be used to help make the final risk-reduction decisions.

Matrix Method

The matrix method was introduced in the two-phase separator example (Table 21-3). Also refer to Chapter 18 "Managing the Justifying Recommendations" for the pros and cons of different types of matrices.

Numerical Criteria Method

The risk criteria established using this approach is based on a maximum tolerable risk per scenario and a variety of consequence categories, such as:

- Human injury (normally expressed in terms of mortality)
- Environmental impact
- Property damage dollar loss
- Loss of production dollar loss
- Releases of hazardous materials
- Fire
- Explosion

For example, an organization may establish the tolerable risk criteria as a maximum frequency (per year or per 1000 hours) of a single fatality.

Number of IPL Credits

This method specifies the number of IPL credits for scenarios of certain consequence levels and frequency (see Table 21-11). Hence, the tolerable criteria are not shown explicitly in this method. This method typically assigns a PFD of 1×10^{-2} to 1 IPL credit. The number of credits assigned to a scenario depends on the severity and frequency of the event. Table 21-11 focuses on human injury and fatality; a similar approach can be applied to other types of consequences such as production loss and environmental impact. In order to account for the various types of consequences, the LOPA calculation needs to take into account the adjustment factors such as enabling event probabilities and conditional modifiers in the frequency calculation.

Table 21-11: IPL Credit Requirements (CCPS, 2001) ***N.B: The following table of values illustrates the methodology***

	Number of IPL Credits Required*	
Adjusted Initiating Event Frequency**	**Consequence Category IV One Fatality**	**Consequence Category V Multiple Fatalities**
Frequency $\geq 1\times10^{-2}$	2	2.5
1×10^{-2} > Frequency $\geq 1\times10^{-3}$	1.5	2
1×10^{-3} > Frequency $\geq 1\times10^{-4}$	1	1.5
1×10^{-4} > Frequency $\geq 1\times10^{-6}$	0.5	1
1×10^{-6} > Frequency	0	0.5

* Adjusted Initiating Event Frequency includes adjustments to the initiating event frequency for $P^{ignition}$, P^{person} present and $P^{fatality}$.

** An IPL Credit is defined as a reduction in event frequency of 1×10^{-2}.

Documenting LOPA

Table 21-12 is a typical template for documenting LOPA. The risk matrix method [make reference to risk matrix in example]is used to establish the tolerable risk criteria. The definitions of column headers are given in Table 21-13.

Table 21-12: Typical LOPA Template

Node: 1. Reboiler EX-103

Consequence		Initiating Event (event per yr)		Enabling Event or Conditions		Conditional Modifiers		Unmitigated Event (event per yr)			Independent Protection Layers			Mitigated Event (event per yr)			Actions Required
Des.	S	Des.	Freq	Des.	Prob.	Des.	Prob.	Freq.	L	RR	Des.	Types	PFD	Freq.	L	RR	
Overpressure potential for distillation column and reflux drum with possibility for leakage, rupture, injury, or fatalities.	4	1. Tube rupture	1.0E-4	1. Probability of defective material of construction can lead to tube rupture.	1.0E-1	1. Probability of personnel affected	1.0	5.0E-6	3	6	1. PV-106 opens to flare	BPCS	5.0E-1	2.5E-9	1	4	4. Add high pressure interlock to restrict actuating air to steam supply valve TV-126, thereby closing control valve.
						2. Probability of fatal injury	5.0E-1				2. PSV	Pressure Relief Device	1.0E-2				
											3. Add high pressure interlock to close TV-126	SIS	1.0E-1				
				2. Operator inadvertently closes cooling water valve.	1.0E-1	2. Probability of taking the wrong action	3.0E-1				2. PSV	Pressure Relief Device	1.0E-2				
											3. PV-106 opens to flare	SIS	1.0E-1				
Rupture of column and damage to flare system due to air in column.	5	1. Startup after annual shutdown	1.0	1. Rapid introduction of feed into distillation unit	5.0E-1	1. Probability of fatal injury	5.0E-1	8.0E-3	5	11	1. PSV (likely to be only partially effective)	Pressure Relief Device	1.0E-1	8.0E-6	1	5	5. Nitrogen purging (including N2 facility) to be incorporated into the design of the light ends distillation unit.
				2. Failure to purge the system of air	8.0E-1	2. Probability of personnel affected	8.0E-1				2. Nitrogen purging prior to column startup.	Other IPL	1.0E-2				
						3. Probability of ignition	5.0E-2										

Table 21-13: Definitions of column headers in LOPA template

Consequence	*Des.* - Description of the final consequence without taking into account the existing safeguards. *S* - The severity ranking of the consequence.
Initiating Event	*Des.* – Description of the initiating event (or cause) together with any assumptions made to establish the initiating event frequency. *Freq.* – Initiating event frequency (typically in "event per year" or event hour")
Enabling Event or Conditions (if applicable)	*Des.* – Description of the enabling event or conditions together with the assumptions used for the values specified in "*Prob.*" column. *Prob.* – Probability that the specified enabling event or conditions would take place.
Conditional Modifiers (if applicable)	*Des.* – Description of the conditional modifiers together with the assumptions used for the values specified in "*Prob.*" column. *Prob.* – Probability used to model the outcome of the consequence.
Unmitigated Event	*Freq.* - This is the event frequency **without taking into account** the existing IPLs. It is the product of the initiating event frequency, the enabling event or conditions probability (s) (if applicable) and the conditional modifiers probability (s) (if applicable). It is typically in "event per year" or event hour". *L* – The likelihood ranking based on the unmitigated event frequency. *RR* – The risk ranking established based on the likelihood ranking, *L* and the severity ranking, *S*, of the consequence.
Independent Protection Layers	*Des.* – Description of the IPL. *Types* – Type of IPL, such as BPCS, Process Design, Operator's Action, SIS, Pressure Relief Device, Other IPL *PFD* – Probability of failure on demand of the IPL.
Mitigated Event	*Freq.* - This is the event frequency **taking into account** the existing IPLs. It is the product of the initiating event frequency, the probability(s) of enabling event or conditions (if applicable), the probability (s) of the conditional modifier(s) (if applicable) and PFDs of existing IPLs. It is typically in "events per year" or events per hour". *L* – The likelihood ranking based on the mitigated event frequency. *RR* – The risk ranking established based on the likelihood ranking, *L* and the severity ranking, *S*, of the consequence.
Action Required	Define the required actions / recommendations. See Chapter 18 "Managing and Justifying Recommendations" for details on how to document the required actions / recommendations.

Benefits of using LOPA

- Requires less time and resources than for a QRA but is more rigorous than HAZOP by itself.
- Many process safety systems are over-engineered for safety with additional costs and have unnecessary complexity. LOPA helps focus the resources on the most critical safety systems.
- Acts as a decision making tool, helps make judgments quicker, resolves conflicts and provides a common base for discussing risks of a scenario.
- Reduces subjectivity while providing clarity and consistency for risk assessment.
- Improves scenario identification by pairing of the cause and consequence from PHA studies.
- Helps to compare risks based on a common ground if it is used throughout a plant.
- Helps decide if the risk is As Low As Reasonably Practicable (ALARP) for compliance to regulatory requirements or standards.
- Identifies operations, practices, systems and processes that do not have adequate safeguards.
- Provides basis for specification of IPLs as per IEC 61511.
- Helps to decide which safeguards to focus on during operation, maintenance and related training.
- Support compliance with process safety regulations - including OSHA PSM 1910.119, Seveso II regulations and IEC 61511.

Disadvantages of LOPA

- LOPA requires more time to reach a risk-based decision than qualitative methods such as HAZOP and What-if.
- Compare to qualitative PHA methods, LOPA requires more time and effort to learn.
- LOPA requires failure rate data to support the methodology. Such data can be difficult to find.
- LOPA is not appropriate for handling complex scenarios such as where multiple shutdown components are linked by a single event such as fire or toxic release requiring complete facility shutdown.
- LOPA is not a hazard identification tool. It relies on other tools like HAZOP to identify hazardous scenarios but provides a semi-quantitative risk evaluation

SUGGESTED READING (URLs current at time of publication)
"Layer of Protection Analysis: Simplified Process Risk Assessment" by AIChE, CCPS, 1st edition, 2001 www.aiche.org/pubcat/seadtl.asp?ACT=S&Keyword=ON&Title=ON&ISBN=ON&Pubnum=ON&srchText=LOPA
"Safety Integrity Level Selection – Systematic Methods Including Layer of Protection Analysis" by Edward M. Marszal, Dr. Eric w. Scharpf, published by ISA, 2002 www.isa.org/Template.cfm?Section=Books1&template=/Ecommerce/ProductDisplay.cfm&ProductID=4517
"Use Layer of Protection Analysis (LOPA) to Comply with Performance-based Standards" by K Bhimavarapu, Control Engineering, February 1, 2002 www.manufacturing.net/ctl/index.asp?layout=article&articleid=CA188586

CHAPTER 22

Quantitative Risk Assessment

ASSESSING & MANAGING RISK

Risk Management strives to reduce hazards to acceptable levels of risk. In the context of hazardous materials it means ensuring the risks are acceptable or tolerable to employees and the public. Before we can manage such risks we must first understand them. This involves conducting a ***risk analysis***, a quantitative evaluation of the consequences of incidents and their likelihood.

Having evaluated the risk, judgements need to be made about risk acceptability. This process is referred to as ***risk appraisal***. Its outcome is the identification of appropriate risk criteria. These vary according to the risk receptors, which could be employees, the public, the environment or the financial impact to the corporation. In contrast to the risk analysis, an objective and technical (engineering) activity, the risk appraisal involves subjective value judgements. In a ***risk assessment***, the results of the risk analysis are compared against the risk criteria in order to draw conclusions about risk acceptability.

Managing risk implies decision-making (see Figure 22-1). If the estimated risk to the risk receptor of interest is acceptable, there is no need to reduce risk. However, a risk management plan ought to include a mechanism for monitoring risk to ensure that the situation does not deteriorate over time. If the risk is not acceptable, then risk reduction options should be evaluated and the most feasible option implemented. This component of risk management is called ***risk control***. In the context of reducing risk it is important to point out that no activity is totally free from risk. Risk can never be totally eliminated, but it can usually be reduced to "acceptable" levels.

RISK ANALYSIS

The first step in a risk analysis is to identify hazards - the ***hazard identification*** phase (Process Hazards Analysis). Depending on the plant or process system a large number of release scenarios can be identified. Risk analyses typically identify only a limited number for evaluation purposes via some sort of identification and screening process. These should cover the range of potential impacts. Thus "Worst possible" and "Worst Credible" type scenarios need to be identified.

Given that a potential release has been identified, the hazard identification phase should also include consideration of post-release influencing factors. For example:

- Is the release a Vapor or Liquid?
- Could we be faced with a toxic release situation, fire or explosion hazard?
- Is the point of release in a confined location?
- Is available mitigation successful?
- Is shielding possible or sheltering?
- Is the wind blowing towards the hypothetical receptor of interest?
- What are the meteorological conditions (stability class; wind speed; ambient temperature)?

Risk management is the process of evaluating risk reduction alternatives with respect to economic considerations. Risk cannot be managed unless it has been assessed both before and after risk control measures have been evaluated.

The value of risk control lies in the overall level of risk reduction that can be achieved with risk control measures in place.

The main steps in the process are:

1. Identify the potential hazards; also identifying potential safeguarding (i.e., Risk Control) features that could be introduced.
2. Estimate the consequences.
3. Estimate the frequencies of the various consequences.
4. Determine the risk without considering any new safeguarding (i.e., Risk Controls) features.
5. Determine the risk with proposed new safeguarding.
6. Evaluate economic impacts of proposed new safeguarding.
7. Compare improvement in risk versus economic impact. Optimize as far as possible.
8. Propose modifications (if needed) to reduce risk to achieve acceptable standards.

Figure 22-1 also shows the risk management framework, which integrates hazard identification, assessment and control.

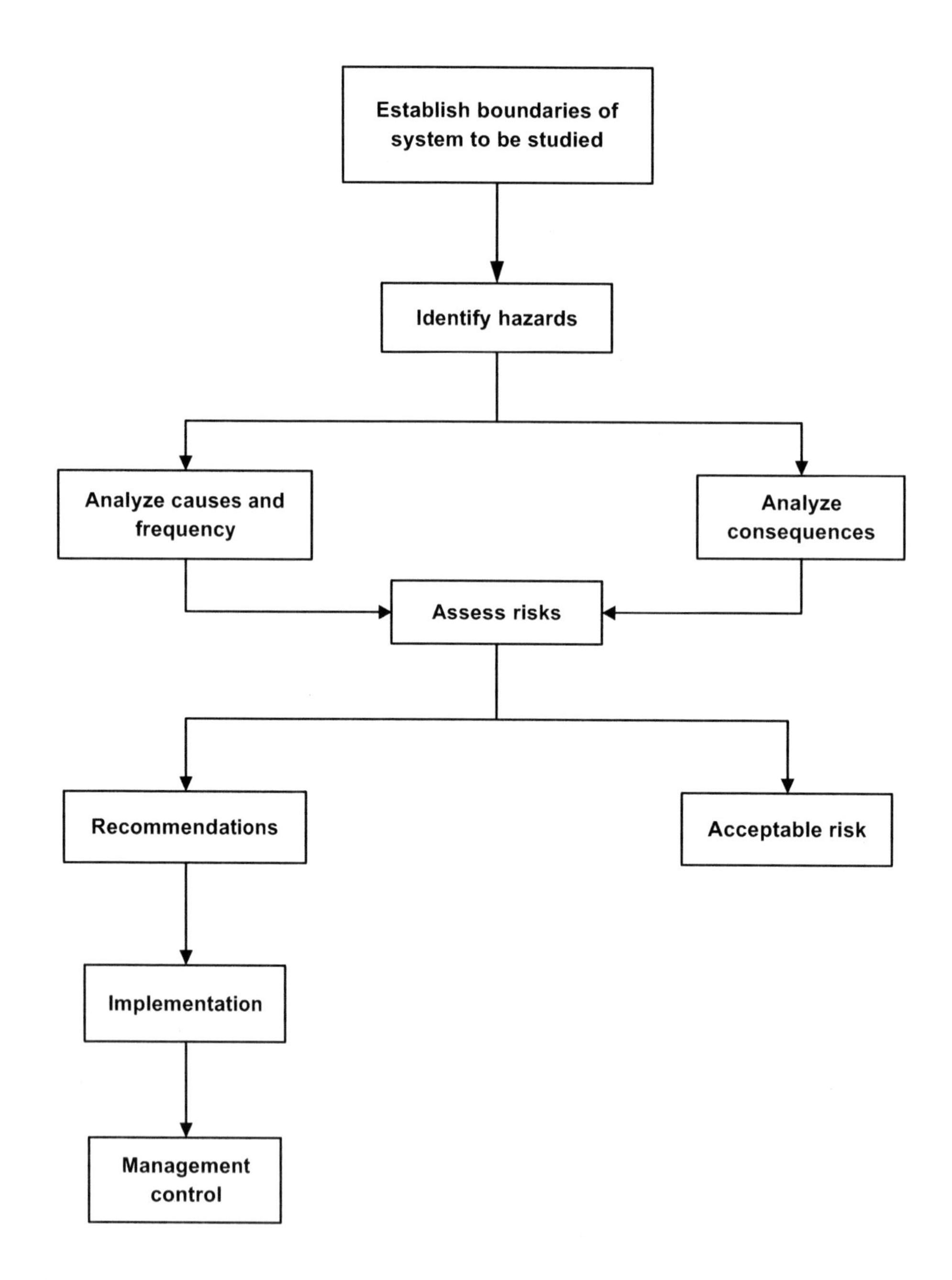

Figure 22-1: Risk Management Framework

FOR A CHEMICAL RELEASE HAZARD (e.g., chlorine, anhydrous ammonia etc. release)

A chemical hazard scenario can be defined in terms of a release scenario and post release influencing factors. For typical releases, the frequency of a chemical hazard scenario can be described by the following equation:

$$F_{HS} = F_{RE} \cdot P_{WD} \cdot P_{ME} \cdot P_S \cdot P_{MI}$$

Where,

F_{HS} = frequency of hazard scenario (yr^{-1})

F_{RE} = frequency of release scenario (yr^{-1})

P_{WD} = probability of wind direction

P_{ME} = probability of meteorological conditions

P_S = probability of failing to take shelter (optional)

P_{MI} = probability of failure for mitigation measures

FOR AN EXPLOSION HAZARD

An explosion hazard scenario can be defined in terms of a release scenario and post release probabilities of (a) an explosion occurring and (b) that mitigation measures fail. For typical releases, the frequency of an explosion hazard scenario can be described by the following equation:

$$F_{HS} = F_{RE} \cdot P_E \cdot P_{MI}$$

Where,

F_{HS} = frequency of explosion hazard scenario (yr^{-1})

F_{RE} = frequency of release scenario (yr^{-1})

P_E = probability that explosion occurs following release

P_{MI} = probability of failure for mitigation measures

FOR A FIREBALL HAZARD

A fireball hazard scenario can be defined in terms of a release scenario and post release probabilities of (a) a fireball occurring and (b) that individual is unable to shelter and (c) that mitigation measures will fail. For typical releases, the frequency of a fireball hazard scenario can be described by the following equation:

$$F_{HS} = F_{RE} \cdot P_F \cdot P_S \cdot P_{MI}$$

Where,

F_{HS} = frequency of fireball hazard scenario (yr^{-1})

F_{RE} = frequency of release scenario (yr^{-1})

P_F = probability that fireball occurs following release

P_S = probability of failing to take shelter (optional)

P_{MI} = probability of failure for mitigation measures

FOR A POOL FIRE HAZARD

A pool fire hazard scenario can be defined in terms of a release scenario and post release probabilities of (a) a pool fire occurring and (b) that individual is unable to shelter and (c) that mitigation measures will fail. For typical releases, the frequency of a pool fire hazard scenario can be described by the following equation:

$$F_{HS} = F_{RE} \cdot P_{PF} \cdot P_S \cdot P_{MI}$$

Where,

F_{HS} = frequency of fireball hazard scenario (yr^{-1})

F_{RE} = frequency of release scenario (yr^{-1})

P_{PF} = probability that pool fire occurs following release

P_S = probability of failing to take shelter (optional)

P_{MI} = probability of failure for mitigation measures

CALCULATION OF TOTAL RISK

The last step in a risk analysis is to combine the consequence information with the frequency information. Mathematically, individual risk is calculated by the following equations:

$$R_{HSj}(x) = F_{HSj} \times P_{DIj}(x)$$

$$R_T(x) = \sum_{j}^{N} R_{HSj}(x)$$

Where,

$R_T(x)$ = Total risk at a distance x from the hazard source - this represents the annual probability of fatality or serious injury to a hypothetical receptor.

$R_{HSj}(x)$ =Risk due to hazardous scenario j at a distance x from the hazard source

$P_{PIj}(x)$ = Probability of fatality or serious injury - i.e. the consequence, of hazardous event j at a distance x from the hazard source

F_{HSj} = Frequency of event j (yr^{-1})

N = Number of hazardous scenarios evaluated for risk

Note: The number and type of hazards to be concomitantly considered will also have to be evaluated, since risk is an additive function.

RISK MEASUREMENT

Risk can be measured in terms the following over a specific time period:

- Numbers of injuries caused,
- Numbers of deaths caused, and
- Damage incurred (financial)

When the focus is on safety, the relevant parameters are death and injury. However, since injuries are hard to quantify and cover a wide spectrum, they are less easily quantifiable than mortality. Therefore, mortality has become a standard method of measuring risk.

RISK ESTIMATION & ACCEPTABILITY CRITERIA

Types of Risk

Individual Risk The risk posed to an individual who is exposed to a hazardous activity. For example, the annual risk of death due to smoking for an individual smoker is about one death per year per 330 individuals who smoke or 3 x 10^{-3} deaths/year/individual.

Societal Risk The risk posed to a societal group who are exposed to a hazardous activity. It is the summation of individual risk for actual persons within the vicinity in question (and considers population distribution). Thus,

$$S_{T,i} = \sum_{i}^{T} S_i$$

$$S_i = F_i C_i$$

Where:

$S_{T,i}$ = overall societal risk

S_i = societal risk for individual i

F_i = frequency of hazard i (events/year)

C_i = consequences of hazard i (deaths/year)

Chronic Risk Typically, it is the environmental risk arising through releases into water, soil and the atmosphere. Chronic = long term effects, such as carcinogenic conditions (cancers)

Acute Risk Typically, rapid or short term caused by exposure to fire, blast or highly toxic chemicals.

Acute = short-term effects, usually burns, damage to body or death

Usually expressed as percentage chance of mortality.

Property Risk Damage to property and usually divided into:

- Onsite risk to plant
- Offsite risk to surrounding area

Voluntary Risk Is risk that you choose to expose yourself to, such as:

- Rock climbing
- Skiing
- Motor racing
- Sky diving

It can also be considered as voluntary risk when an employee chooses to work in a work place and knowingly submits him/herself to work of a hazardous nature, e.g., fire fighters.

Involuntary Risk Is risk that is imposed either on an individual or community that is deemed beyond their control or without their knowledge or agreement, e.g., effects of plant explosions/releases from a chemical plant in the neighborhood.

Environmental Risk Is risk that is imposed on the environment such that use by humans, plants, or animals is curtailed in a detrimental manner.

Comparative Risk

Comparative risk is the frequency of occurrence of specified consequences from a representative selection of hazardous events associated with an activity under alternative situations. Risk estimates include a range of assumptions plus uncertainties in data. Therefore, greater confidence can be placed in risk-based decisions when made on a comparative basis, since errors that are introduced will be common to each of the alternatives.

Uncertainty in Risk Estimation

Risk estimates are based upon:

- Available failure rate data
- Modeled data (mathematical estimates)
- Best estimates

Such data has error bands associated with it (upper/lower). Main concern is with underestimation of risk; therefore, upper bound error band is of greater importance.

RISK APPRAISAL

Risk Appraisal is the process of identifying risk criteria and is required for risk-based decision making

Public risk criteria may be applied to individuals living near hazardous facilities (individual risk criteria) or the society at large (societal risk criteria); in this latter case f/n curves may also be a consideration. One basis of acceptability applied by previous (MIACC) criteria is as follows.

- A negligible level of risk is deemed to be equal to 1% of the accidental death rate for the general population (considers all accidents - motor vehicles, fires, drownings, etc.). This is the additional risk to members of the public living near hazardous facilities. This level is about equal to an annual probability of fatality of one-in-a-million (or 1×10^{-6}).
- An unacceptable level of risk is considered to be a factor of 100 greater than the acceptable risk level - or about one hundred-in-a-million (or 1×10^{-4}).

Occupational risk to employees is deemed to be "voluntary". This assumes that they are aware of the risks present in their working environment and are trained to deal with such emergencies. In Canada, there are no guidelines or legislative requirements on occupational risk. However, it is generally considered that there ought to be about an

order-of-magnitude difference between voluntary and involuntary risks. Therefore, a negligible occupational risk would be about 1×10^{-5} per year (annual probability of death or serious injury). An unacceptable level of risk would be about 1×10^{-3} per year. In between, occupational risks are considered tolerable only if emergency procedures are in place and if cost-effective measures to reduce risk have been implemented.

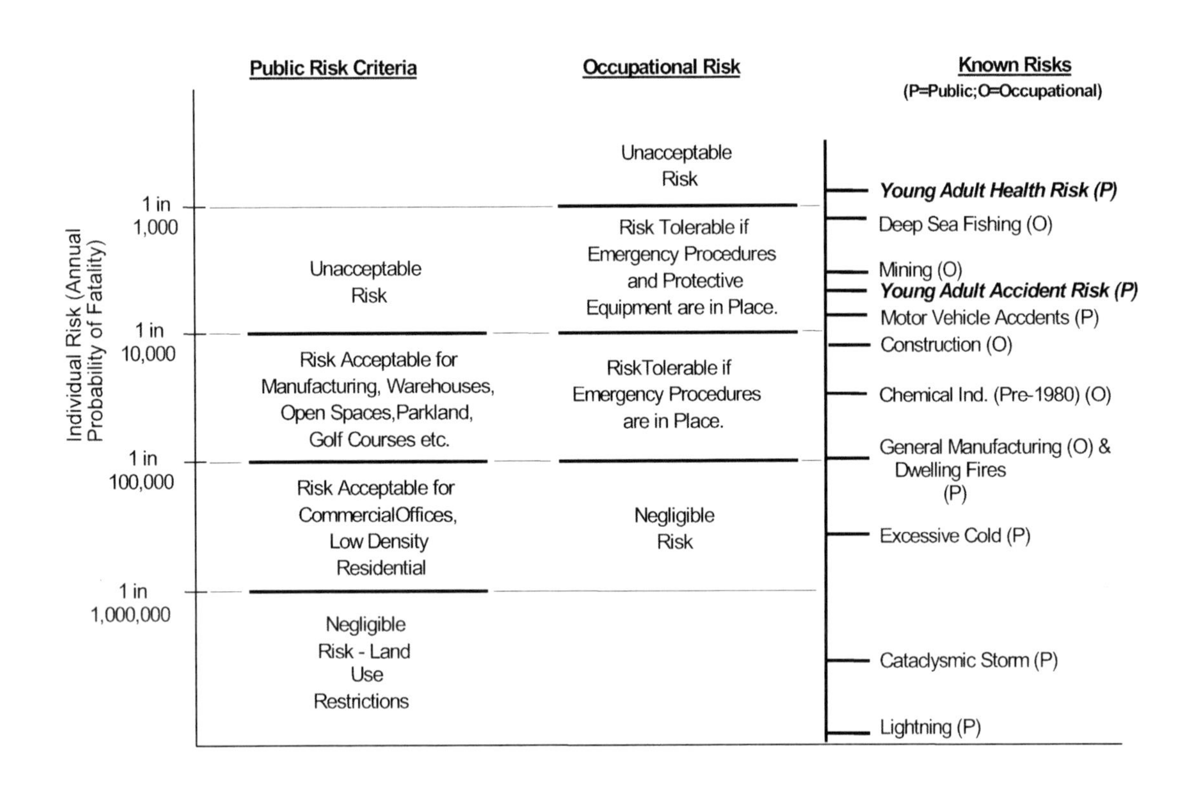

Figure 22-2 (a) Comparison of Public (MIACC) and Occupational Risk Criteria

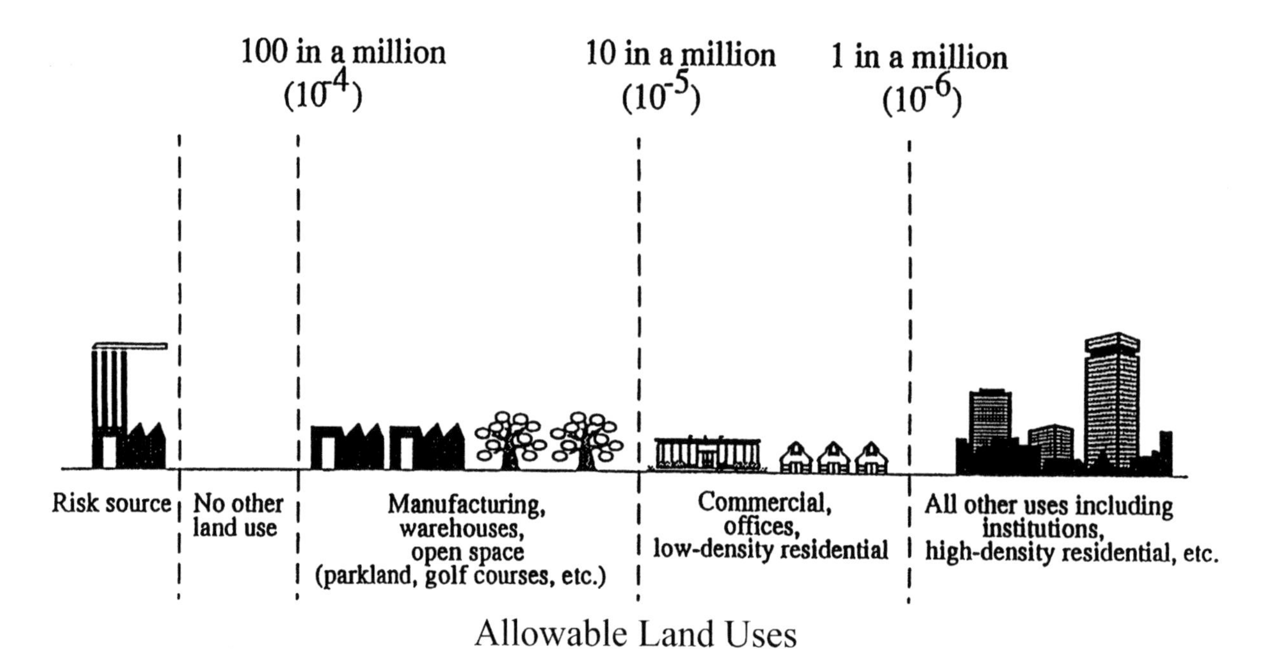

Figure 22-2(b): MIACC's Risk Acceptability Criteria

Note: (Former) Major Industrial Accident Council of Canada (MIACC) established these guidelines in 1994

RISK ASSESSMENT RESULTS AND LAND USE PLANNING

Based upon Figure 22(b) the MIACC criteria shown are specific to land use planning within Canada. UK Health and Safety Executive (HSE) have published "Risk Criteria for Land-user Planning in the Vicinity of Major Industrial Hazards," recommending that:

1. For new developments with less than 25 residents, 10^{-5} deaths/year as a maximum level of individual risk

2. For new developments with less than 75 residents, 10^{-6} deaths/year as a maximum level of individual risk

3. For schools, hospitals and old peoples' homes where persons are exposed to an involuntary risk, 10^{-7} deaths/year as maximum level of individual risk

RISK ASSESSMENT AND EMERGENCY RESPONSE PLANNING

A Risk Assessment identifies situations where accident prevention, an emergency response plan or where land use restrictions may be required. The output of a risk analysis is a risk-distance plot usually based on fatalities.

Depending on the findings a detailed ERP may be created. This will typically address the ERPG levels 1, 2 and 3.

Note: Distance to the ERPG-3 level (the maximum airborne concentration below which it is believed that nearly all individuals could be exposed for one hour without experiencing or developing life-threatening health effects). Distance to the ERPG-2 level (the maximum airborne concentration below which it is believed that nearly all individuals could be exposed for one hour without experiencing irreversible or other serious health effects or symptoms that could impair their abilities to take protective action).Distance to

the ERPG-1 level (the maximum airborne concentration below which it is believed that nearly all individuals could be exposed for one hour without experiencing other than mild transient adverse health effects or perceiving a clearly objectionable odor).

Consideration on how to handle community awareness and responsible planning for emergencies is an extremely sensitive issue. The following is a graph depicting Individual Risk Vs. Distance for an LPG Terminal.

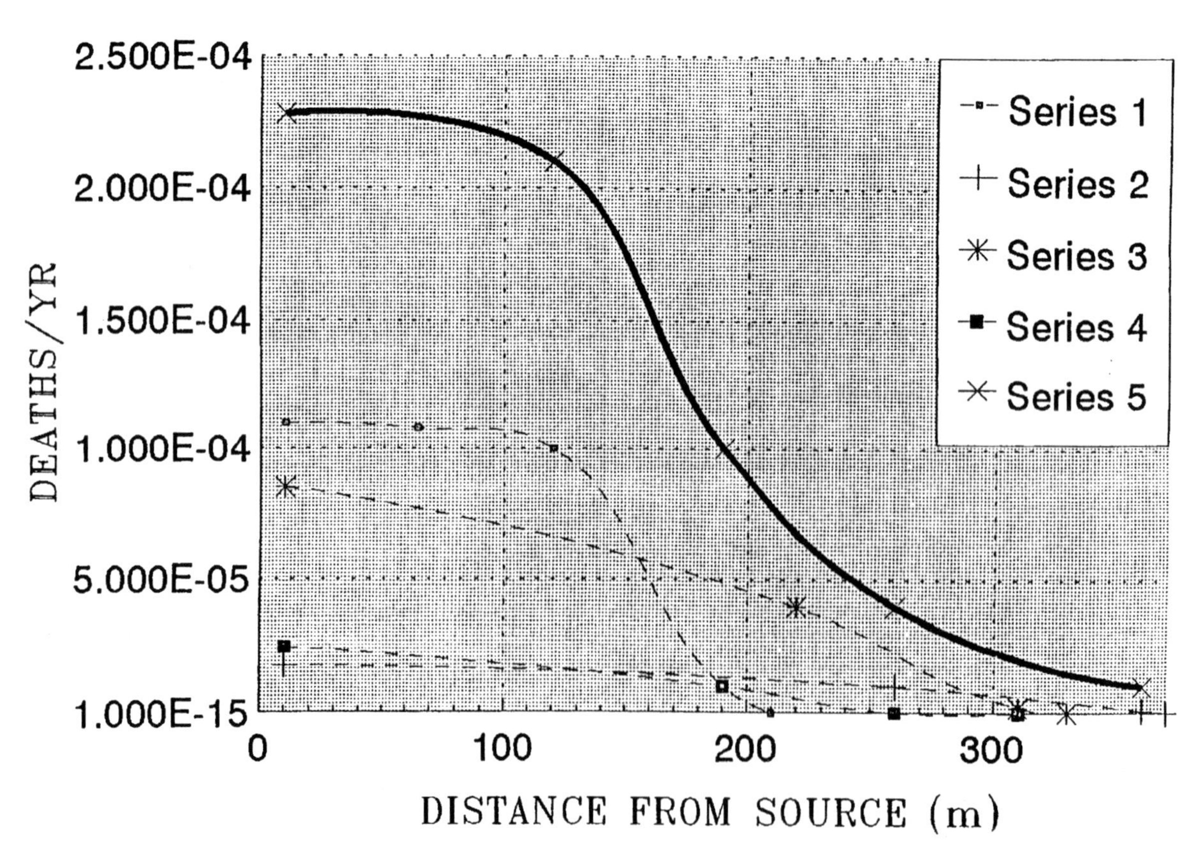

Scenarios include overfilling, hose rupture, derailment, and vandalism.
Thick line on graph represents the sum of dotted lines.

Figure 22-3: Individual Risk vs. Distance for LPG Terminal

Risk Acceptability Criteria

Risk analysts rely principally on risk acceptability criteria established by government agencies or studies, which have specifically addressed the issue. Foremost agencies include:

- Health & Safety Executive (HSE), UK.
- Ministry of Housing, Physical Planning & Environment, the Netherlands.

Comparative Common Risks

Normal everyday risk can serve as a basis for gauging risk. Caution needs to be exercised when making these comparisons since people are willing to accept higher risk levels when the risks are voluntary (*e.g., hanging gliding*) as opposed to risks that are involuntary (*e.g., living near a dangerous facility*).

The following pages contain tables depicting Comparative Risk Data, Table 22-1(a), Risks Estimated to Increase the Probability of Death in Any Year by One Chance in a Million, Table 22-1(b) and How People See It, Table 22-2.

Comparative Risk Data

Table 22-1(a): Example of Mortality Statistics (Ref: Mortality Statistics for USA, 1974 and revised, 2000, Chemical Manufacturer's Association)

Hazard	Total Number of Deaths	Individual Chance of Death per Year[a]
Heart disease	757,075	3.4×10^{-3}
Cancer	351,055	1.6×10^{-3}
Work accidents	13,400	1.5×10^{-4}
All accidents	105,000	4.8×10^{-4}
Motor vehicles	46,200	2.1×10^{-4}
Homicides	20,465	9.3×10^{-5}
Falls	16,300	7.4×10^{-5}
Drowning	8,100	3.7×10^{-5}
Fires, burns	6,500	3.0×10^{-5}
Poisoning by solids or liquids	3,800	1.7×10^{-5}
Suffocation, ingested objects	2,900	1.3×10^{-5}
Firearms, sporting	2,400	1.1×10^{-5}
Railroads	1,989	9.0×10^{-6}
Civil aviation	1,757	8.0×10^{-6}
Water transport	1,725	7.8×10^{-6}
Poisoning by gases	1,700	7.7×10^{-6}
Pleasure boating	1,446	6.6×10^{-6}
Lightning	124	5.6×10^{-7}
Hurricanes	93	4.1×10^{-7}
Tornadoes	91	4.1×10^{-7}
Bites and stings	48	2.2×10^{-7}

[a] These statistics are based on continuous exposure of the total U.S. population in 1974 or other years for which data were available.

Table 22-1(b): Risks Estimated to Increase the Probability of Death in Any Year by One Chance in a Million

Activity	Cause of Death
Smoking 1.4 cigarettes	cancer, heart disease
Drinking .5 liter of wine	cirrhosis of the liver
Spending 1 hour in a coal mine	black lung disease
Spending 3 hours in a coal mine	accident
Living 2 days in New York or Boston	air pollution
Traveling 6 minutes by canoe	accident
Traveling 10 miles by bicycle	accident
Traveling 300 miles by car	accident
Flying 1000 miles by jet	accident
Flying 6000 miles by jet	cancer caused by cosmic radiation
Living 2 months in Denver	cancer caused by cosmic radiation
Living 2 months in average stone or brick building	cancer caused by natural radioactivity
One chest X ray taken in a good hospital	cancer caused by radiation
Living 2 months with a cigarette smoker	cancer, heart disease
Eating 40 tablespoons of peanut butter	liver cancer caused by aflatoxin B
Drinking Miami drinking water for 1 year	cancer caused by chloroform
Drinking 30 12 oz cans of diet soda	cancer caused by saccharin
Living 5 years at site boundary of a typical nuclear power plant	cancer caused by radiation
Drinking 1000 24-oz soft drinks from plastic bottles	cancer from acrylonitrile monomer
Living 20 years near a polyvinyl chloride plant	cancer caused by vinyl chloride (1976 standard)
Living 150 years within 20 miles of a nuclear power plant	cancer caused by radiation
Living 50 years within 5 miles of a nuclear power plant	cancer caused by radiation
Eating 100 charcoal-broiled steaks	cancer from benzopyrene

Source: Adapted from Wilson, R., "Analyzing the Daily Risks of Life." *Technology Review*, 81, 1979, pp. 40–46.

Note: These data are based on simple extrapolations from population averages. Some data are based on actuarial statistics (e.g., coal mine accidents) and others are based on theoretical models (e.g., cancers from chlorinated water).

Table 22-2: Risk: How People See It (Ref: Dun's Review by Dun & Bradstreet)

	Activity (Estimated Deaths per Year)	Risk Ranking by Group		
		League of Women Voters	College Students	Business & Professional Club Members
1.	Smoking (150,000)	4	3	4
2.	Alcoholic beverages (100,000)	6	7	5
3.	Motor vehicles (50,000)	2	5	3
4.	Handguns (17,000)	3	2	1
5.	Electric power (14,000)	18	19	19
6.	Motorcycles (3,000)	5	6	2
7.	Swimming (3,000)	19	30	17
8.	Surgery (2,800)	10	11	9
9.	X-rays (2,300)	22	17	24
10.	Railroads (1,950)	24	23	20
11.	General (private) aviation (1,300)	7	15	11
12.	Large construction (1,000)	12	14	13
13.	Bicycles (1,000)	16	24	14
14.	Hunting (800)	13	18	10
15.	Home appliances (200)	29	27	27
16.	Fire fighting (195)	11	10	6
17.	Police work (160)	8	8	7
18.	Contraceptives (150)	20	9	22
19.	Commercial aviation (130)	17	16	18
20.	Nuclear power (100)	1	1	8
21.	Mountain climbing (30)	15	22	12
22.	Power mowers (24)	27	28	25
23.	High school & college football (23)	23	26	21
24.	Skiing (18)	21	25	16
25.	Vaccinations [b]	30	29	29
26.	Food coloring [b]	26	20	30
27.	Food preservatives [b]	25	12	28
28.	Pesticides [b]	9	4	15
29.	Prescription antibiotics [b]	28	21	26
30.	Spray cans [b]	14	13	23

[b] Death estimates not available.

RISK CONTROL (RISK MITIGATION)

Table 22-3:Typical Risk Control Measures

Measure	Active or Passive in Order to be Effective	Applicable to New Facilities	Applicable to Existing Facilities
Good access roads	Passive	Yes	Maybe
Increase buffer zones	Passive	Yes	No
Improve facility layout/spacing	Passive	Yes	No
Additional containment for hazardous materials (e.g. double walls)	Passive	Yes	No
Add fireproofing	Passive	Yes	Probably
Bury critical cabling	Passive	Yes	Maybe
Bury fire mains to protect from blast	Passive	Yes	Maybe
Blast protection for critical buildings	Passive	Yes	Maybe
Safe location of control center	Passive	Yes	No
Provide safe havens	Passive	Yes	Yes
Additional spill containment	Passive	Yes	Maybe
Add access and escape routes	Passive	Yes	Maybe
Add crash barriers	Passive	Yes	Yes

Table 22-4: Process Specific Measures

Measure	Active or Passive in Order to be Effective	Applicable to New Facilities	Applicable to Existing Facilities
Change process to less aggressive conditions, e.g. lower pressures	Passive	Yes	No
Change process chemistry	Passive	Unlikely	No
Reduce process inventories	Passive	Yes	No
Better controls, alarms, interlocks	Active	Yes	Maybe
Better pressure relief systems	Passive (normally)	Yes	Maybe
Additional isolation valves, etc.	Active	Yes	Maybe
Control ignition sources	Passive	Yes	Maybe
Automated shutdown systems	Active	Yes	Maybe
Better scheduling of hazardous activities	Passive	Yes	Yes
Better operating and maintenance practices	Active	Yes	Yes
Sparing of critical equipment	Active	Yes	Maybe
Corrosion/erosion monitoring	Active	Yes	Yes
Critical piping mods, to reduce stress	Passive	Yes	Yes

Table 22-5: Emergency Measures

Measure	Active or Passive in Order to be Effective	Applicable to New Facilities	Applicable to Existing Facilities
Emergency response plan	Active	Yes	Yes
Add fire monitors	Active	Yes	Maybe
Add fire detection	Active	Yes	Yes
Add deluge systems	Active	Yes	Maybe
Add water curtains, stream curtains	Active	Yes	Maybe
Add safety console to plot hazardous releases	Active	Yes	Maybe
Better protective equipment for personnel	Active	Yes	Yes
Additional training programs	Active	Yes	Yes
Better construction practices	Active	Yes	Less applicable
Introduce Process Safety Management systems	Active	Yes	Yes

RELATIONSHIP BETWEEN EVENTS (INCIDENTS) AND EFFECTS (IMPACTS)

A consequence can be divided into two parts:

- The event or incident itself, e.g., fireball, and
- The effect or impact caused, e.g., death, injury, or damage.

These can be related by what is known as a probit equation. A probit is a PROBability unIT, Pr, and has the form:

$$Pr = a + b\,\{\ln(V)\}$$

Where Pr is the probit value, V is the causative variable and a and b are probit constants based on specific exposures.

Examples of causative variables include:

Peak Overpressure

Impulse

Effective Exposure Time

Effective Radiation Intensity

Concentration

A form of probit equation frequently used for chemical exposure is:

$$Pr = a + b\{\ln(C^n\, t)\}$$

Where:

- a, b, and n are parameters dependent upon the toxic or harmful nature of the hazard. n lies usually between 0.6 and 3.
- C is the concentration or exposure dosage, usually in parts per million.
- t is the exposure time, usually in minutes.

In cases where the exposure concentration may vary the term $C^n t$ is replaced by the integral

$\Sigma C^n_i \Delta t_i$

Once the probit unit has been evaluated, it can be related to percentage (%) mortality by the following table (Table 22-6): Transformation of Percentages to PROBITs in Toxicity Calculations (Ref: Finney, 1971 – extracted from Lees, F.P., Loss Prevention in the Process Industries, Vol. 1, pg. 9/73, 1996)

Table 22-6: Transformation of Percentages to Probits in Toxicity Calculations (Ref: Finney, 1971)

%	0	1	2	3	4	5	6	7	8	9
0	-	2.67	2.95	3.12	3.25	3.36	3.45	3.52	3.59	3.66
10	3.72	3.77	3.82	3.87	3.92	3.96	4.01	4.05	4.08	4.12
20	4.16	4.19	4.23	4.26	4.29	4.33	4.26	4.39	4.42	4.45
30	4.48	4.50	4.53	4.56	4.59	4.61	4.64	4.67	4.69	4.72
40	4.75	4.77	4.80	4.82	4.85	4.87	4.90	4.92	4.95	4.97
50	5.00	5.03	5.05	5.08	5.10	5.13	5.15	5.18	5.20	5.23
60	5.25	5.28	5.31	5.33	5.36	5.39	5.41	5.44	5.47	5.50
70	5.52	5.55	5.58	5.61	5.64	5.67	5.71	5.74	5.77	5.81
80	5.84	5.88	5.92	5.95	5.99	6.04	6.08	6.13	6.18	6.23
90	6.28	6.34	6.41	6.48	6.55	6.64	6.75	6.88	7.05	7.33
%	0	0.1	1.2	0.3	0.4	0.5	0.6	0.7	0.8	0.9
99	7.33	7.37	7.41	7.46	7.51	7.58	7.58	7.65	7.88	8.09

The following are probit correlations for a Fire and Explosion exposures (Ref: Eisenberg, Lynch and Breeding, 1975 – extracted from Lees, F.P., Loss Prevention in the Process Industries, Vol. 1, pg. 9/64, 1996)

Table 22-7: Probit Correlations for Fire and Explosion Exposures (Ref: Eisenberg, Lynch and Breeding, 1975)

Hazard	Injury or Damage	Causative Variable	Probit Parameter a	Probit Parameter b
Fire	Deaths from thermal radiation	$t_e I_e^{4/3}/10^4$	-14.9	2.56
Explosion	Deaths from Lung Hemorrhage	P^o	-77.1	6.91
	Eardrum Ruptures	P^o	-15.6	1.93
	Deaths from Impact	J	-46.1	4.82
	Injuries from Impact	J	-39.1	4.45
	Injuries from Flying Fragments	J	-27.1	4.26
	Structural Damage	P^o	-23.8	2.92
	Glass Breakage	P^o	-18.1	2.79

Where:

t_e	=	Effective time duration, in seconds
I_e	=	Effective radiation intensity, in Watts/square meter
P^o	=	Peak overpressure, in Newtons/square meter
J	=	Impulse, in Newtons/square meter

For toxic releases the following probits (Table 22-8) are taken from Louvar, J.F. and Louvar, B.D., Health & Environmental Risk Analysis: Fundamentals with Applications (1998) and * Lees, F.P., Loss Prevention in the Process Industries, Vol. 2, pg. 18/60 (1996)

Table 22-8: Parameters used in Probit Equation for Toxic Releases

Material	a	b	n
Acrolein	-9.93	2.05	1.0
Acrylonitrile	-7.81	1.00	1.3
Allyl alcohol	-4.22	1.00	1.0
Ammonia	-16.14	1.00	2.0
Benzene	-109.78	5.30	2.0
Bromine	-10.50	1.00	2.0
Carbon Disulfide	-46.56	4.20	1.0
Carbon Monoxide	-7.25	1.00	1.0
Carbon Tetrachloride	-6.29	0.41	2.5
Chlorine	-13.22	1.00	2.3
Ethylene Oxide	-6.19	1.00	1.0
Formaldehyde *	-12.24	1.30	2.0
Hydrogen Chloride	-6.20	1.00	1.0
Hydrogen Cyanide	-9.68	1.00	2.4
Hydrogen Fluoride *	-35.87	3.354	1.0
Hydrogen Sulfide	-11.15	1.00	1.9
Methyl Bromide	-5.92	1.00	1.0
Methyl Isocyanate	-0.34	1.00	0.7
Nitrogen Dioxide	-17.95	1.00	3.7
Parathion	-2.84	1.00	1.0
Phosgene	-27.20	5.10	1.0
Phosphamidon	-3.14	1.00	0.7
Phosphine	-2.25	1.00	1.0
Propylene Oxide	-7.42	0.51	2.0
Sulfur Dioxide	-1.22	1.00	2.4
Tetraethyl Lead	-1.50	1.00	1.0
Toluene	-6.79	0.41	2.5

Example:

Suppose a group of people is subjected to chlorine vapors as follows:

200 ppm for 150 minutes

100 ppm for 50 minutes

50 ppm for 20 minutes

What is the percentage of deaths likely arising from these exposures?

From the above, the following equation applies:

$$Pr = -13.22 + 1.00\{\ln \Sigma (C^{2.3}\ t)\}$$

Concentration, ppm	**Exposure time, minutes**	**$C^{2.3}$ t**
200	150	29407645
100	50	1990536
50	20	161682
	$\Sigma (C^{2.3}\ t)$	31414363
	$\{\ln \Sigma (C^{2.3}\ t)\}$	17.263
	$Pr = -13.22 + 1.00\{\ln \Sigma (C^{2.3}\ t)\}$	4.043
	% Mortality (from Table of Transformation)	17%

TRUE RISK VERSUS POTENTIAL RISK

Although the risk equation,

$$\textbf{Risk} = \textbf{Consequence} \times \textbf{Frequency}$$

exists there is a considerable variance between levels of calculated risk depending upon:

- to what extent risk mitigation (control) measures are in place
- the perception that only the target, per se, should figure in the consequence calculations

Potential risk is calculated initially independent of the level of risk mitigation available. However, mitigation could reduce the consequences and/or decrease the frequency of the event. By applying risk mitigation (control) measures the aim is to reduce the risk to levels deemed as acceptable when compared with specific risk criteria (e.g. mortality statistics, f/n curves).

However this does not represent what we might term "true risk" because other important considerations are frequently omitted, these are, namely:

i) the dwell factor, which considers what percentage of the time there is exposure to risk by the target (recipient) which actually occurs

ii) the potential effect of sheltering in place being effective, in the event of an incident (which may, or may not have been considered in the original calculation)

iii) the potential for escape in the event of an incident

iv) the potential for rescue in the event of an incident

However, since rescuers e.g. fire fighting crews, are also frequently exposed to high levels of risk, in effecting a potential rescue, this should also be factored into the calculation.

We will illustrate these considerations by way of applying a simple calculation of a single event type is scenario. Consider the filling of an ammonia storage vessel from a road tanker and the potential for a leak in the area due to a ruptured line. We shall assume that without risk mitigation the individual mortality level is 10^{-3} deaths per annum and that with mitigation this is reduced to 10^{-4} deaths per annum.

In this case, suppose that the dwell factor, i.e. person likely to be present is 20%, the effect of sheltering is 40% effective, the potential for escape is 80% the potential for rescue is 30% and the chance of a rescuer being killed in a rescue attempt is 20%. Thus we have a "true risk" value of -

$$= 10^{-4} \times \frac{20}{100} \times \frac{(100-40)}{100} \times \frac{(100-80)}{100} \times \frac{(100-20)}{100} \times \frac{(100+20)}{100}$$

$$= 10^{-4} \times 0.2 \times 0.6 \times 0.2 \times 0.8 \times 1.2$$

$$= 10^{-4} \times 0.02 = \underline{\underline{2 \times 10^{-6} \ deaths / annum}}$$

Thus the true risk is around three orders of magnitude below that of the potential risk. This also reflects on the reason why there tend to be more near misses as opposed to incidents for although the frequency of occurrence is unchanged the consequences are reduced.

Note: In the event that the user takes sheltering into account as shown, it should not be repeated as P_s, probability of failing to take shelter as shown for the frequency calculations for chemical release hazards, fireball hazards or pool fire hazards as shown in pages 22-25, 26.

FAULT TREE ANALYSIS (FTA)

The fault tree technique was introduced in 1962 at the Bell Telephone Laboratories in connection with a safety evaluation of the launching system for the intercontinental Minuteman missile. The Boeing Company improved the technique and introduced computer programs for solving fault trees. The technique has subsequently been very widely used in the nuclear power industry. The FTA technique is described in the Industrial Electro-technical Commission (IEC) standard 1025 (1990).

FTA breaks down an accident or an event into its contributing causes, provided that they can be identified as discrete, specific and definable. The event is deemed to be the top event. The result of the FTA is a combination of failures and sub-events that are sufficient to result in the top event. A fault tree may be analyzed to obtain the minimum cut sets. A cut set is a set of primary events or underdeveloped faults, which can give rise to the top event. A minimum cut set is one that does not contain within itself another cut set. The complete set of minimum cut sets is the set of principal fault modes for the top event.

Symbols used in Fault Trees

The following are sets of symbols used to represent logic gates and events:

Logic Gate Symbols

Gate Symbol	Gate Name	Causal Relation
	AND	Output event occurs if all input events occur simultaneously
	OR	Output event occurs if any one of the input events occurs

Event Symbols

Event Symbol	Event Name	Meaning
	CIRCLE	Basic event with sufficient data
	DIAMOND	Undeveloped event
	RECTANGLE	Event represented by a gate
	TRIANGLE	Transfer symbol

The following simple example of Fault Tree Analysis shows how the top event – an automobile accident – might be caused by either driver error or brake failure due to improper maintenance or faulty brake components.

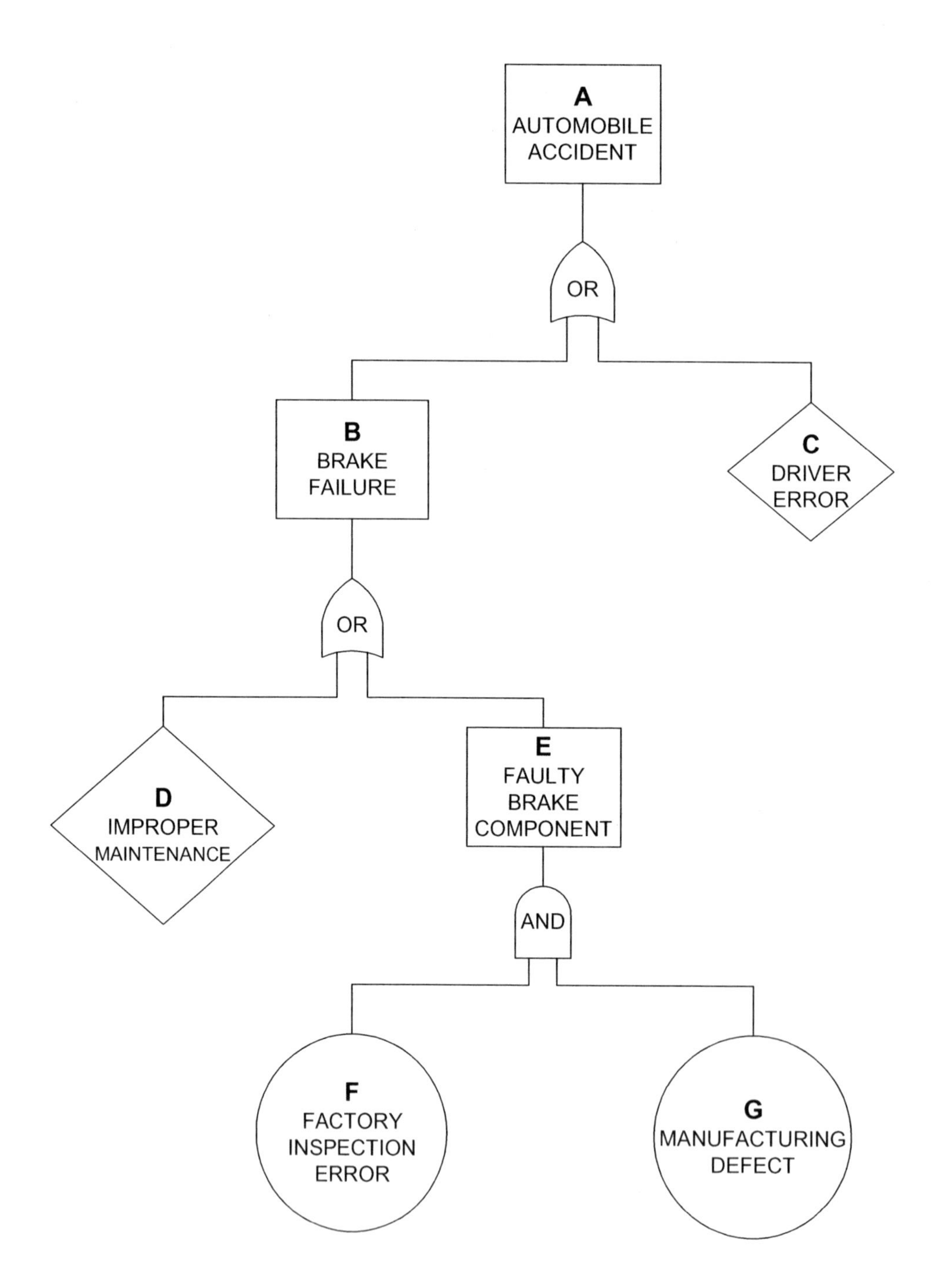

Figure 22-4: Simple Fault Tree of Accident Due to Faulty Brake Component

It should be noted, however, that this is an overly simplistic example and clearly does not mean to suggest that all motor accidents can be represented as such, because there are very many potential causes and contributors to motor accidents.

In the example shown, there are two minimum cut sets, namely DFC and DGC.

When it comes to the numerical evaluation of fault trees, we handle probabilities and event frequencies. Strict adherence must be paid to ensuring that the fault trees are correct by the rules of dimensional analysis. For example, the units of probability are dimensionless, whereas failure rates are on a per-unit time basis.

When we consider an OR gate, , if we define the probability of failure of A as P(A) and likewise, that of B as P(B), then the probability of A or B failing is $P(A \cup B) = P(A) + P(B)$ with the constraint that $P(A \cup B)$ cannot exceed a value of unity. Thus for n failures, $P(1 \cup 2 \cup \ldots n) = P(1) + P(2) + P(3) + \ldots P(n)$ with the overall constraint of unity.

When we consider an AND gate, , if we define the probability of failure of A as P(A) and likewise, that of B as P(B), then the probability of A and B failing together simultaneously is $P(A \cap B) = P(A) \times P(B)$. Thus for n failures, $P(1 \cap 2 \cap \ldots n) = P(1) \times P(2) \times P(3) + \ldots P(n)$.

From a dimensional analysis standpoint:

1) Failure rates may be summated as in OR gates, but they cannot be multiplied together.

2) Failure rates may be multiplied by probabilities, as in AND gates.

3) When the top event is probabilistic, all other events must also be probabilistic.

When evaluating a fault tree, there are two particularly important features, namely (a) the frequency (or probability) of the top event and (b) the contribution made by minimum cut sets to see what the main contributors to the top event may be.

FAILURE RATES

Q: Where do we get failure rate data from?

A: From a number of possible sources, namely:

- Historical data on failures experienced at various industrial facilities;
- Failure rate data provided by equipment vendors (if available);
- Published data, e.g., CCPS reliability data, IEEE, etc.;
- Failure rates derived by considering contingent or dependent factors.

Table 22-9: Typical failure rate data for process plant instruments (reference: Lees, "Loss Prevention in the Process industries," Second Edition, Tables 13.6/14:

Instrument	Failure (faults/year)
Control valve	0.60
Solenoid valve	0.42
Pressure measurement	1.41
Flow measurement (fluids)	1.14
Level measurement (liquids)	1.70
Thermocouple temperature measurement	0.52
Radiation pyrometer	2.17
Controller	0.29
Pressure switch	0.34
Flame failure detector	1.69
pH meter	5.88
Gas-liquid chromatograph	30.6
Electrical conductivity meter (liquids)	16.7
Impulse lines	0.77

Table 22-10: Typical failure rate data for process plant instruments (reference: Lees, "Loss Prevention in the Process industries," Second Edition, Tables 13.6/14:

Loop failure (by type of loop)	Faults/year
Pressure Indicating Controller (PIC)	1.15
Pressure Recording Controller (PRC)	1.29
Flow Indicating Controller (FIC)	1.51
Flow Recording Controller (FRC)	2.14
Level Indicating Controller (LIC)	2.37
Level Recording Controller (LRC)	2.25
Temperature Indicating Controller (TIC)	0.94
Temperature Recording Controller (TRC)	1.99

Table 22-11: Typical failure rate data for process plant instruments (reference: Lees, "Loss Prevention in the Process industries," Second Edition, Tables 13.6/14:

Loop failure (by element in loop)	% faults
Sensing/sampling	21
Transmitter	20
Transmission	10
Receiver (e.g., indicators, recorders)	18
Controller	7
Control valve	7
Other	17

Failure rates and probabilistic failures from contingent and dependent factors:

In certain instances, a failure rate or specific probability cannot be simply looked up in a book or obtained from an established data source. In these cases, the failure, probability or contribution may have to be derived by considering contingent factors. Consider, for instance, the case of a release of a flammable vapor cloud of propane gas and the potential for ignition due to passing rail trains on a nearby track. What are the chances of a train igniting such a release?

For ignition to occur due to a passing train we must have an AND gate situation, namely that while the vapor cloud is in the immediate region of the tracks, a train also passes by. Let us suppose that a dispersion calculation had established that the vapor cloud would be in the area for approximately 10 minutes before dispersing below the flammable limit and that, on average, there is one vapor cloud release in 5 years. Also, there are 8 trains a day that pass through the area and they each take, on average, 6 minutes to pass through the area. What is the probability that the vapor cloud will be ignited by the passing train?

Consider, first, the probability, P(Cloud), of even having a vapor cloud in the area in question. On an annual basis there are 1/5 = 0.2 releases. Each release lasts 10 minutes, thus in one year the probability of there being a vapor cloud over the tracks is:

$P(\text{Cloud}) = 0.2 \times (10/60) / (365 \times 24) = 0.0000038 = 3.8\text{ E-06}$

On an annual basis there are 8 × 365 = 2920 trains. As each train is in the region for 6 minutes, the probability, P(Train), of a train being in the region at any one time is:

$P(\text{Train}) = 2920 \times (6/60) / (365 \times 24) = 0.0333 = 3.33\text{ E-02}$

Thus, for both to occur over the same time period, causing ignition of the cloud by the train, we have P(Ignition of Cloud by Train) = $P(\text{Cloud} \cap \text{Train})$,

and $P(\text{Cloud} \cap \text{Train}) = P(\text{Cloud}) \times P(\text{Train}) = (3.8\text{ E-06}) \times (3.33\text{ E-02}) = 1.26\text{ E-07}$ in any one year, which means that the event frequency for this is once every 7.9 million years. It may be assumed that the rarity of such an event can allow us to ignore such an event. However, if there were say 10 times more releases, 10 times more trains, and the vapor hung around in a trough in the area of the track, causing it to linger for 10 times longer, then the probability increases 1000 fold to 1.26 E-04, which is once every 7.9 thousand years. It can also be readily seen that in deriving these failure probabilities we are creating mini fault trees to assist us.

Summary Of Basic Probability Relations:

Logic Gate	Meaning	Boolean Algebra Relation	Probability Relations
A (AND gate) B C	**AND** (i.e., both B and C needed for event A)	A = BC	P(A) = P(B).P(C)
A (OR gate) B C	**OR** (i.e., either B or C needed for event A)	A = B + C	P(A) = P(B) + P(C) – P(B).P(C) but since P(B).P(C) is usually very small, this reduces to P(A) = P(B) + P(C)

Summary Of Relations Involving Frequencies And/Or Probabilities, Where P(..) Denotes Probability And F(..) Denotes Frequency:

Gate	Inputs	Output
OR A (OR gate) B C	• P(B) OR P(C) • F(B) OR F(C) • F(B) OR P(C)	• P(A) = P(B) + P(C) – P(B).P(C) ≅ P(B) + P(C) • F(A) = F(B) + F(C) • Not permitted; reformulate
AND A (AND gate) B C	• P(B) AND P(C) • F(B) AND F(C) • F(B) AND P(C)	• P(A) = P(B).P(C) • Not permitted; reformulate • F(A) = F(B).P(C)

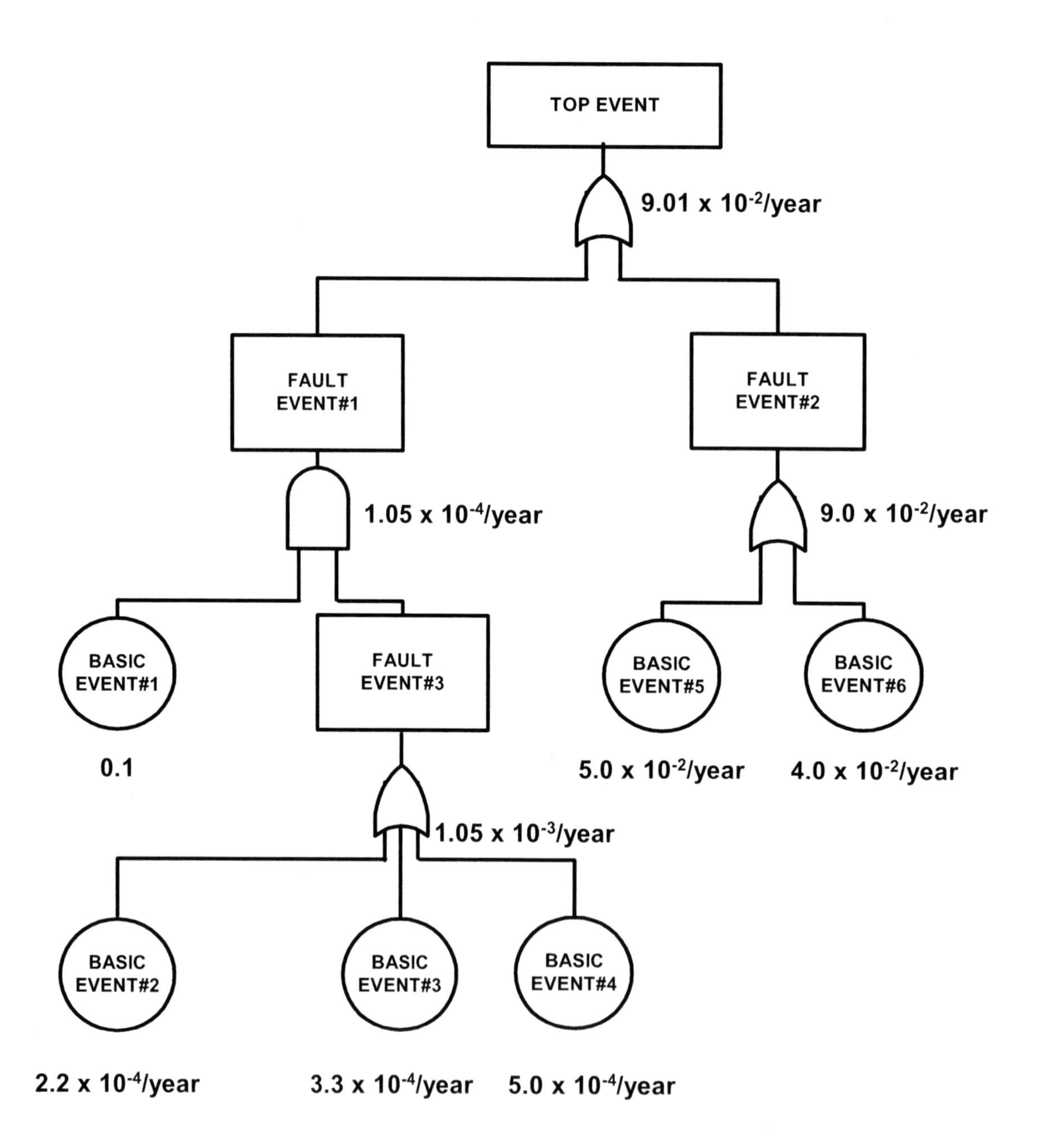

Figure 22-5: Example Of A Fault Tree Construction

The above diagram shows a typical fault tree construction in terms of:

- The Top Event;
- Fault Events;
- Basic Events.

Basic Event # 1 is expressed as a probability; it is dimensionless and without units.

Basic Events # 2, 3, 4, 5 & 6 are expressed as frequencies with dimensions of faults per annum.

Once (a) the basic events are specified and (b) the AND and OR gates are assigned, the values of Fault Events 1 & 2 as well as the Top Event are thus fixed, and there are no degrees of freedom left. Only by changing either the structure of the tree itself and/or the values of the basic events can we change the failure rate value for the top event.

When a fault tree has been constructed, the values for individual basic events can be assessed to understand their relative contributions. Furthermore, when mitigating risk, the failure of the mitigation devices can be assessed and therefore, if necessary, further mitigation can be incorporated until an acceptable level of risk is achieved.

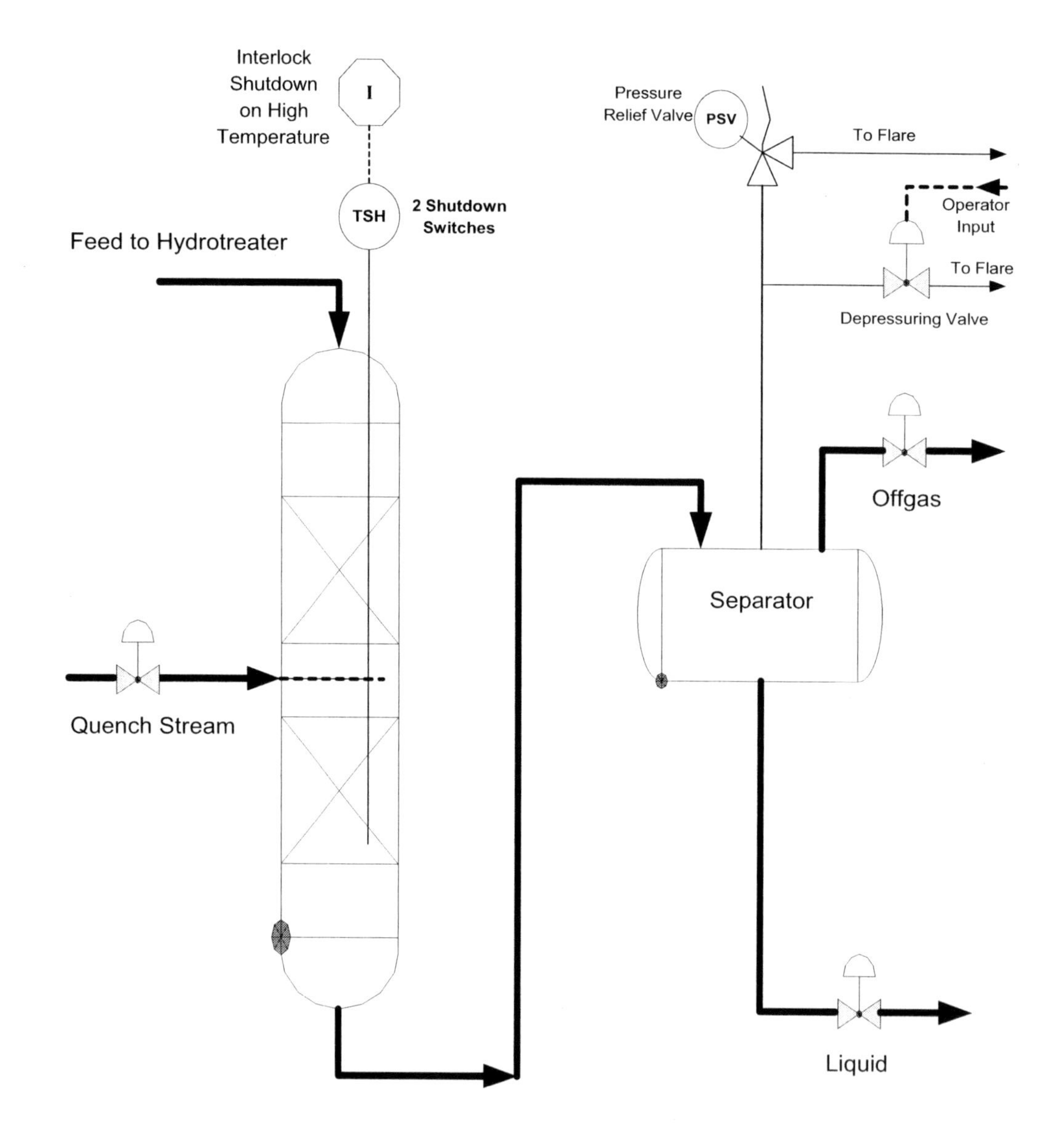

Figure 22-6: Example of Fault Tree Applied to a Hydrotreater Reactor

Let us consider the case of a hydrotreater, with a trickle bed design, working normally at about 1800 psig. The potential exists for a runaway reaction due to an exothermic reaction, which could result either from a hot spot developing in the reactor bed or a loss of quench hydrogen. Should the temperature of the reactor bed increase, then at some point this should trigger a shutdown via the interlocks of the high temperature sensors. Should these fail, then the pressure will rise, leading to overpressure and the relief valve lifting on the separator. (With hydrotreaters, PSVs are not located on the reactors themselves, only on the downstream separator.) In addition to the PSV lifting, the plant operator should (remotely) open the depressuring valve to flare. If these safeguards do not function, there is every possibility that the reactor will experience a runaway condition, leading to over-temperature and overpressure, and it will eventually rupture.

Table 22-12: Table of Basic Events for Hydrotreater:

Basic Event	Value Assigned	Source
Hot spot in Reactor	Once every 5 years, i.e., 0.2/yr	Order of Magnitude Value – experience based
Quench Fails (Closed)	0.3/yr	50% of time for control valve failure
Temp. Switch 1 fails	Probability of 0.52	Assumed to be non-repairable until annual plant shutdown occurs
Temp. Switch 2 fails	Probability of 0.52	Assumed to be non-repairable until annual plant shutdown occurs
PSV fails to open	Probability of 0.00175	From unreliability calculation due to 0.2 E-06 failures per annum
Operator fails to depressure	Probability of 0.1	Order of magnitude for failure due to human error

From the fault tree constructed, it can be seen that a fairly high top event failure rate of 1.37 E-02/yr is obtained, i.e., once every 73 years. From inspection of these numbers, it can thus be easily deduced that having many more temperature sensors in the reactor bed would substantially reduce this top event frequency.

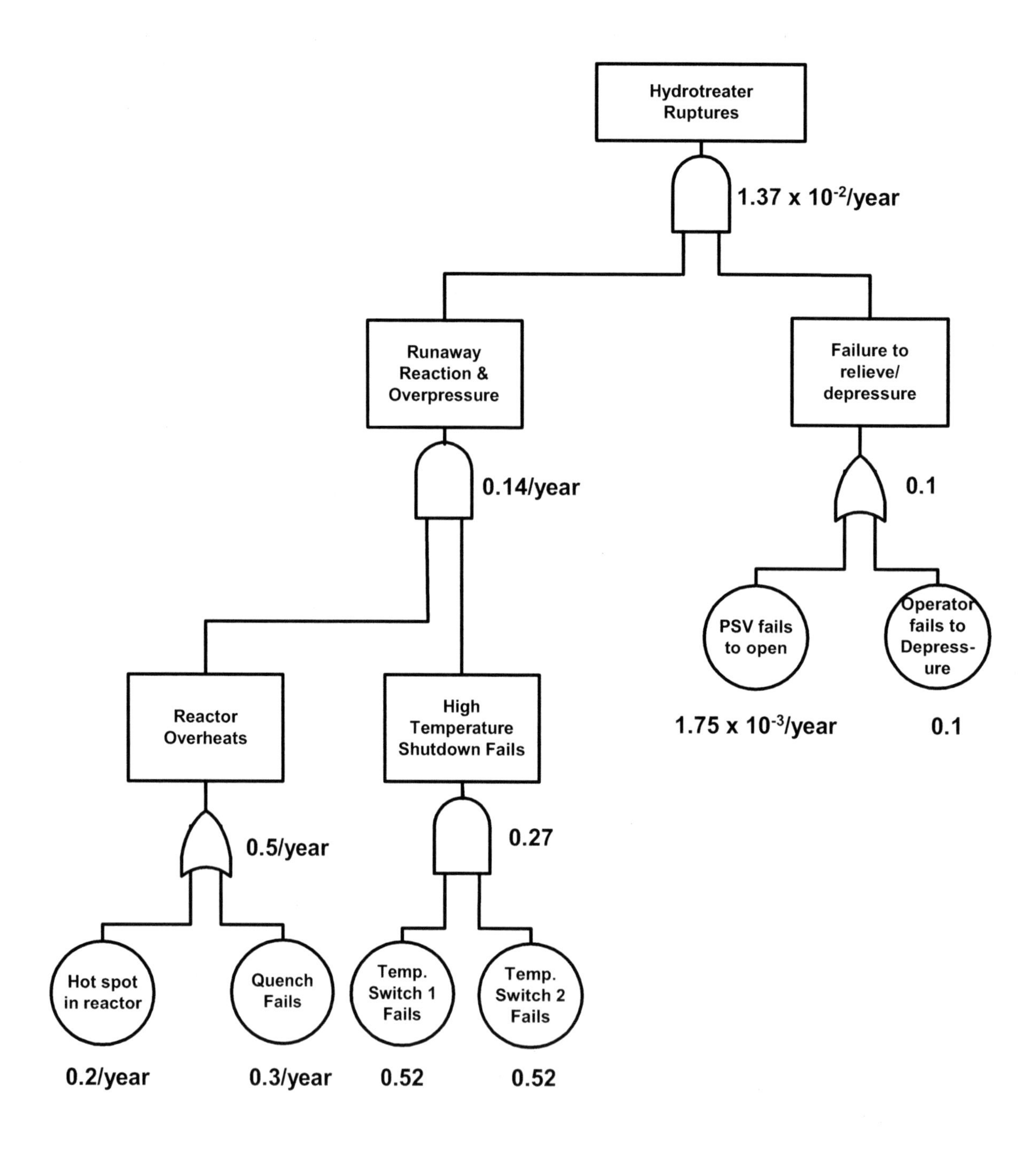

Figure 22-7: Fault Tree Analysis Example for Hydrotreater Rupture

EVENT TREE ANALYSIS

With an Event Tree, as shown below in Figure 22-8, the Basic Event, a Large LPG Release is depicted on the left side of the diagram together with its frequency of occurrence. The next Event, the Wind Disperses Vapour, is considered: there is an assigned probability of, say 0.7, that this will occur (i.e., *Yes*) and thus there is also an assigned probability of, 0.3, by deduction, that this will not occur (i.e., *No*). This pattern of Event consideration, for "Ignition Source Ignites Vapour", for "Vapour Cloud Detonates", and for "Fire Impacts Rail Car" are all evaluated in a similar manner. The result is the Outcome, expressed as Events such as Flash Fire, No Impact, Vapour Cloud Explosion (VCE), Boiling Liquid Vapour Cloud Explosion (BLEVE) and Fireball are shown, together with their individual expected frequencies. It also follows that the sum of these individual event Outcome frequencies must equal the Basic Event frequency, namely that of the original Large LPG Release frequency.

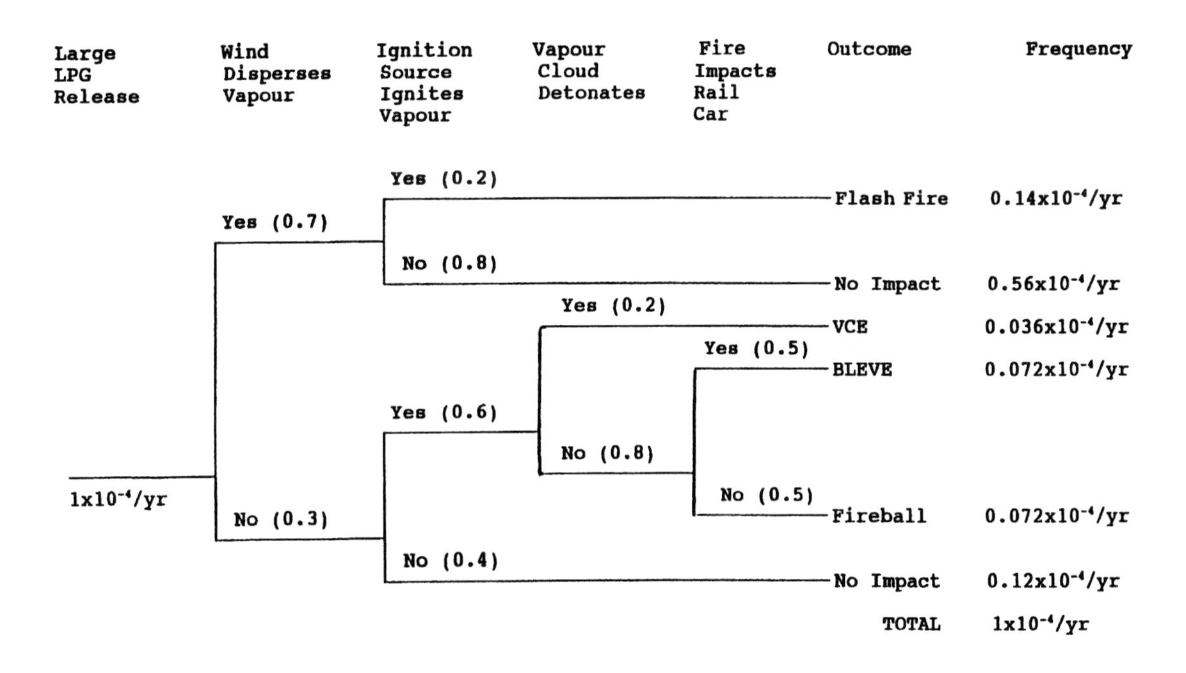

Typical Event Tree for a Large LPG Release

Figure 22-8

FAILURE RATE ESTIMATION AND RELIABILITY DATA

Failure Rates Estimation

- Established data sources, *e.g. CCPS*
- Historical records and plant information logs
- Fault/Event tree constructs

Failure rates are expressed as individual failure rates per annum or per hour.

The following pages contain:

- A graph showing the Rupture Rate Leaks for Piping, Failure Rate Vs. Line Size (Figure 22-9).
- A Sample Reliability Data Sheet (Table 22-13).

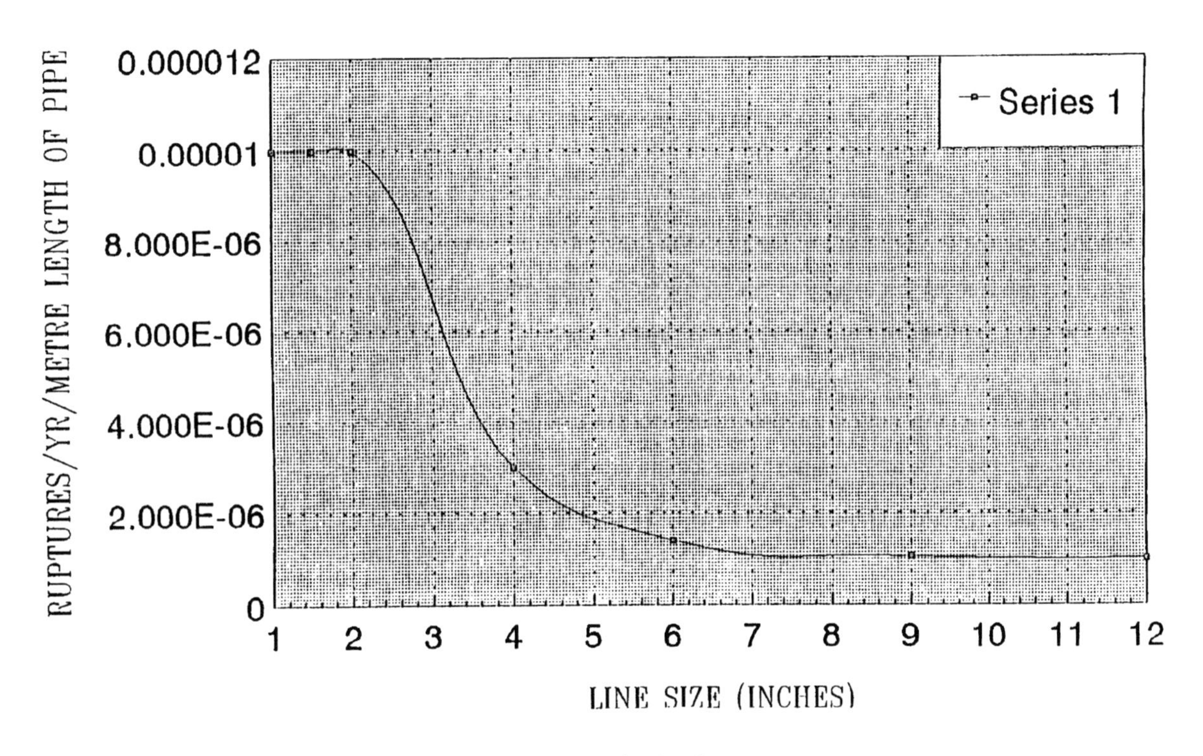

Valve Rupture Leaks: 0.0001/yr/valve
Pump Rupture Leaks: 0.0003/yr/pump

Figure 22-9: Rupture Rate Leaks for Piping (Ref: Cox, A.W., Lees, F.P., and Ang, M.L., Classification of Hazardous Locations, Inst. of Chemical Engineers, 1992)

Table 22-13: Sample Reliability Data Sheet

DATA ON SELECTED PROCESS SYSTEMS AND EQUIPMENT	
Taxanomy No. 3.5.2	Equipment Description VALVES-MANUAL
Operating Mode	Process Severity UNKNOWN

Population	Samples	Aggregated time in service (10^6 hrs)		No. of Demands
		Calendar time	Operating time	

Failure mode	Failures (per 10^6 hrs)			Failures (per 10^6 hrs)		
	Lower	Mean	Upper	Lower	Mean	Upper
CATASTROPHIC						
a. Leakage 0 - 10%	0.0141	0.152	0.501	0.0141	0.291	1.06
b. Leakage > 10%						
c. Rupture						
d. Normally Open/Fails Open						
e. Normally Closed/Fails Closed						
f. Normally Open/Fails Plugged						
g. Normally Closed/Fails Open						
DEGRADED						
INCIPIENT						
a. Wall Thinning						
b. Embrittlement						
c. Cracked or Flawed						
d. Internal Leakage						

Equipment Boundary

Data Reference No. (Table 5.1): 8, 8.1, 8.2, 8.3, 8.7, 8.12

Ref: CCPS, Guidelines for Chemical Process Quantitative Risk Analysis, 1989

INTRODUCTION TO CONSEQUENCE ANALYSIS

Consequence analysis relates to the impacts caused by exposure to hazards, whether chronic or acute.

Chronic (long term) usually relates to:

- Carcinogenic effects
- Deteriorative disease

Acute (short term) usually refers to:

- Death or injury
- Damage to property

Environmental can refer to:

- Human health impacts (usually chronic)
- Flora, fauna impacts

Main Concern is with Acute Effects

- Fire
- Explosion
- Toxic release, e.g., H_2S
- Property damage

Major Plant Hazards

Fires:

Pool fires

Fireballs

Jet fires

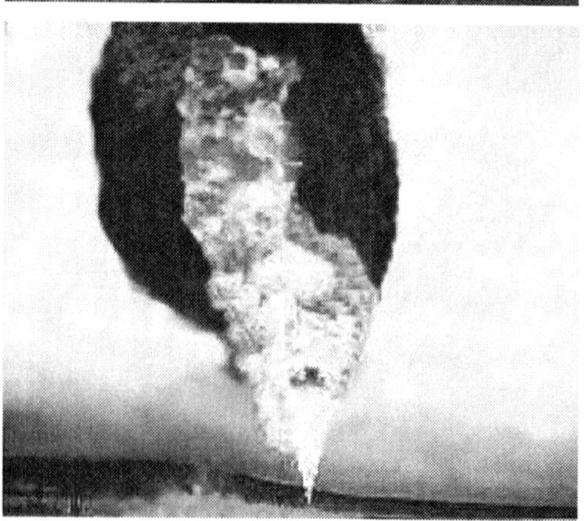

Explosions:

Boiling Liquid Expanding Vapor Explosion (BLEVE)

Vapor Cloud Explosions (VCE); Confined and Unconfined and Condensed Phase

Toxic Releases:

Toxic VCEs, vapor/liquid releases – contaminates environment (air/water/soil)

Radiation Hazards:

For example: nuclear sources

Hazards

Primary Elements in the system that are inherently hazardous, e.g.

- Presence of hazardous materials, e.g. HF, H_2S
- Possibility of runaway reactions
- High temperature/high pressure processes, e.g. hydroprocessors
- Ignition sources
- Possibility of human error
- Possibility of mechanical failure
- Large inventory – storage tanks, large vessels, extensive piping systems

Secondary Result of primary hazards, e.g.

- Fire
- Explosion
- Release of toxic material
- Toxic products of combustion
- Asphyxiation
- Impact blows from missiles

Tertiary Result of secondary hazards – “domino” effect, e.g.

- Fire spread
- Secondary explosion
- Impact blows from missiles
- Loss of control

Domino potential is strongly influenced by

1. Plant size
2. Plant layout
3. Plant spacing
4. Inventories

CONSEQUENCE MECHANISMS

Typical Release Mechanisms

Typical release points

- Hairline crack
- Gasket failure on flange
- Pipe/vessel/tank ruptures
- Pump/compressor seal leaks/failures
- Hose and disconnected failures
- Valve seal failure
- Releases from vents, drains, safety valves, loading arms

Typical causes

- Overstressed lines
- Overstressed vessel nozzles
- Two phase flow forces
- Line freeze-up
- Water hammer in line
- Inadequately supported lines

- Vessels/tanks/lines overpressured
- Excess corrosion/erosion
- Lack of maintenance
- Low temperature embrittlement
- Hydrogen blistering
- Stress corrosion cracking
- Impact forces
- Thermal expansion

Consequences of Explosion, Fire, Toxic Release

Table 22-14: Primary, Secondary and Tertiary Consequences of Loss of Containment

	PRIMARY CONSEQUENCES Release Event	SECONDARY CONSEQUENCES Forces Manifested	TERTIARY CONSEQUENCES Final Impact
Loss of Containment	**Explosion** - mechanical, vapor cloud, BLEVE, condensed phase	blast, overpressure, thermal radiation, flame contact, missiles	injury, death, property damage, environmental damage, possible domino effects
	Fire - pool, flash, jet fire	thermal radiation, flame contact	same as above
	Toxic Release - to the environment	toxics dispersed in vicinity	injuries, death, environmental damage

FIRE & EXPLOSION EFFECTS

The following graph portrays a propane BLEVE fireball, based on 100,000 kg of propane.

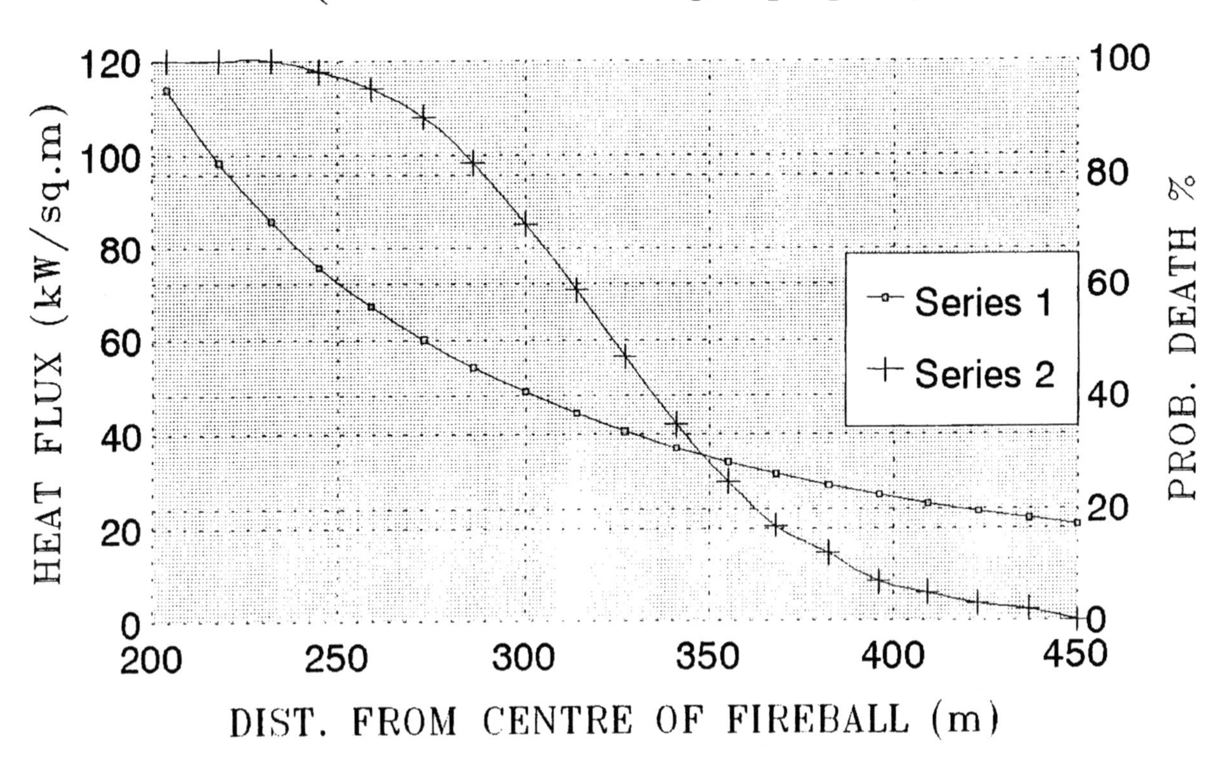

Fireball heat flux: 350 kW/sq.m
Max. fireball diameter: 273 m
Duration: 16.4 s

Figure 22-10: Heat Flux & Probability Death vs. Distance from Center of Fireball

Types of Explosions

Mechanical (i.e., No chemical change)

- Gas freely expands and performs work of expansion.
- Usual serious consequence local to explosion.

  Example: *Explosion caused by rupture of air receiver under pressure.*

Condensed phase (Chemical change)

- Supply own expansion gases.
- Can operate a vacuum.
- Supply own energy.
- Associated with ***extremely*** high local pressures ("Chapman-Jouget") in the region of thousands of pounds per square inch.

  Example: *High explosives such as TNT, nitro glycerin, picric acid, guncotton.*

Vapor cloud (VCE) (Chemical change, i.e., Combustion)

- Flammable substance present in sufficient quantity.
- Rapid combustion (detonation) in air.
- Needs ignition source.

  Example: *Propane release and forming a vapor cloud, which subsequently ignites, and combusts with explosive violence.*

Dust explosion (Chemical Change)

- Combustible material in a finely divided state.

  Example: *Coal dust. (Many problems for the coal mining industry.)*

Factors Making VCEs Explosions More Likely

- Any degree of confinement
- Low auto-ignition temperature, *e.g., H_2S*
- High flame speeds, *e.g., Hydrogen*
- Large releases
- Wide explosion limits
- Local turbulence, *e.g., high velocity jet*

Problems With Modeling Vapor Cloud Explosions

- Not as simple to model as TNT explosions.
- TNT model severely overestimates near field effects.
- VCEs rarely exceed 10 to 30 PSI at center.
- Explosion yield is not a reliable parameter.
- VCEs not well understood: Continuing research.

Important Note

1. **Both** over pressure **and** impulse forces are very important for predicting the real effects of blast.
2. **Both** overpressure **and** vacuum phases exist.

Explosion Modeling Methods

TNT Model

- Not reliable in near field, overestimates effects.
- Reasonable agreement in far field.

Baker-Strehlow Method

- Blast waves considered to be generated by constant velocity and accelerating flames propagating in a spherical geometry.
- Can overestimate overpressures significantly.

Wiekema Piston Blast Model (TNO)

Needs knowledge of reactivity of substance to predict effects.

Multi-Energy Model (TNO)

- Divides forces generated into areas depending on degree of confinement.
- Hard to apply.

Hardest Factors to Estimate

- Amount of material in vapour cloud contributing to the explosion.
- Degree of entrainment of liquids (usually taken as equivalent to the mass vapor fraction flashing off).
- Period of release before explosion occurs (could be anywhere from around 20 sec. to around 20 min.).
- Degree of confinement (Multi-Energy Model).
- Yield based on TNT equivalency: Could be anywhere from 1 to 10% (or more!).

TNT Equivalency

$$M_{TNT} = \frac{M_c \times H_c \times Y}{2000}$$

Where:

M_{TNT} = TNT equivalent, in lb.

M_C = Mass of vapor cloud, lb.

Y = Explosive yield (expressed as a fraction)

H_C = Net heat of combustion of cloud material, BTU/lb.

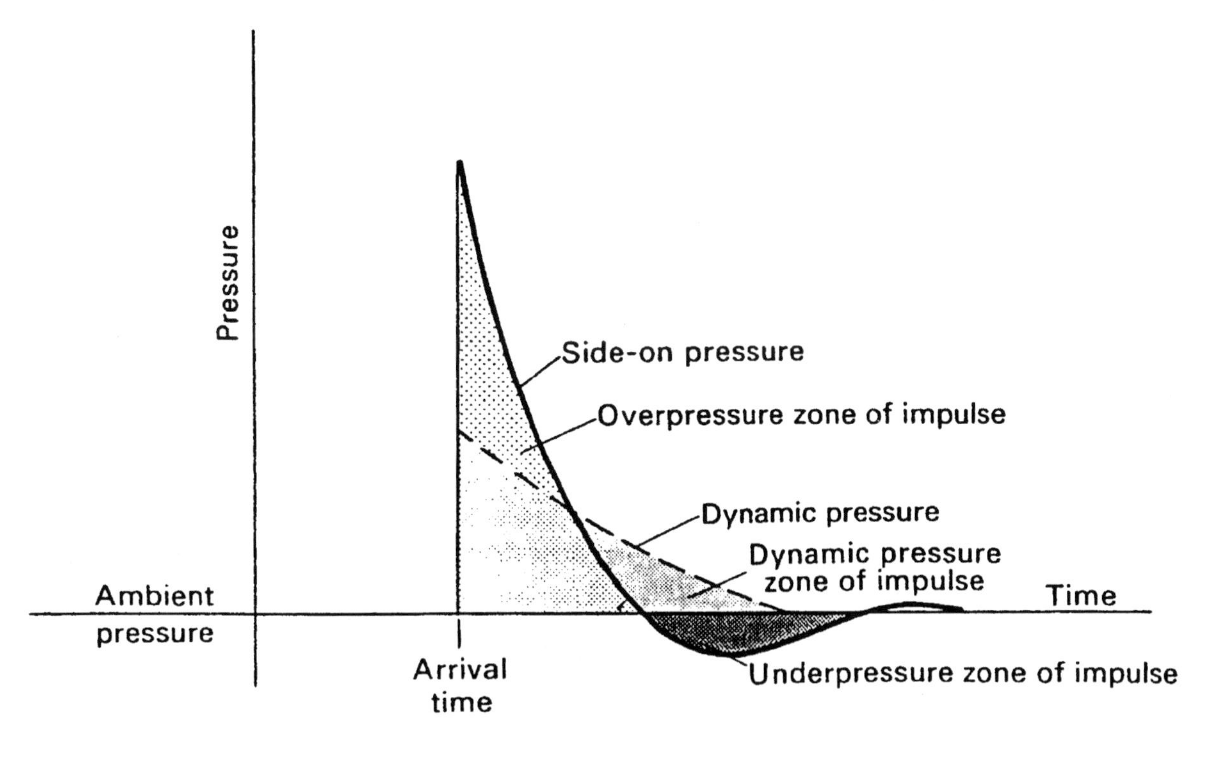

Pressure change experienced at a point given distance from an explosion
Ref: "Safety in Process Plant Design", G.L. Wells, Halsted Press, 1980

Figure 22-11

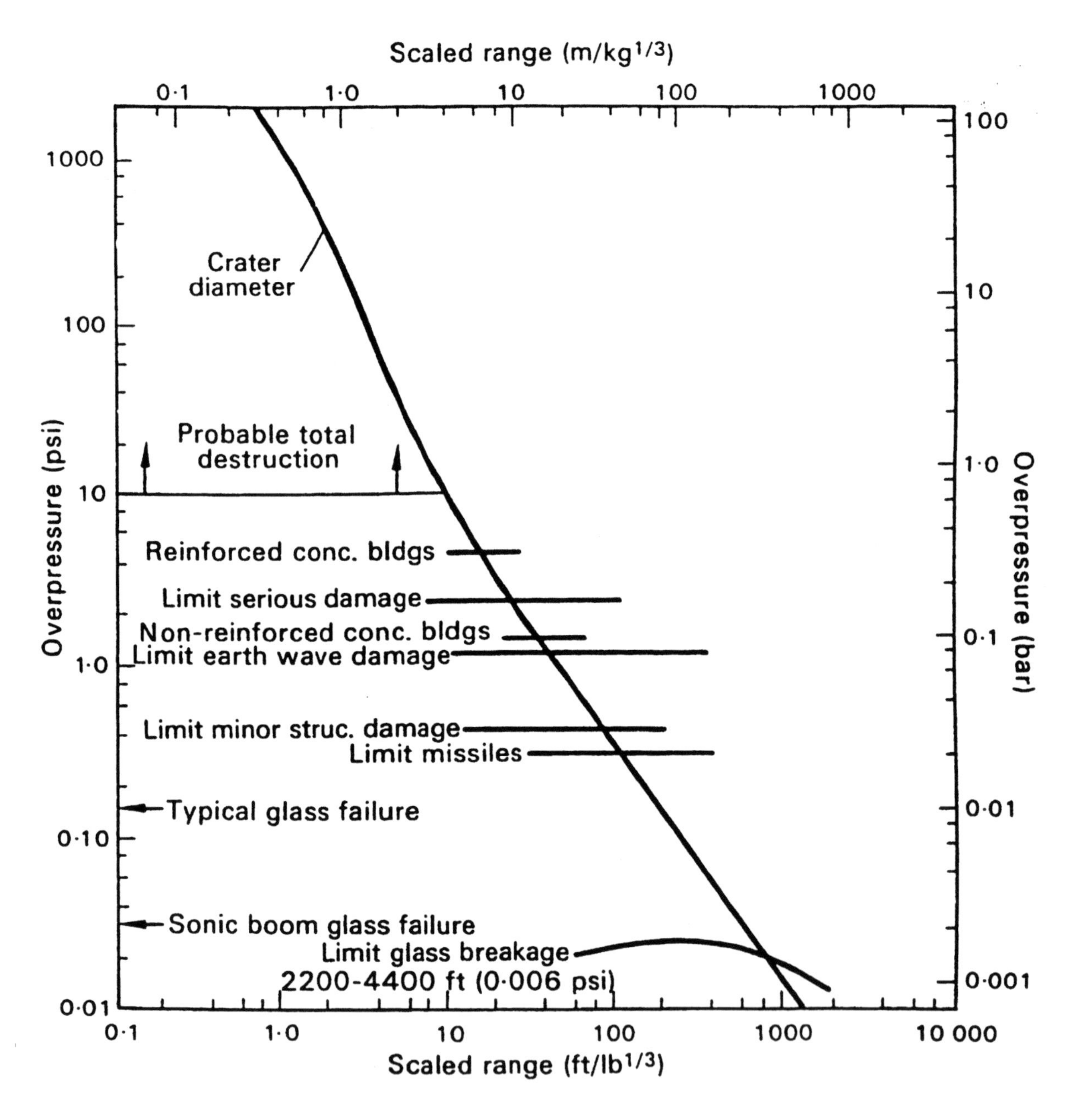

Overpressure scaled-distance plot showing typical levels for blast damage
Ref: "Safety in Process Plant Design", G.L. Wells, Halstead Press, 1980

Figure 22-12

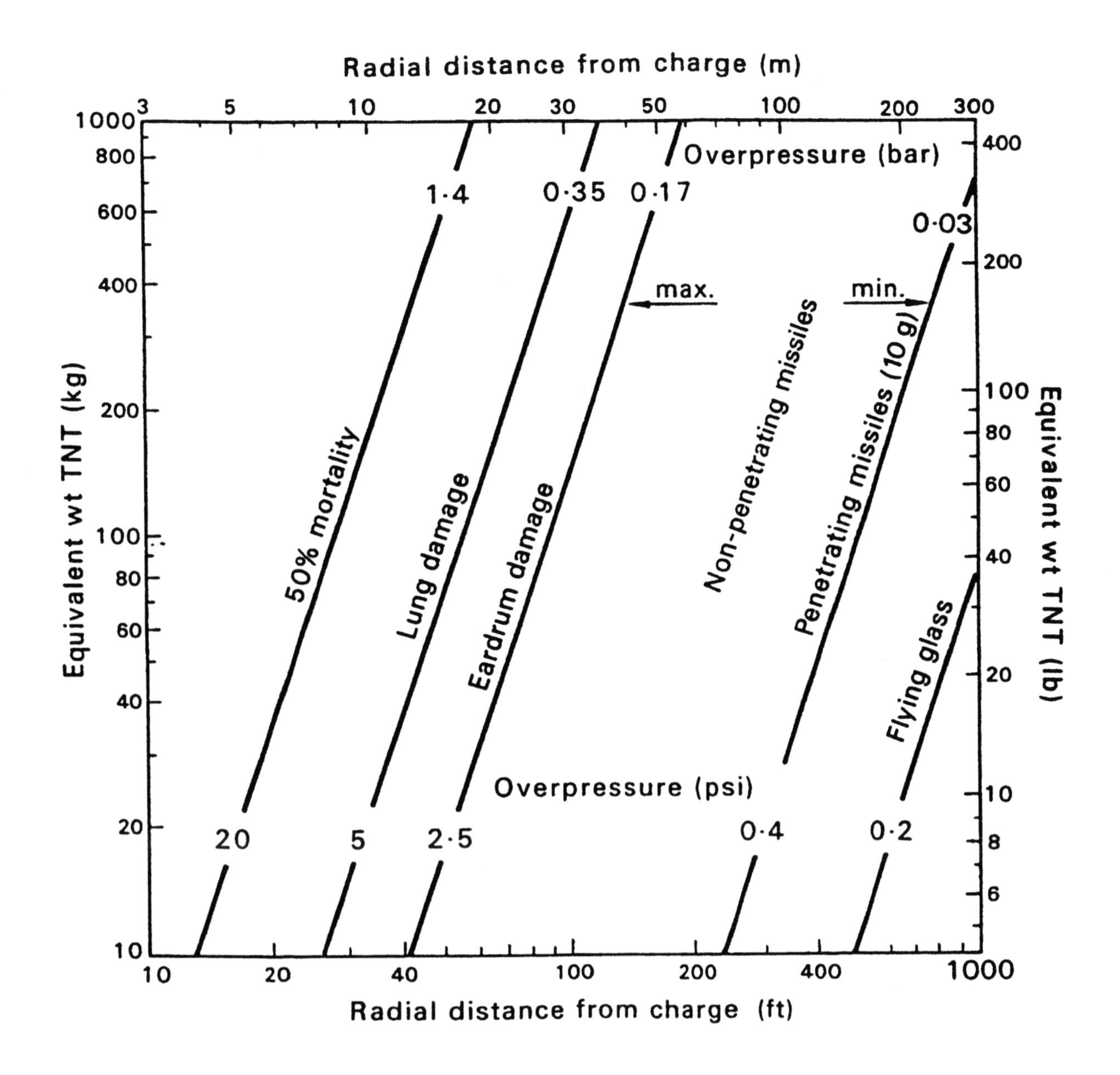

Typical Blast Injuries

Ref: "Safety in Process Plant Design", G.L. Wells, Halstead Press, 1980

Figure 22-13

Damage Caused by Overpressure Effects of an Explosion

Table 22-15: Damage Caused by Overpressure Effects of an Explosion (Ref: Stephens, M. M., Minimizing Damage to Refineries, U.S. Dept. of the Interior, Office of Oil & Gas, February 1970.)

Equipment	Overpressure PSI																							
	0.5	1.0	1.5	2.0	2.5	3.0	3.5	4.0	4.5	5.0	5.5	6.0	6.5	7.0	7.5	8.0	8.5	9.0	9.5	10.0	12.0	14.0	16.0	18.0	20.0
Control house steel roof	A	C	D				N																		
Control house concrete roof	A	E	P	D			N																		
Cooling tower	B			F			O																		
Tank: cone roof		D				K							U												
Instrument cubicle			A			LM						T													
Fire heater				G	I					T															
Reactor: chemical				A				I					P					T							
Filter				H					F										V		T				
Regenerator						I				IP					T										
Tank: floating roof						K							U												D
Reactor: cracking							I							I							T				
Pine supports							P					SO													
Utilities: gas meter									Q																
Utilities: electric transformer									H						I					T					
Electric motor										H								I							V
Blower										Q										T					
Fractionation column											R			T											
Pressure vessel horizontal												PI						T							
Utilities: gas regulator												I								MQ					
Extraction column													I							V	T				
Steam turbine															I						M	S			V
Heat exchanger															I			T							
Tank sphere																I						I	T		
Pressure vessel: vertical																					I	T			
Pump																					I		Y		

A. Windows and gauges break
B. Louvers fall at 0.3 – 0.5 psi
C. Switchgear is damaged from roof collapse
D. Roof collapses
E. Instruments are damaged
F. Inner parts are damaged
G. Brick cracks
H. Debris-missile damage occurs
I. Unit moves and pipes break
J. Bracing fails
K. Unit uplifts (half-filled)
L. Power lines are severed
M. Controls are damaged
N. Block walls fail
O. Frame collapses
P. Frame deforms
Q. Case is damaged
R. Frame cracks
S. Piping breaks
T. Unit overturns or is destroyed
U. Unit uplifts (0.9 filled)
V. Unit moves on foundations

Typical Ignition Sources

- Hot surfaces
- Electrical equipment sparking
- Motor Vehicles
- Electrostatic discharges
- Open flame heaters (Boilers, Fired heaters)
- Welding operations
- Mechanical sparking (Friction)
- Lighting
- Solar effects
- Smoking
- Lightning

To cause ignition, a minimum ignition energy level must be exceeded:

- 0.25 millijoules for deflagration
- 2.5×10^9 millijoules for detonation

Note: A deflagration can lead to detonation due to an accelerating wave front when confinement exists.

The following pages contain:

- A table depicting Explosion Effects (Table 22-16).
- A graph showing a Propane Vapor Cloud Explosion, Overpressure vs. Distance (Figure 22-14)

Table of Explosion Effects

Table 22-16: Explosion Effects at Various Overpressures

Overpressure* (psig)	Expected Damage
0.03	Occasional breaking of large windows already under stress.
0.04	Loud noise (143 dB); sonic boom glass failures.
0.10	Breakage of small windows under strain.
0.15	Typical pressure for glass failure.
0.30	Some damage to house ceilings; 10% window glass breakage.
0.40	Limited minor structural damage.
0.50 – 1.0	Windows usually shattered; some window frame damage.
0.7	Minor damage to house structures.
1.0	Partial demolition of houses; made uninhabitable.
1.0 – 2.0	Corrugated metal panels fail and buckle. Housing wood panels blown in.
1.0 – 8.0	Range for slight to serious injuries due to skin lacerations from flying glass and other missiles.
1.3	Steel frame of clad building slightly distorted.
2.0	Partial collapse of walls and roofs of houses.
2.0 – 3.0	Non-reinforced concrete or cinder block walls shattered.
2.3	Lower limit of serious structural damage.
2.4 – 12.2	Range for 1-90% eardrum rupture among exposed populations.
2.5	50% destruction of home brickwork.
3.0	Steel frame building distorted and pulled away from foundation.
3.0 – 4.0	Frameless steel panel building ruined.
4.0	Cladding of light industrial buildings ruptured.
5.0	Wooded utility poles snapped.
5.0 – 7.0	Nearly complete destruction of houses.
7.0	Loaded train wagons overturned.
7.0 – 8.0	8-12 in. thick non-reinforced brick fail by shearing of flexure.
9.0	Loaded train boxcars demolished.
10.0	Probable total building destruction.
15.5 – 29.0	Range for 1-99% fatalities among exposed populations due to direct blast effects.

* These are the peak pressures formed in excess of normal atmospheric pressure by blast and shock waves. Source: Lees, F.P, Loss Prevention in the Process Industries, Vol. 1, Buttersworths, London and Boston, 1980.

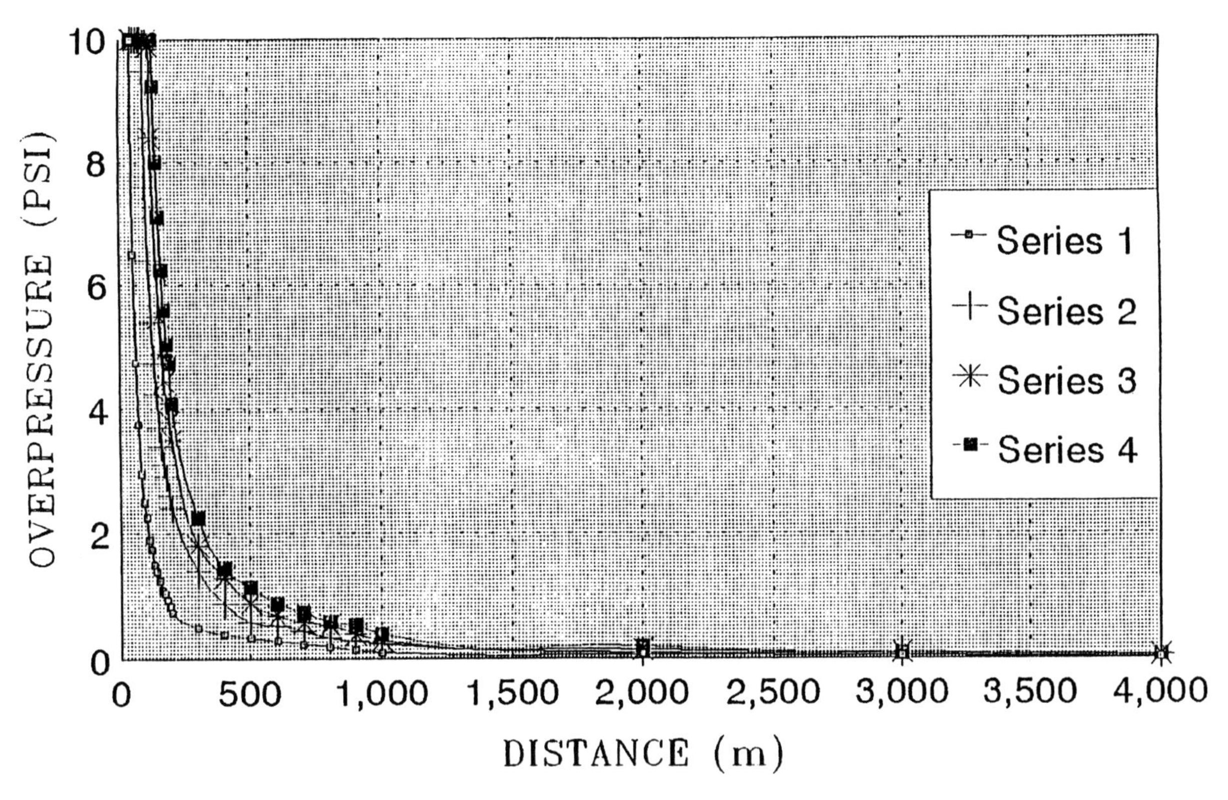

57% flash (inc. entrainment) of liquid propane
Series 1: 10,000 kg liquid release
Series 2, 3, 4: 100000, 200000, and 300000 kg, respectively

Figure 22-14: Overpressure vs. Distance for Propane Vapor Cloud Explosion

CONSEQUENCE ANALYSIS CALCULATIONS

The calculation methodologies for consequence analyses can be both complex and highly specialized. This text does not include a major review of methodologies but, rather, (a) provides reference sources for the reader and (b) provides a small sample of calculation methodologies.

The following are a list of typical useful reference sources that can provide significant assistance in computing the effects and impacts from toxic, fire and explosion hazards:

- "Guidelines for Chemical Process Quantitative Risk Analysis" " by AIChE, CCPS, 2000
- "Guidelines for Evaluating the Characteristics of Vapor Cloud Explosions, Flash Fires and BLEVE's" by AIChE, CCPS, 1994
- "Methods for the Calculation of Physical Effects", Part 1 and 2, Third Edition, 1997, 'Yellow Book', Committee for the Prevention of Disasters, by TNO, Netherlands
- "Classification of Hazardous Locations" by A.W. Cox, F.P.Lees, M.L.Ang, published by IChemE, 1990
- "Dow's Chemical Exposure Index Guide", 1st edition, 1994
- "Loss Prevention in the Process Industries" by F.P.Lees, published by Butterworth-Heinemann, 1996. (Volumes 1,2, 3)
- "Handbook for Chemical Hazard Analysis Procedures", plus "ARCHIE", (Automated Resource for Chemical Hazard Incident Evaluation), developed jointly by the U.S. Department of Transportation, the Federal Emergency Management Agency, and the Environmental Protection Agency ,1989
- "Process Plant Layout", by J.C.Mecklenburgh [see Appendix B], ISBN 0608131474
- "Safety in Process Plant Design", by G.L.Wells, Halsted Press, 1980

(Website URL's are provided under Suggested Reading at the end of this Chapter)

CONSEQUENCE ANALYSIS – Specific Situations

Pool Fires:

Pool Spreading

1. For <u>bunded</u> conditions, use area of bund or enclosing walls.
2. For <u>continuous</u> spills, the liquid will spread and increase the burning area until the total burning rate is equal to the spill rate, i.e., equilibrium exists between spill rate and burning rate.
3. For <u>instantaneous</u> spills, the unconfined pool fire grows in size until a barrier is reached or until all the fuel is consumed.

Rate of burning in kg/m²s

$$\frac{dm}{dt} = \frac{0.001 H_c}{C_p (T_b - T_a) + H_{vap}}$$

Where:
H_c = heat of combustion in J/kg
C_p = specific heat of liquid in J/kg°K
T_b = normal boiling point, °K
T_a = ambient temperature, °K
H_{vap} = enthalpy of evaporation, J/kg
Note:
Typically, dm/dt is 0.05 for gasoline to 0.12 for LPG

Also, when Tb ≤ Ta, $\frac{dm}{dt} = 0.001 \frac{H_c}{H_{vap}}$

Pool diameters for continuous spills

Pool diameter in meters: $D = 2\left[\frac{\dot{V}_L}{\pi \dot{y}}\right]^{\frac{1}{2}}$

Where:

$\dot{V}_L$ = continuous liquid spill rate, m^3/s

$\dot{y}$ = liquid burning rate, m/s and $\dot{y} = \frac{1}{\rho_l} \times \frac{dm}{dt}$

Where:
ρ_l = liquid density, kg/m^3

Instantaneous spills

Maximum diameter in meters: $D_{max} = 2\left[\frac{V_L^3 g}{\dot{y}^2}\right]^{\frac{1}{8}}$ and $t_{max\,sec} = 0.6743\left[\frac{V_L}{g \dot{y}^2}\right]^{\frac{1}{4}}$

Where:
V_L = volume spilled, m^3

g = 9.81 m/s^2 and for spills on water, use $g' = \frac{g(1-\rho_l)}{\rho_w}$

Note:

At anytime: $\left(\frac{D}{D_{max}}\right) = \frac{\sqrt{3}}{2}\left(\frac{t}{t_{max}}\right)\left[1+\left(\frac{2}{\sqrt{3}}-1\right)\left(\frac{t}{t_{max}}\right)^2\right]$

Flame height

Visible flame height in meters: $H = 42D\left[\frac{dm/dt}{\rho_a\sqrt{gD}}\right]^{0.61}$

Where:
ρ_a = ambient air density, kg/m^3 (typically, around 1.2)

Thermal flux received

Two methods are

1. Point Source Model – simplest to use

Radiation flux received, kW/m^2, at distance x is $Q_x = \frac{\tau Q_R}{4\pi x^2}$

Where:
Q_R = total heat radiated, kW, and

$$Q_R = \left(\frac{dm}{dt}\right)\times(pool\_area)\times H_c \times\left[\frac{\%\_radiative/combustion\_output}{100}\right]$$

x = distance from point source to target, m (assume flame is cylindrical)
τ = atmospheric transmissivity, dimensionless, and $\tau = 2.02(P_w x)^{-0.09}$
P_w = water pressure, Pa (conservative value, τ = 1)

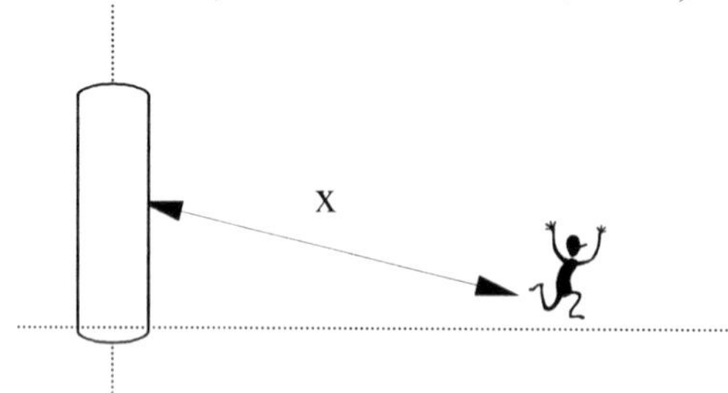

Table 22-17: Radiation Fraction of Combustion Energy for Hydrocarbon Pool Fires

Hydrocarbon	Pool Size (m)	% Radiative Output/ Combustion Output
Methanol	1.22	17.0
LNG on land	18.0 0.4-3.05 1.8-6.1 20.0	16.4 15.0-34.0 20.0-25.0 36.0
LNG on water	8.5-15.0	12.0-31.0*
LPG on land	20.0	7.0
Butane	0.3-0.76	19.9-26.9
Gasoline	1.22-3.05 1.0-10.0	40.0-13.0* 60.1-10.0
Benzene	1.22	36.0-38.0
Hexane	-	40
Ethylene	-	38

* Smaller diameter fires were associated with higher radiative outputs.

2. Solid Flame Model – more accurate but difficult to use. (It is not described here)

Flame tilt

θ = angle of flame tilt
$\cos(\theta) = 1$, for $u^* < 1$
$\cos(\theta) = (1/u^*)^{1/2}$, for $u^* \geq 1$

Where:
$u^* = u_w/u_b$ = (wind speed)/(characteristic buoyancy velocity), and
$u_b = [(mgD)/(\rho_v)]^{1/3}$

Where:
m = fuel burn rate, kg/s
ρ_v = fuel vapor density at ambient conditions

Table 22-18: Effects of thermal radiation

Incident Flux		Impact
kW / m^2	**BTU / hr / ft^2**	
37.5	11,890	Sufficient to cause damage to process equipment (100% lethality).
25	7,926	Minimum energy required to ignite wood at long exposures (non-piloted.
12.5	3,963	Minimum energy for piloted ignition of wood, melting of plastic (100% lethality).
9.5	3,000	Exposure must be limited to a few seconds, sufficient for escape only. Pain threshold reached after 8 seconds, second degree burns after 20 seconds.
6.31	2,000	Personnel could tolerate for up to 1 minute without shielding but with appropriate clothing.
4	1,268	Sufficient to cause pain if cover is not reached in 20 seconds. First degree burns.
1.6	507	Tolerable for long periods.

Fireballs:

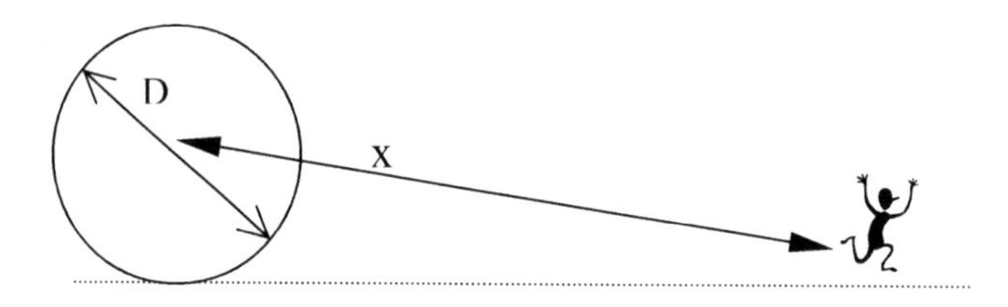

Where: $Q_x = \dfrac{\tau \times E \times D^2}{4x^2}$

Q_x = radiation received at target, kW/m^2

E = surface emitted flux (max. ~350 kW/m^2)

τ = atmospheric transmissivity

x = distance in meters

D = fireball diameter in meters = 6.48 x $M^{0.325}$

Where:

M = mass, kg

Duration of the fireball: $t_{sec} = 0.825 \times M^{0.26}$

Jet Fire Characteristics

In order to compute potential knock-on or impingement effects from a jet fire the following simplified equation may be used:

$$F_{len} = D_{jet}\ (1050/C_{cl}) \sqrt{\frac{M_a}{M_f}}$$

where:

F_{len} = Flame length in meters

D_{jet} = Diameter of jet in meters

C_{cl} = Lean limit concentration in volume %

M_a = Molecular weight of air (= 29)

M_f = Molecular weight of fuel

One of the concerns of a jet fire is "Torching" where a flame impacts adjacent equipment/ pipework and thus locally weakens metal leading to potential destruction. Flange jet leaks are especially important, e.g.

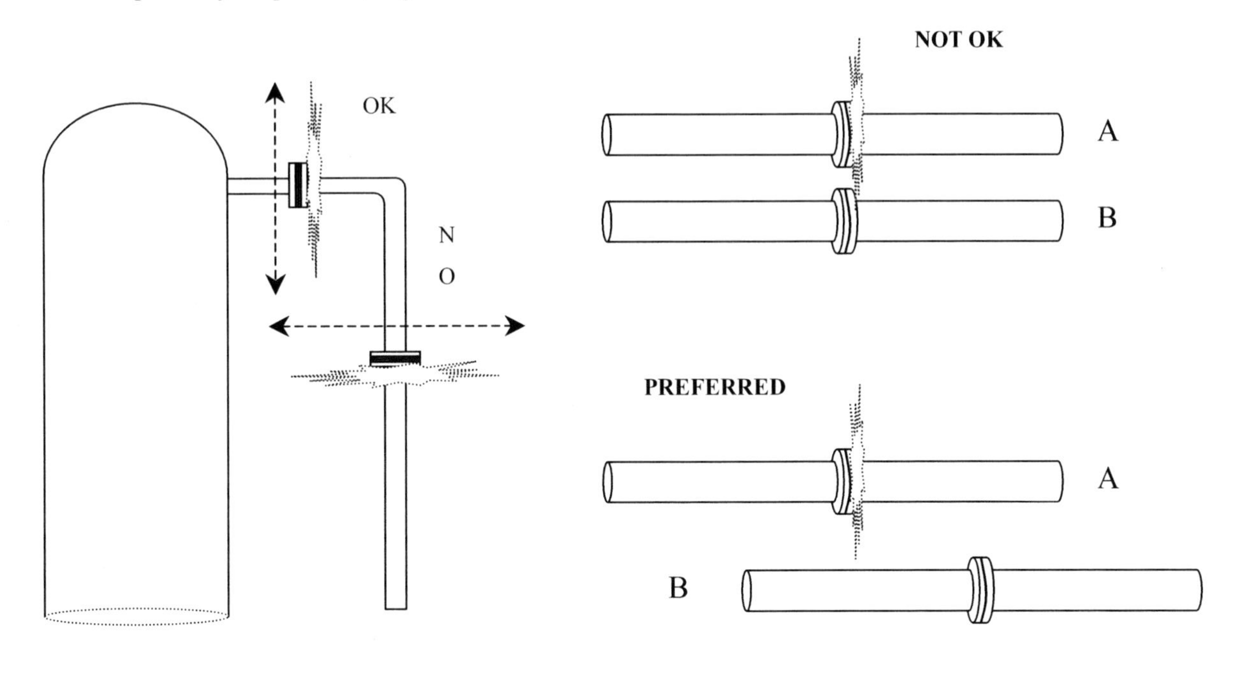

Note:

- Torching can impact adjacent flanges. A staggered arrangement of flanges is preferred.
- Hydrogen leaks burn with invisible flames.

Vapor Cloud Explosions:

- Releases do not normally explode when below 5 tonnes.
- Maximum size considered being around 50 tonnes.

For calculation the explosive effect of TNT equivalent, refer to page 22-55. For materials that can flash at atmospheric condition, the following equation can be used to determine the flash fraction:

$$m_v = 1 - exp\left(-\frac{C_P}{\Delta H_v} \Delta T \right)$$

where:

m_v = Vapor mass fraction or quality

C_p = Specific heat capacity at constant pressure, in kJ/kg.K

ΔH_v = Latent heat of vaporization of liquid, in kJ/kg

ΔT = Ambient temperature – boiling point at normal temperature and pressure, in K

Because flash evaporation is a fairly violent process and vaporization is accompanied by considerable liquid, it is common to assume that the mass of the spray is equal to that of the material vaporised.

For:

x = distance to given overpressure, ft, and

$$x = M_{TNT}^{1/3} \exp\left(3.5031 - 0.7241 \ln(O_p) + 0.0398 (\ln O_p)^2\right)$$

O_p = overpressure, psi

Note – Model based on condensed phase explosives, i.e., near field estimates are higher. However, estimates in the distant field are reasonably accurate using this model.

Table 22-19: Yield Factors for Explosive Vapors and Gases

Substances with Yield Factors of Y = 0.03	
Acetaldehhyde	3-Methyl-Butene-1
Acetone	Methyl-Butyl-Ketone
Acrylonitrile	Methyl Chloride
Amyl Alcohol	Methyl-Ethyl-Ketone
Benzene	Methyl Formate
1,3-Butadiene	Methyl Mercaptan
Butene-1	Methyl-Propyl-Ketone
Carbon Monoxide	Monochlorobenzene
Cyanogen	N-Amyl Acetate
1,1-Dichloroethane	Naphthalene
1,2-Dichloroethane	N-Butane
Di-Methyl Ether	N-Butyl Acetate
Dimethyl Sulphide	N-Decane
Ethane	N-Heptane
Ethanol	N-Hexane
Ethyl Acetate	N-Pentane
Ethylamine	N-Propanol
Ethyl Benzene	N-Propyl Acetate
Ethyl Chloride	O-Dichlorobenzene
Ethyl Cyclohexane	P-Cymene
Ethyl Formate	Petroleum Ether
Ethyl Propionate	Phthalic Anhydride
Furfural Alcohol	Propane
Hydrocyanic Acid	Propiaonaldehyde
Hydrogen	Propylene
Hydrogen Sulphide	Propylene Dichloride
Iso-Butyl Alcohol	P-Xylene
Isobutylene	Styrene
Iso-Octane	Tetrafluoroethylene
Iso-Propyl Alcohol	Toluene
Methalamine	Vinyl Acetate
Methane	Vinyl Chloride
Methanol	Vinylidene Chloride
Methyl Acetate	Water Gas
Substances with Yield Factors of Y = 0.06	
Acrolein	Ethylene
Carbon Disulphide	Ethylene Nitrite
Cyclohexane	Methyl-Vinyl-Ether
Di-Ethyl Ether	Phthalic Anhydride
Di-Vinyl Ether	Propylene Oxide
Substances with Yield Factors of Y = 0.19	
Acetylene	Isopropyl Nitrate
Ethylene Oxide	Methyl Acetylene
Ethyl Nitrate	Nitromethane
Hydrazine	Vinyl Acetylene

GAS VESSEL EXPLOSIONS: Mechanical Explosions Without Combustion

Energy release: $E = \frac{P_1 V_1}{\gamma - 1}\left[1 - \left(\frac{P_a}{P_1}\right)^{\frac{\gamma-1}{\gamma}}\right]$

Where:

P_a = ambient pressure

P_1 = burst pressure

V_1 = volume of vessel

Note:

E can be re-expressed, in energy terms, as TNT equivalent.

For compressed liquefied gases:

E.g. liquid N_2, compute internal energy charge at burst (can simulate).

$U = H - \Delta(PV)$

Table22-20: Effects of Explosions on People

Overpressure (psi)	Casualty Probability (fatalities/injuries)
< 1	0
1 – 3	0.10
3 – 5	0.25
5 – 7	0.70
> 7	0.95

"Risk Assessment and Risk Management for the Chemical Process Industry", edited by H.R. Greenberg, J.J. Cramer, Published by van Nostrand Reinhold, 1991, (ISBN 0-442-23438-4)

FACTORS INCREASING LIKELIHOOD OF EXPLOSIONS

- Wide flammability limits.
- High flame speeds, e.g. hydrogen.
- Low autoignition temperatures, e.g. H_2S.
- Ignition sources, e.g. fired heaters.
- Quiescent atmospheric conditions (little wind – low dispersion).
- Confined spaces (low dispersion).

Based on 12 recorded releases ("Unconfined Vapor Cloud Explosions" by K. Gugan):

- Approximately 40% result in explosions (mean).
- Upper limit 60% result in explosions (large releases).
- Lower limit 30% result in explosions (small releases).

VAPOR DISPERSION MODELING (GAUSSIAN DISPERSION)

Many models available, depending on conditions. Simplest and most widely used model is the Pasquill-Gifford model, based on Gaussian Dispersion.

(1) For continuous releases:

$$C_{@(x,y,z)} = \frac{G}{2\pi\sigma_y\sigma_z}\left\{\exp\left(\frac{-y^2}{2\sigma_y^2}\right)\left[\exp\left(\frac{-(z-h)^2}{2\sigma_z^2}\right)+\exp\left(\frac{-(z+h)^2}{2\sigma_z^2}\right)\right]\right\}$$

Where:
C = concentration, kg/m3 @ location (x, y, z)
G = emission rate, kg/s
h = effective height of emission, m (stack height plus plume rise)
x = downwind distance from source, m
y = crosswind distance from center plane, m

z = vertical distance from center plane, m

U_z = wind velocity at height z, m/s, and

$$U_z = U_{10}\left(\frac{z}{10}\right)^P \quad \text{(cont'd)}$$

Where:
U_{10} = wind velocity @ 10 m height
z = height above ground level, m
P = power coefficient

Note:
- σ_y, σ_z are predicted from formulas (recommended by Briggs).
- U_z is never considered @ < 0.5 m/s

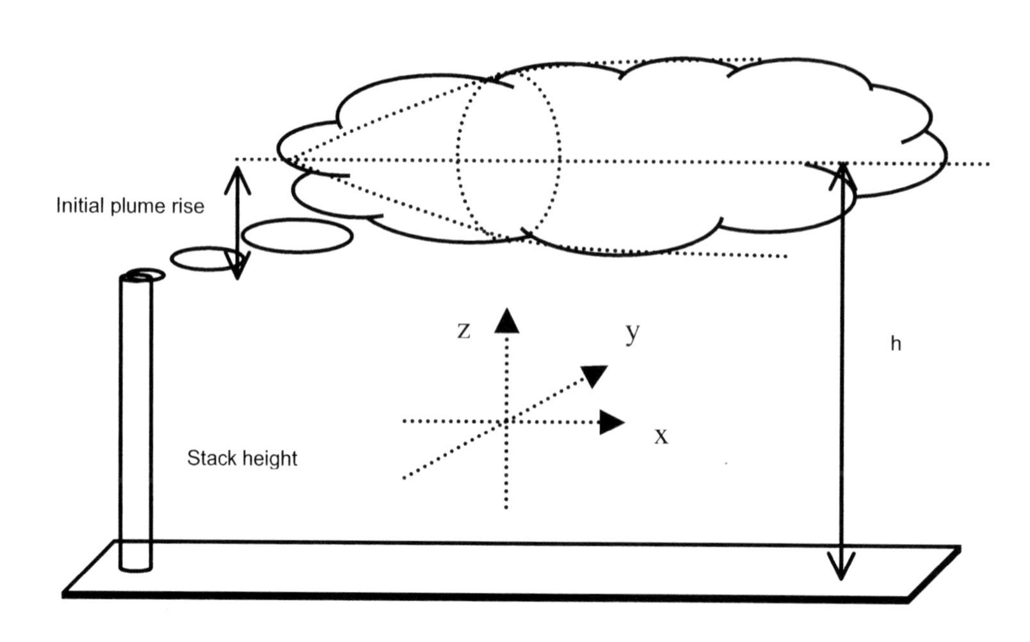

Note:

There are many methods of evaluating initial plume rise, shape, etc depending upon buoyancy and other factors. A first approximation is to treat the plume as a jet-mixing zone where exit vapors loose momentum.

Cude formula: $H_J = d_s \times \frac{v_{se}}{v_w} \times \sqrt{\frac{0.07641}{\left(\frac{MW_g}{10.73}\right) \times \left(\frac{P_s}{T_{sg}}\right)}}$

H_j = jet-mixing zone height, ft

d_s = stack tip diameter, ft

v_{se} = stack exit velocity, ft/s

v_w = wind velocity, ft/s

MW_g = molecular weight of gas

P_s = pressure at stack tip, psia

T_s = stack gas temperature, °R

Table 22-21: Pasquill Stability Meteorological Conditions

Surface Wind Speed (m/s) @10 m height	Daytime Isolation			Nighttime Conditions	
	Strong	Moderate	Slight	Thin Overcast or ≥ Low 4/8 Cloudiness	≤ 3/8 Cloudiness
< 2	A	A – B	B		
2 – 3	A – B	B	C	E	F
3 – 4	B	B – C	C	D	E
4 – 6	C	C – D	D	D	D
> 6	C	D	D	D	D

Where:

A: Extremely unstable conditions	**D: Neutral conditions**
B: Moderately unstable conditions	**E: Slightly stable conditions**
C: Slightly unstable conditions	**F: Moderately stable conditions**

Table 22-22: Wind Speed Correction Coefficients

Stability	Power Law: Atmospheric Coefficient (p)	
	Urban	Rural
A	0.15	0.07
B	0.15	0.07
C	0.20	0.10
D	0.25	0.15
E	0.40	0.35
F	0.60	0.55

Formulas Recommended by Briggs (1973) for:

- $\sigma_y(x)$, and
- $\sigma_z(x)$, $[10^2 < x < 10^4$ m].

Table 22-23: Formulas for σ_y and σ_z

Pasquill Type	σ_y, m	σ_z, m
Open-Country Conditions		
A	$0.22x(1 + 0.0001x)^{-1/2}$	$0.20x$
B	$0.16x(1 + 0.0001x)^{-1/2}$	$0.12x$
C	$0.11x(1 + 0.0001x)^{-1/2}$	$0.08x(1 + 0.0002x)^{-1/2}$
D	$0.08x(1 + 0.0001x)^{-1/2}$	$0.06x(1 + 0.0015x)^{-1/2}$
E	$0.06x(1 + 0.0001x)^{-1/2}$	$0.03x(1 + 0.0003x)^{-1}$
F	$0.04x(1 + 0.0001x)^{-1/2}$	$0.016x(1 + 0.0003x)^{-1}$
Urban Conditions		
A – B	$0.32x(1 + 0.0004x)^{-1/2}$	$0.24x(1 + 0.001x)^{1/2}$
C	$0.22x(1 + 0.0004x)^{-1/2}$	$0.20x$
D	$0.16x(1 + 0.0004x)^{-1/2}$	$0.14x(1 + 0.0003x)^{-1/2}$
E – F	$0.11x(1 + 0.0004x)^{-1/2}$	$0.08x(1 + 0.00015x)^{-1/2}$

(2) For puff emissions:

$$C = \frac{M}{(2\pi)^{3/2}\sigma_x\sigma_y\sigma_z}\left\{\exp\left(\frac{-(x-ut)^2}{2\sigma_x^2} - \frac{y^2}{2\sigma_y^2}\right)\left[\exp\left(\frac{-(z-h)^2}{2\sigma_z^2}\right) + \exp\left(\frac{-(z+h)^2}{2\sigma_z^2}\right)\right]\right\}$$

Assume:

$\sigma_x \cong \sigma_y$

and:

σ_y = half those for continuous plumes

σ_z = same as those for continuous plumes

ATMOSPHERIC DISPERSION – HEAVY GAS DISPERSION

Dispersion will largely depend upon whether a specific vapor is heavier or lighter than air. For example, Chlorine and Sulfur Dioxide vapors have molecular weights greater than 29 and, at ambient temperatures, are denser than air. Anhydrous Ammonia, although having a molecular weight of 17 can produce a colder-than-ambient vapor and/or produce aerosol clouds which are initially heavier-than-air. A so-called Heavy Gas Dispersion Model is required to model the transport and diffusion of these releases. As indicated, these clouds are heavy relative to the surrounding atmosphere by virtue of having a greater molecular weight. When the release is a flashing liquid, a cloud is heavier by virtue of experiencing local auto-refrigeration effects. As a result of slumping and spreading radially due to gravity, heavy gas clouds, in contrast to clouds that are neutrally buoyant or lighter than air, may spread against the direction of the wind and take much longer to disperse since atmospheric turbulence cannot readily penetrate into the cloud. Therefore, relative to neutrally buoyant clouds, they have the potential to travel large distances without dispersing to "safe levels". If the buoyancy of the vapor cloud is neutral then a Gaussian plume dispersion model may be appropriate. However, if the cloud is dense and heavier than air, then a heavy gas model such as SLAB, may be required. SLAB is one of a number of dense gas models which include:

- ADAM
- ALOHA™
- DEGADIS
- EMGRESP
- HARMI-II (UF_6)

- HGSYSTEM/ UF_6
- PLUME (UF_6)
- TOXIGAS
- WHAZAN

SLAB was developed by the Lawrence Livermore National Laboratory. It is a widely used, well validated and quality assured heavy gas dispersion model. SLAB is useful in that it can realistically model aerosol evaporation within a cloud which is a real occurrence for flashing liquid releases (such as liquid releases of pressurized liquefied gases such as chlorine, anhydrous ammonia and sulfur dioxide. The input data required by SLAB consists of the following:

- release type, e.g., flashing jet or Vapor
- orientation of release, e.g., vertical jet or horizontal jet
- physical and thermodynamic properties of chlorine and sulfur dioxide
- release parameters, e.g. release rate, duration, release height, flash fraction
- meteorological parameters, e.g., stability, wind speed, surface roughness, ambient temperature, relative humidity
- concentration averaging time

As before, for the Gaussian dispersion modeling, the meteorology controls the dispersion of an emission that has been discharged into the atmosphere. The specific meteorological parameters required by dispersion models are (i) atmospheric stability, (ii) wind speed, (iii) ambient temperature and (iv) ambient relative humidity. The more important parameters are referenced by the Pasquill Stability categories previously described. In addition, dispersion modeling requires accounting for surface roughness and elevation changes. Surface roughness acts as a kind of friction factor affecting the flow of a cloud. In dispersion modeling this is accounted for by a surface roughness length, (Zo). The greater the surface roughness length, the more turbulence that is generated in the

atmosphere and the greater the dispersion. This results, all other factors being equal, in a shorter distance traveled before a cloud disperses to safe levels.

The effects of elevation changes can be important but are difficult to assess unless evaluated in a wind tunnel. Current heavy gas dispersion models, have been developed for dispersion over flat terrain.

QUANTITATIVE RISK ASSESSMENT DUE TO RELEASE OF TOXIC VAPORS

A Consequence Analysis can model the health effects from exposure to toxic vapor clouds.

The hazard identification stage, as a result of a Process Hazards Analysis, will identify a number of toxic vapor release scenarios. For each of these scenarios, the following are determined:

- release rate and release duration [see Chapter 17, referring to Loss of Containment Calculations]
- downwind dispersion characteristics [see above, in relation to Dispersion Modeling]
- health effects [see earlier section in this chapter, on probit analysis]

Health effects are due to primarily by inhalation but may also be via eye and skin contact, especially if the material is intensely hygroscopic (such as with Anhydrous Ammonia, Hydrogen Chloride and Hydrogen Fluoride). A consequence analysis contributes to a quantitative risk assessment (QRA) by predicting the quantitative likelihood of fatality or serious injury to an unprotected individual. This information is particularly useful in helping to define the area around a hazardous facility that is covered by an emergency response plan (ERP).

For each release scenario, following the above steps allows the prediction the probability of death or serious injury, P_{Death} , as a function of separation distance (x) from the release point, as shown below:

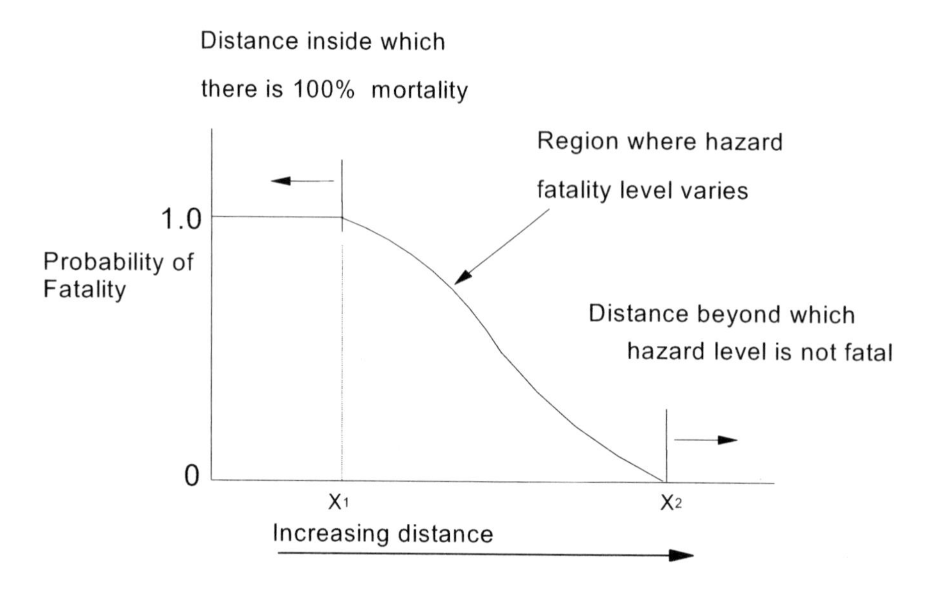

Figure 22-15: Probability of Fatality versus Distance from Release

The area that needs to be covered by an ERP should be conservatively based on the following:

- a worst credible case (or Alternative Case, as defined by RMP) release scenario
- conservative meteorological conditions
- an appropriate concentration level and exposure time, which are believed not to impair escape/evacuation in one hour, and in conformance with ERPG definitions.

SPECIFIC RELEASE SCENARIOS

As a result of the process hazards analyses it is possible to identify hazardous release situations and thence to model specific release scenarios in terms of:

- Case numbers for each individual release scenario
- Release scenario description
- Release source
- Source temperature in °F
- Release rate in kg/s
- Release duration in seconds
- Total release, in kg
- Vapor fraction
- Approx. release height in m.

Release scenarios are calculated based on:

- Hole or aperture size at point of release
- Whether release is liquid, gas or flashing liquid
- Fluid properties of releasing fluid at saturated conditions
- Saturation pressure at source temperature
- Extent of heat input (e.g., for fire condition).

USE OF CONSEQUENCE ANALYSIS

Consequence analyses should typically address:

1. Distance within which exposure is rapidly fatal to all exposed - i.e., within a few breaths.
2. Distance at which the probability of death is 50%. At this distance, the "average" person would be affected
3. Distance to the lethality limit. At this distance, the most sensitive person in a societal group would be affected
4. Distance to the ERPG-3 level (the maximum airborne concentration below which it is believed that nearly all individuals could be exposed for one hour without experiencing or developing life-threatening health effects).
5. Distance to the ERPG-2 level (the maximum airborne concentration below which it is believed that nearly all individuals could be exposed for one hour without experiencing irreversible or other serious health effects or symptoms that could impair their abilities to take protective action).
6. Distance to the ERPG-1 level (the maximum airborne concentration below which it is believed that nearly all individuals could be exposed for one hour without experiencing other than mild transient adverse health effects or perceiving a clearly objectionable odor).

With respect to these estimated impact areas it is important to note the following:

- They apply to outdoor receptors. People who are indoors and stay indoors may experience lesser impacts. Sealed buildings may offer some protection against a passing hazardous cloud, because a contaminant can only slowly infiltrate a building; this will depend on the building ventilation rate [the infiltration rate will likely be higher in winter when greater quantities of air enter due to combustion/heating air requirements].

- They apply along the cloud centerline only. As a receptor moves in the perpendicular direction away from the axis of cloud travel, concentration levels decrease.
- It is a normal practices in consequence analysis modeling to make conservative assumptions in order to be on the safe side. The consequence analyses assume constant atmospheric and terrain conditions. In reality, factors such as wind speed and surface roughness can change with cloud travel. Thus conservative estimates are required.

SUGGESTED READING (Note: URLs current at date of publication)
"Guidelines for Evaluating the Characteristics of Vapor Cloud Explosions, Flash Fires and BLEVE's" by AIChE, CCPS, 1994 www.aiche.org/pubcat/seadtl.asp?Act=C&Category=Sect4&Min=20
"Methods for the Calculation of Physical Effects", Part 1 and 2, Third Edition, 1997, 'Yellow Book', Committee for the Prevention of Disasters, by TNO, Netherlands www.tno.nl:1201/compass?scope=Yellow+book&browse-category=English&search-category=ROOT&ui=sr&chunk-size=&page=1&taxonomy=TNO
"Guidelines for Chemical Process Quantitative Risk Analysis" " by AIChE, CCPS, 2000 www.aiche.org/pubcat/seadtl.asp?ACT=C&Category=Sect4
"Guidelines for Consequence Analysis of Chemical Releases" by AIChE, CCPS, 1999 www.aiche.org/pubcat/seadtl.asp?Act=C&Category=Sect4&Min=20
"Guidelines for Process Equipment Reliability Data, with Data Table" by AIChE, CCPS, 1989 www.aiche.org/pubcat/seadtl.asp?Act=C&Category=Sect4&Min=30
"Quantified risk assessment: Its input to decision making" published by UK Health & Safety Executive, 1980 http://www.hsebooks.co.uk/homepage.html
"A Basic Approach for the Analysis of Risks from major Toxic Hazards" by R.P.Pape and C.Nussey, pages 367 to 388, IChemE Symposium Series No.93 www.icheme.org/framesets/aboutusframeset.htm
"Risk Analysis of Six Potentially Hazardous Industrial Objects in the Rijnmond Area A Pilot Study", Rijnmond Public Authority, February 1982, ISBN 90-277-1393-6 http://www.wkap.nl/home/search/?view=Search+results
"Classification of Hazardous Locations" by A.W. Cox, F.P.Lees, M.L.Ang, published by IChemE, 1990 www.icheme.org/framesets/aboutusframeset.htm
"Use risk analysis as a safety tool" by G.Mani, Hydrocarbon Processing, September 1994, pages 87 to 94 www.hydrocarbonprocessing.com/contents/publications/hp/

"Occupant response shelter evacuation model" by Electrowatt Engineering (UK) Ltd. for HSE, UK, Contract Research Report 162/1998 www.hse.gov.uk/research/crr_pdf/1998/crr98162.pdf
"The implications of major hazard sites in close proximity to major transport routes" by W.S. Atkins for HSE, UK, Contract Research Report 163/1998 www.hse.gov.uk/research/crr_pdf/1998/crr98163.pdf
"Dow's Fire & Explosion Index Hazard Classification Guide", 7th edition, published by AIChE, 1994 www.forensicdna.com/Bookstore/arson.html
"Dow's Chemical Exposure Index Guide", 1st edition, 1994 http://www.NetStoreUSA.com
"Hazard Identification, Analysis and Control" by H.Ozog, Chemical Engineering, February 18, 1985, pages 161 to 170 www.che.com/
"Probabilistic Risk Assessment in the CPI" by P.Guymer et al., Chemical Engineering Progress, January 1987, pages 37 to 45 www.cepmagazine.org/
"Risk criteria for land-use planning in the vicinity of major industrial hazards", published by HSE, UK, through HMO, 1989 http://www.hsebooks.co.uk/homepage.html
"Risk-based Land Use Planning Guidelines", by MIACC, 1995 www.ptsc-program.org/order_publications.htm
"Eliminating Potential Process Hazards" by Trevor Kletz, Chemical Engineering, April1, 1985, pages 59 to 68 www.che.com/
"Process Safety Analysis, An Introduction" by Bob Skelton, IChemE, 1997 www.icheme.org/framesets/aboutusframeset.htm
"Estimating Hazard Distances from Accidental Releases", by A.Kumar, Chemical Engineering, August 1998, pages 121 to 128 www.che.com/
"Quantify BLEVE Hazards", by R.W.Prugh, Chemical Engineering Progress, February 1991, pages 66 to 72 www.cepmagazine.org/

"Quantitative Evaluation of Fireball Hazards", by R.W.Prugh, Process Safety Progress,April, 1994, pages 83 to 91

www.aiche.org/safetyprogress/

"Assessment of Mathematical Models for Fire and Explosion Hazards of Liquefied Petroleum Gases", by W.P.Crocker and D.H.Napier, Journal of Hazardous Materials, 20 (1988), pages 109 to 135, Elsevier Science

www.elsevier.nl/inca/publications/store/5/0/2/6/9/1/

"Prediction of thermal hazards from fireballs", by K.Satyanarayana et al., J.Loss Prev. Process Ind., 1991, Vol 4, October

www.elsevier.nl/inca/publications/store/3/0/4/4/4/

"Thermal Radiation Hazards from Hydrocarbon Pool Fires", by K.S. Mudan, Prog. Energy. Combust. Sci., 1984, Vol 10, pages 59 to 80, Pergamon Press

http://www.elsevier.com/inca/search

"Overpressure Monograph", I.Chem.E., 1989

www.icheme.org/framesets/aboutusframeset.htm

"Ammonia Toxicity Monograph", I.Chem.E., 1989

www.icheme.org/framesets/aboutusframeset.htm

"Chlorine Toxicity Monograph", I.Chem.E., 1989

www.icheme.org/framesets/aboutusframeset.htm

"Thermal Radiation Monograph", I.Chem.E., 1989

www.icheme.org/framesets/aboutusframeset.htm

"Explosion Hazards and Evaluation", by W.E.Baker et al., Elsevier Scientific Publishing Company, 1983

http://isbn.nu/0444420940/price/2

"Estimating the Flammable Mass of a Vapor Cloud" by AIChE, CCPS, 1999

www.aiche.org/pubcat/seadtl.asp?ACT=C&Category=Sect4

"Loss Prevention in the Process Industries" by F.P.Lees, published by Butterworth-Heinemann, 1996. (Volumes 1,2, 3)

www.aiche.org/pubcat/seadtl.asp?Act=C&Category=Sect4&Min=50

www.bhusa.com/gulf/us/subindex.asp?maintarget=bookscat%2Fsearch%2Fresults%2Easp&country=United+States&ref=&mscssid=GKTMNF4S2L2C8K5B017248LP4MJXFWVF

"Assessment of Uncertainties in Risk Analysis of Chemical Establishments – The ASSURANCE Project – Final Summary Report", by K.Lauridsen et al., Riso-R-1344(EN), Riso National Laboratory, Denmark, May 2002

www.risoe.dk/rispubl/SYS/syspdf/ris-r-1344.pdf

"Risk analysis and risk policy in the Netherlands and the EEC" by J.M.Ale, J.Loss Prev. Process Ind., 1991, Vol 4 January, pages 58 to 64 http://www.elsevier.nl/locate/jlp
"Handbook for Chemical Hazard Analysis Procedures", plus "ARCHIE", (Automated Resource for Chemical Hazard Incident Evaluation), developed jointly by the U.S. Department of Transportation, the Federal Emergency Management Agency, and the Environmental Protection Agency ,1989 http://hazmat.dot.gov/risk_tools.htm
"Process Plant Layout", by J.C.Mecklenburgh [see Appendix B], ISBN 0608131474 http://www.addall.com/Browse/Detail/0608131474.html
"Safety in Process Plant Design", by G.L.Wells, Halsted Press, 1980 http://www.amazon.com/exec/obidos/ISBN%3D0470269073/vermontsiriA/002-4893859-2409650
"Risk Assessment and Risk Management for the Chemical Process Industry", edited by H.R. Greenberg, J.J. Cramer, Published by van Nostrand Reinhold, 1991, (ISBN 0-442-23438-4) http://www.wiley.com/remsearch.cgi?query=Greenberg&field=author&x=9&y=9
"Atmospheric Dispersion Modeling Resources ©", 2nd edition,U.S. Department of Energy, 1995 (Website) http://www.bnl.gov/scapa/admr.pdf
"Risk = Hazard + Outrage", by V.T. Covello et al. (Website) http://www.psandman.com/articles/cma-0.htm
"Manual of Industrial Hazard Assessment Techniques", 1985, published by World Bank

APPENDIX I

Deriving Deviations from First Principles

Introduction

For the users of structured hazards analysis techniques such as Hazard and Operability (HAZOP) analysis and Failure Modes and Effects Analysis (FMEA) there are distinct differences and alternative applications for their usage. Although HAZOP is very well established and practiced, its exact basis is rather more empirical and subject to considerable variations. This situation has significant disadvantages including inconsistency of methodologies and lack of agreed criteria for assessment. In the field this has lead to a number of different approaches so that repeatability, a hallmark of the scientific method, is largely absent.

The problem is the inability to derive Deviations for HAZOP from first principles. This has lead to experience and empirically based methods filling the gap. These methods are more subjective and dependent upon those executing the analyses. Therefore considerable diversity in results is very frequent. The question is thus one of "who's or what" method should be used for a particular application. This clearly leads to the demand for a basic root methodology that removes it from the vagaries of subjectivism.

This paper addresses the situation by proposing Component Functional Analysis (CFA) as the root methodology. It also shows how Deviations (for HAZOP) may be derived through

the concomitant identification of malfunctions and counter functions from CFA using FMEA. It shows how both FMEA and HAZOP are closely related through consideration of functional envelopes.

Critique of Current Methods of Structured Hazards Analysis

The main forms of process hazards analysis (PHA) include Guide Word HAZOP, FMEA and What If analysis. The first two methods (HAZOP and FMEA) are highly structured while What If has little structure and is largely experience based.

FMEA, the most structured of the two remaining methods, is widely used as MIL - STD 1629 in the defense, aircraft and automobile industries. The technique is relatively straight forward and considers the failure modes of specific components.

Guide Word HAZOP, unlike FMEA, is dependent upon applying Deviations (and Disturbances) from the Design Intent. This method is somewhat more complex than FMEA and is very widely used in the process industries.

Although highly structured from an applications standpoint Guide Word HAZOP has the underlying weaknesses that the Deviations (for HAZOP) are assumed to be self evident, ***which indeed, they are not***. With Guide Word HAZOP the deviations are obtained by the application of Guide Words to Properties (which includes Parameters and Activities). However the methodology for determining which Properties should be selected is not obvious. Ellis Knowlton (Chemetics) has developed an approach based on the application of Guide Words to Materials, Activities, Sources and Destinations: this requires imagination and skill on the part of the user while the CCPS (ref. “Guidelines to Hazards Evaluation Procedures”, 2nd edition) has endorsed the Parametric Deviation methodology. Knowlton's method is most suitable for batch processes while the Parametric Deviation methodology is more suitable for continuous processes.

One of the major problems with Guide Word HAZOP, is that Deviations (for HAZOP) may be either over or under specified. This leads to wasted efforts or oversights respectively. For example, it may appear to be obvious, for a process line, to examine High, Low, Reverse Flow, High, Low Pressure, High, Low Temperature and so forth. But what is the basis for this? Unfortunately, applying a non-structured basis to a highly structured methodology leads to somewhat arbitrary results as the basis of a supposedly complete analysis.

Under these circumstances who can dictate that one set of deviations is right and another wrong? Furthermore if a specific methodology for deriving Deviations is correct, should not those systems with the greatest potential for failure also be associated with the greatest number/ range of Deviations? There is some good evidence, as this paper suggests, that equipment, such as pumps (and compressors) are underassessed using the current HAZOP methodology while other equipment is being over-assessed.

Component Functional Analysis

With any form of structured analysis the greater the number of assumptions and simplifications the greater the tendency to compromise the methodology and its effectiveness. By way of an analogy, computers, which are not programmed to take short cuts, are effective simply because the level of detail needed is not a constraint in program execution. Insisting upon thoroughness, when it is affordable, is understandable. Settling for less, because greater effort is required, may be a false economy.

There is an oversimplification of the basic issues in structured process hazards analysis. This has lead to overlooking of some very basic principles, namely the components themselves, their contributional functionality and their failure to perform, in the functional sense.

Unlike HAZOP, Component Functional Analysis (CFA) with FMEA requires no simplifying assumptions. It requires a basic understanding of the systems being analyzed.

The analysis requires breakdown into parts (i.e., nodes, subsystems). Typically these are line(s), vessel(s), heat exchanger(s), pump(s), compressor(s) and so forth.

The CFA requires the parts to be broken down into components. By way of example, consider a line passing from one point to another as typically having isolation valves, flow transmitter, control valve, piping, drain, vent etc. as components each having specific functions. Hence the line should be understood not simply as "line" but rather as an integrated group of components, each with their own purpose, i.e., functions.

The corollary to knowing the function(s) of each component is that diametrically opposite considerations, i.e., malfunctions, can be specified. In addition to malfunctions, components may have functions that are mutually antagonistic; when this occurs these are identified as counter functions. Following the listing of the malfunctions and counter functions the conventional process of specifying Causes, Consequences, Safeguards and Recommendations follows.

An example, as per the attached sketch, shows a transfer line from a caustic storage vessel, automatically neutralizing an acid stream in a neutralizing vessel. CFA with FMEA applied is used to derive Deviations. It can be seen that the malfunctions are related to failure modes. It is not necessary to specify the Design Intent because it is implicit within the methodology.

Component Functionality: a Pivotal Benchmark for establishing Failure Modes and Deviations

FMEA depends upon dividing subsystems into components. Failure Modes are then postulated, together with listing of Effects, Safeguards and Recommendations. With HAZOP the system is divided into Nodes which are examined for Deviations from the Design Intent. In the latter case prefixing Guide Words to a Property term, a Parameter or Activity, creates Deviations. However the choice of combinations relies upon experience rather than the application of any well recognized method.

In Tables I-1 and I-2 the application of Component Functional Analysis is used to derive both Functional Failures (for FMEA) and Deviations (for HAZOP) using the example.

The steps are as follows:-

- Specify the subsystem/ node to be analyzed.
- Break the subsystem/ node into components, including the functional envelope of the process material.
- List the functions/functional envelopes of each component listed.
- Prepare a list of malfunctions, diametrically opposite to the functions, for each of the components listed.
- Examine each of the components to find if any of the functions could be antagonistic to one another. List these as counter functions.
- The malfunctions can be considered as Failure Modes for FMEA with the added extra bonus of counter functions (not normally identified with FMEA).
- The malfunctions can be directly translated into Deviations, such as High Flow, High Pressure and so forth.
- Summarize and condense the list of Deviations (the same Deviation may have been repeated a number of times).

Functionality and functional envelopes can also be extended to activities and operations where specific errors are identified through the same route of malfunctions and counter functions.

Use and Advantages of Component Functional Analysis over other methods of Structured Hazards Analysis

CFA applied to FMEA can be used in a number of ways. Firstly, it can be used directly as a thorough method for performing Structured Hazards Analysis wherever FMEA or HAZOP is normally used. Secondly it can be used for deriving Failure Modes for FMEA or Deviations for HAZOP.

There are a number of items of equipment, such as pumps and compressors, where HAZOP fails to identify the key areas of failure. This is of significant concern because it is precisely these types of equipment, involving multiple components, where failures and major problems are most frequently experienced. To demonstrate the effectiveness of CFA Table I-2 shows how it can be applied to a simple Centrifugal Pump in order to derive Functional Failure Modes and Deviations.

Determination of HAZOP Deviations for Parameters and Operations

Functional envelopes can be created for both Parameters and Operations. It is then possible to list Malfunctions and Counterfunctions leading to listing of potential Deviations.

Once the exercise has been performed for ranges of systems it is thus possible to create a library of valid Deviations that can be used on a repetitive basis.

Table I-1: CFA for Acid Neutralizing Process

Node Type: Line from "A" to "B"	Operating Outdoors, range 90°F to -20°F		
Node Components	**Component Functions, Functional Envelope**	**Component Malfunctions & Counter Functions (CF) – Failure Modes**	**Deviations**
Piping from A to B	1. Liquid transfer from A to B 2. Material containment at pressure, temperature	1.1 No transfer 1.2 Two phase transfer with gas blowthrough 1.3 Transfer from B to A 2.1 Loss of containment – leak 2.2 Loss of containment – rupture	1.1. No flow 1.2. Two phase flow 1.3. Reverse flow 2.1. Leak 2.2. Rupture
Isolation valves at A & B	1. Isolate line for maintenance, shutdown	1.1 Line overpressures with heat tracing when blocked in CF 1.2 Valve(s) inadvertently closed	1.1. High pressure 1.2. No flow
Flow transmitter	1. Measure liquid flow rate and transmits hydraulic signal	1.1 Loss of hydraulic signal	1.1. Loss of transmitter signal
Flow/ pH controller	1. Controls liquid flow rate and doses to pH set point	1.1 Flow controller fails control valve open 1.2 Flow controller fails control valve close	1.1. High flow 1.2. Low/ no flow
Control valve	1. Controls flow rate of liquid 2. Maintains seal	1.1 Control valve fails open 1.2 Control valve fails close 2.1 Loss of CV seal	1.1. High flow 1.2. Low/ no flow 2.1. Loss of CV seal
Control valve actuator	1. Operates control valve by I/P conversion	1.1 Loss of air fails CV in locked position	1.1. Loss of instrument air
Bypass valve around CV	1. Manual control when CV is out for maintenance	1.1 Bypass left open or unattended causing excess flow or low flow	1.1. High flow 1.2. Low flow
Block valves in CV set	1. Permits CV to be removed for maintenance	1.1 Leakage when CV is out for maintenance	1.1. Leak
Heat tracing plus insulation of line	1 Maintains line at 60°F to 80°F	1.1 Heats line too much due to loss of temperature control 1.2 Loss of heat tracing	1.1 High temperature 1.2. Low temperature
Process material	1. 50% caustic at 50 psig, 150°F with F.P. of 60°F	1.1.Source pressure too high 1.2.Source pressure too low 1.3.Temperature below F.P.	1.1. High pressure 1.2. Low pressure 1.3. Low temperature

Summary of Deviations

1. No/ low flow
2. High flow
3. Two phase flow
4. Reverse flow
5. High pressure
6. Low pressure
7. High temperature
8. Low temperature
9. Leak
10. Rupture
11. Loss of transmitter signal
12. Loss of instrument air

Table I-2: CFA for Centrifugal Pump

Node Type: Centrifugal Pump	Operating Outdoors, range 90°F to -20°F		
Node Components	**Component Functions, Functional Envelope**	**Component Malfunctions & Counter Functions (CF) – Failure Modes**	**Deviations**
Pump impeller	1. Rotates at constant speed, fixed location 2. Creates differential head, imparts energy to liquid 3. Rotates counter-clockwise 4. Creates low pressure at eye of impeller	1.1. Overspeeding 1.2. Loss of balance for impeller 2.1. Discharge pressure too high 2.2. Discharge pressure too low 3.1. Reverse rotation 4.1. Pump can activate with too low NPSH	1.1. High speed 1.2. Loss of balance, vibration 2.1. High discharge pressure 2.2. Low discharge pressure 3.1. Reverse rotation 4.1. Cavitation, NPSH
Pump casing	1. Material containment 2. Directs liquid through flow path 3. Withstands centripetal forces generated	1.1. Casing leaks 1.2 Casing rupture 2.1. Internal recycling, loss of efficiency 3.1. Cannot withstand high pressures generated	1.1. Leak 1.2. Rupture 2.1. Loss of efficiency 3.1. High pressure
Pump seal	1. Maintains seal between static and moving parts	1.1. Loss of pump seal – leak 1.2. Loss of pump seal - contamination	1.1. Leak 1.2. Contamination
Seal flush medium	1. Lubricates and cools seal	1.1. Loss of flushing medium	1.1. Loss of flushing medium
Pump vent	1. Vents air at pump start	1.1. Vent leaks when operating 1.2. Vent won't vent	1.1. Leak 1.2. No venting
Pump drain	1. Drains pump on shutdown	1.1. Drain leaks when operating 1.2. Drain won't drain	1.1. Leak 1.2. No draining
Pump coupling	1. Maintains transmission between motor and pump	1.1. Loss of flexibility, vibration	1.1. Vibration
Pump shaft	1. Transmits force from motor to impeller via coupling	1.1. Misalignment – vibration 1.2. Shaft breaks	1.1. Vibration 1.2. Shaft breaks
Motor driver	1. Generates energy from electrical supply 2. Maintains constant speed 3. Operates within temperature range 4. Operates within load range	1.1. Loss of power 2.1. Excessive speed 3.1. Motor overheats 4.1. Motor overloads	1.1. Loss of power 2.1. High speed 3.1. High motor temperature 4.1. High motor load
Motor control switch	1. Starts motor 2. Stops motor	1.1. Motor won't start 1.2. Motor won't stop	1.1. No motor start 1.2. No motor stopping
High load motor cutout	1. Stops motor on over load	1.1. Motor overload won't operate when required	1.1. No motor overload trip
Power supply	1. Supplies power to motor and in range	1.1. Excess power to motor 1.2. Loss of power to motor	1.1. More power 1.2. No power
Baseplate	1. Maintains alignment 2. Anchors pump	1.1. Loss of alignment – vibration 2.1. Loss of anchoring	1.1. Vibration 2.1. Vibration
Process material	1. Hydrocarbon liquid with SG of 0.9 and vapor	1.1. Flashing to vapor if eye of impeller falls below 12 psia	1.1. Cavitation, NPSH

	pressure of 12 psia at 60	(CF)	

Summary of Deviations

(1) High speed
(2) Vibration
(3) High discharge pressure
(4) Low discharge pressure
(5) Reverse rotation
(6) NPSH
(7) Leak
(8) Rupture
(9) Loss of efficiency
(10) Contamination
(11) Loss of seal flush
(12) No venting
(13) No draining
(14) (High) vibration
(15) Shaft breaks
(16) No power
(17) High motor temperature
(18) High motor start
(19) No motor start
(20) No motor stop
(21) No overload trip

Acid Neutralizing Process

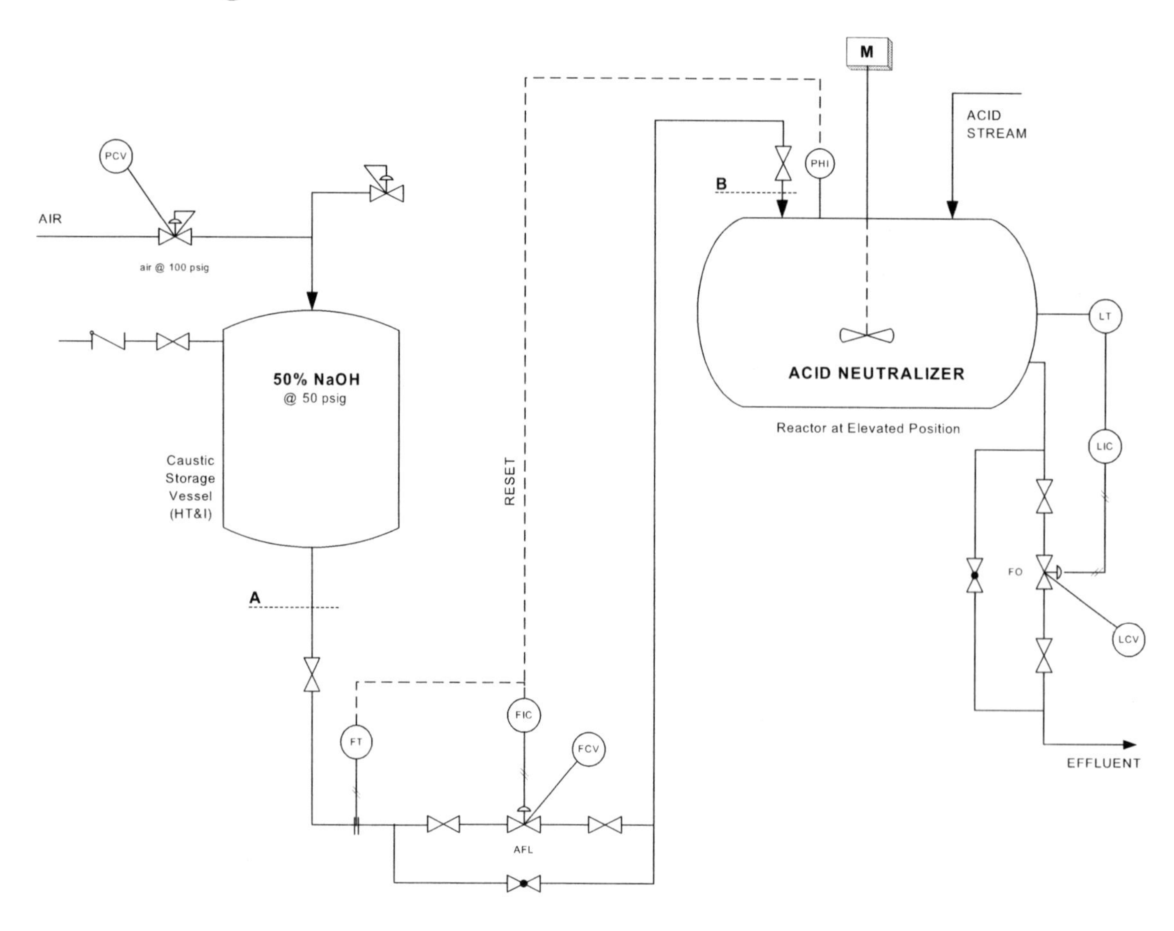

Figure I-1: P&ID of Acid Neutralizing Process

Centrifugal Pump

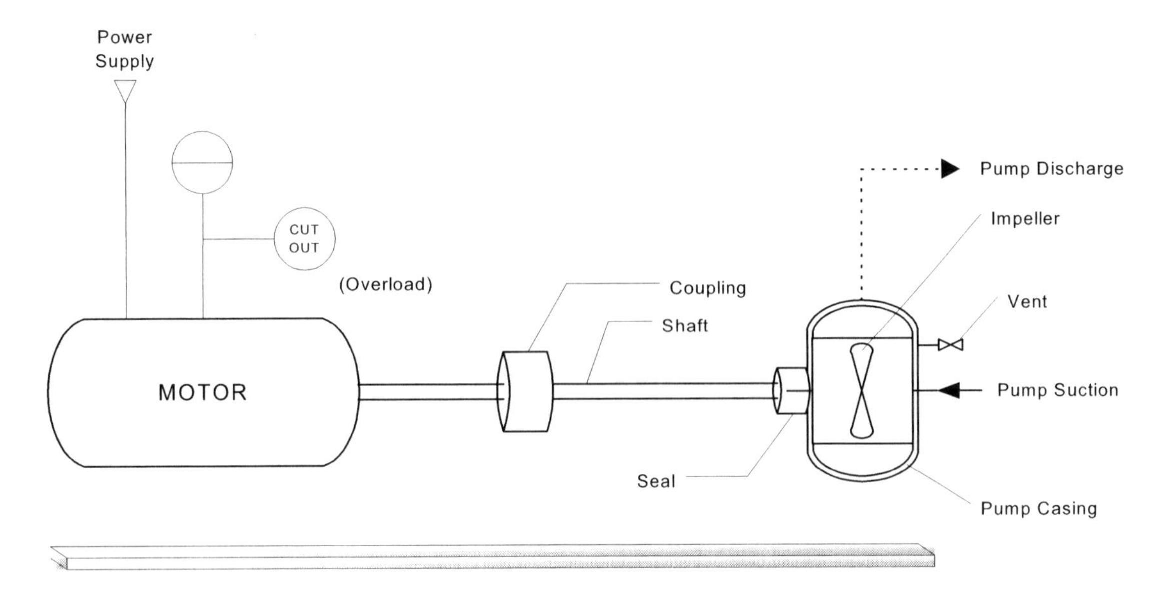

Figure I-2: Schematic Representation of Centrifugal Pump

SUGGESTED READING (Note: URLs current at date of publication)
"The 'Traffic HAZOP' – an approach for identifying deviations from the desired operation of driving support systems", by H.M.Jagtman, (Website) www.tbm.tudelft.nl/webstaf/ellenj/publications/TPM1stdoc_consortiumJagtman.pdf
"Qualitative Techniques", by R.Pudduck, (Website) www.ams.mod.uk/ams/content/docs/stgweb/pages/ssmo/lecture/RINA_19_11_99.htm

APPENDIX II

Different Types of HAZOP

A. Parametric Deviation Based HAZOP

Parametric Deviation Based HAZOP relies on establishing sets of commonly applied deviations by establishing typical parameters/properties/operations and assigning Guide Words (such as High, Low, No, Part of, Other than, As well as, etc.).

The resultant deviations effectively form a "library of deviations" which can be repetitively used, depending on the equipment type being HAZOPed.

It is the most widely used form of HAZOP in the world today.

Advantage of Parametric Deviation Based HAZOP

It has the advantage of giving reasonably consistent results and is simple to use. It also adds a certain degree of quality assurance.

Disadvantage of Parametric Deviation Based HAZOP

It has the disadvantage that certain interactions and special case deviations may be overlooked. In addition, more deviations than are really required may also be processed, thus consuming excessive time and effort.

Furthermore the basis for using such established deviations is experience as opposed to a basic methodology based on logic/ reasoning.

An ultra conservative approach would be to use a very extensive list of deviations, say around 20 (or even more), for every node reviewed. However, such an approach is very time consuming.

This could lead to frustration, boredom and lack of conviction, by team members, that any specific deviation is particularly relevant; this can compromise quality.

The key to efficient HAZOPs is:

1. Making node sufficiently large to minimize repetition
2. Using correct deviations (not too many, not too few)
3. Control of HAZOP sessions (See Chapter 19)

The following table, Table II-1, shows typical Deviations for Various Items of Equipment.

Parametric Deviation based Methodology

Table II-1: Examples of Equipment Types and Assigned HAZOP Deviations

DEVIATION	CENTRIFUGAL COMPRESSOR	CENTRIFUGAL PUMP	COLUMN	FURNACE HEATER	HEAT EXCHANGER	LINE
High Suction Pressure	X					
High Pressure		X		X	X	X
High Discharge Pressure	X					
High Temperature			X	X	X	X
High Discharge Temperature	X					
High Flow				X		X
High Bottoms Level			X			
High Concentration of Impurities			X			
Low Pressure			X	X	X	X
Low Suction Pressure	X					
Low Flow, Low/ No Flow				X		X
Low Temperature			X	X	X	
Low Bottoms Level			X			
Low Tray Level			X			
Reverse/ Misdirected Flow				X		X
Column Flooding			X			
Contaminants Enter Equipment	X			X		X
Leakage	X	X	X	X	X (Tube & Shell)	X
Rupture	X	X	X	X	X (Tube & Shell)	X
Cavitation		X				
Maintenance Hazards			X			
Startup/ Shutdown Hazards			X			
Loss of Performance					X	

B. "Creative Identification of Deviations & Disturbances" Methodology for Performing HAZOPs

Ref: "A Manual of Hazard & Operability Studies – The Creative Identification of Deviations and Disturbances" by R. Ellis Knowlton, published by Chemetics International Ltd., 1992

Example:

Consider a kettle, operating on a batch basis, into which are fed several liquids – X and Y – for mixing and emulsification. Consider that X is being pumped from a supply drum to the kettle via a feed line.

The process intention can be described in the following words: "Transfer X from the supply drum to the kettle via supply pump and feed line." The design intention is analyzed in terms of 5 specific components: Node, Material handled, Activity undertaken, Source, and Destination. See the following table.

Node	Supply Pump and Feed Line
Material	X
Activity	Transfer
Source	Supply Drum
Destination	Kettle

Advantages

- Thorough
- Good for batch type operations

Disadvantages

- Not all deviations may be valid or could be hard to interpret.
- May be cryptic and hard to audit unless very well documented.
- Can be hard to apply to continuous operations.

The deviations are obtained by applying Guide Words to Material, Activity, Source, and Destination, as follows:

Table II-2: Deviations derived by applying Guide Words to Material, Activity, Source, and Destination

	MORE	LESS	REVERSE	AS WELL AS	PART OF	OTHER THAN
Material	More X	Less X	Reverse X	As well as X	Part of X	Other than X
Activity	Transfer more	Transfer less	Reverse transfer	As well as transfer	Part of transfer	Other than transfer
Source	More from supply drum	Less from supply drum	Reverse from supply drum	As well as from supply drum	Part of from supply drum	Other than from supply drum
Destination	More to kettle	Less to kettle	Reverse to kettle	As well as to kettle	Part of a kettle	Other than to kettle

C. Procedural HAZOP

Batch processes are often used to produce various kinds of materials in the chemical industry. Continuous operations are also operated in batch modes, e.g. during startup, shutdown, maintenance, etc.

In such cases the HAZOP can be performed by sequentially analyzing the operating procedures of the particular batch process.

The operating instructions of the batch process are divided and simplified and rewritten, if necessary, so that each instruction represents the design intention.

For e.g., one instruction might be “Fully open valve V-101 to transfer 4500 kg of reactant X to the reactor R-201.” This can be broken into more elemental actions; “Fully open V-101” and “Transfer 4500 kg of X into R-201”. Each can then be combined with Guide Words to establish deviations, as follows.

- Not / Fully open V-101
- As well as / Fully open V-101
- More / Transfer 4500 kg of X into R-201
- Less / Transfer 4500 kg of X into R-201, and so on.

Procedural HAZOP Example

Procedure Description

The start-up operations of the light ends column C-101 is used to illustrate the hazard and operability technique used in batch processes. The following steps illustrate the procedures followed during the startup of the column. (See Figure II-1 for reference)

1. Put cooling water on light ends condenser EX-102 to condense light ends flashing from feed.
2. Open bypass around PV-106 to allow non-condensibles to pass to flare without pressure build-up in light ends stripper, C-101.
3. Set FRC-101 on feed supply to minimum setting (after opening up battery limit valve on feed from feed drum V-101).
4. When level in base of column reaches normal liquid level on LIC-110 crack open manual by pass around TV-126.
5. Observe level in reflux drum V-102 on LIC-107 and close manual bypass around PV-106 ensuring that setpoint of PIC-106 is set for normal design when low level is reached on LIC-107.
6. Start reflux pump P-101 and ensure total reflux with FRC-116 set for design flow.
7. Increase steam on reboiler to design flow by setting TRC-126.
8. When medium high level is almost reached in column bottoms, LIC-119, start bottoms pump P-102. With LV-119 fully closed maintain minimum flow conditions on P-102.
9. Stop feed to column, maintain reflux but do not export distillate or bottoms. Keep reboiler running and maintain check on overheads composition.
10. When overheads material is fully up to specification introduce more feed at reduced flowrate and export distillate and bottoms to maintain equilibrium.
11. Increase feed flow to design flowrate over duration of shift.

Figure II-1: P&ID of Light Ends Process

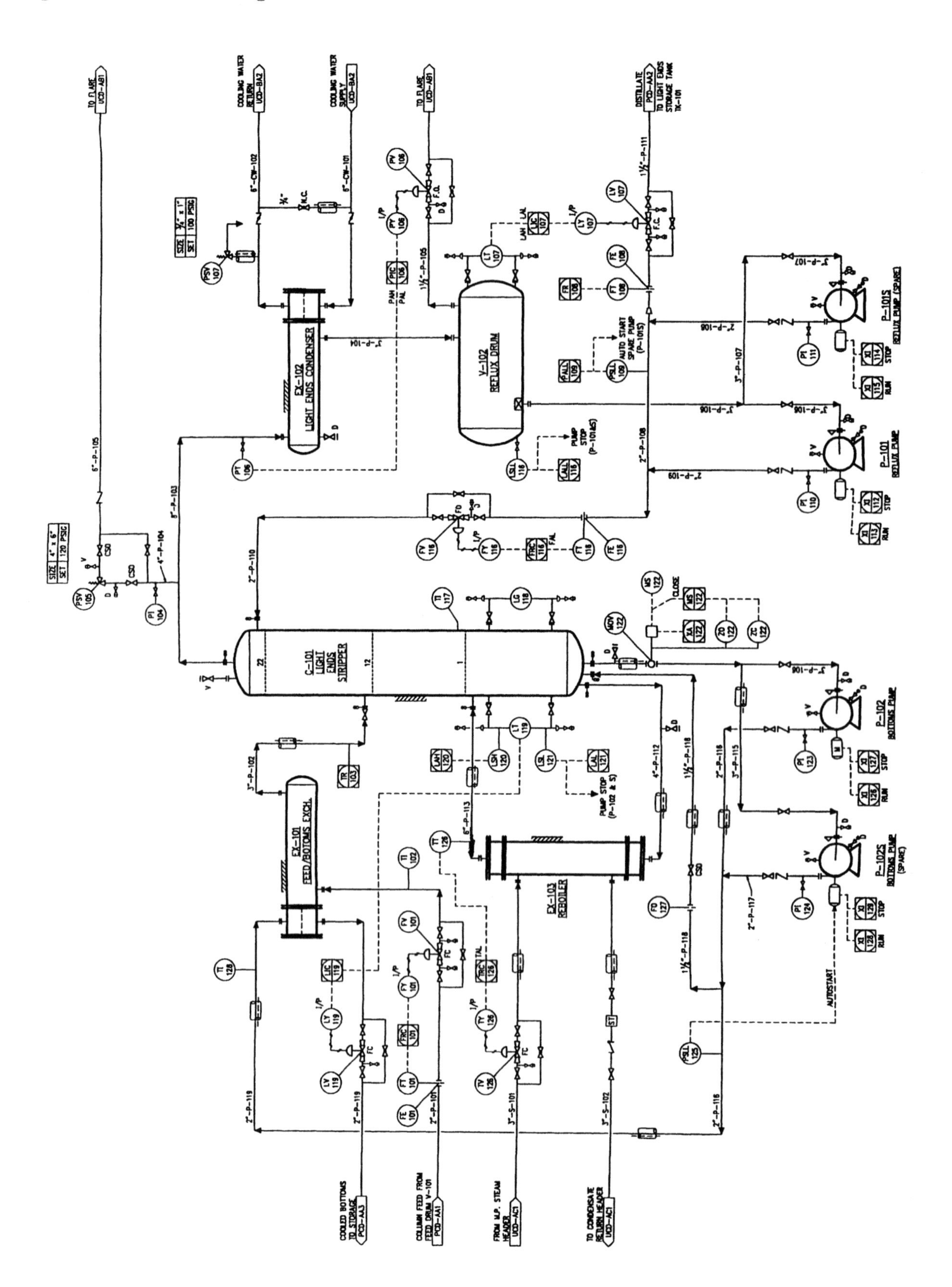

Procedural Step Evaluation

In the above example, each of the startup procedures was considered to be a node and the deviations associated with each of these nodes were evaluated. The procedure is illustrated using the first procedure.

Startup Procedure: Put cooling water on light ends condenser EX-102 to condense light ends flashing from feed.

Assigned Node: Cooling water on light ends condenser EX-102

Deviations:

- **No** cooling water on light ends condenser EX-102
- **Sooner/Later** cooling water on light ends condenser EX-102
- **More** cooling water to EX-102
- **Less** cooling water to EX-102
- **Reverse** cooling water to EX-102
- **Other than** cooling water to EX-102
- **As well as cooling water to EX-102**

Deviation:	No Cooling water on light ends condenser EX-102
Cause:	Frozen pipeline in winter
Consequence:	Can't startup
Safeguards:	¾" Bypass line which is electrically traced
Recommendations:	Ensure that there is a low point drain and a high point vent

Similar evaluation is conducted on each of the above nodes. The above procedural step methodology can be conducted to evaluate batch operations, operating procedures, operating manual instructions etc.

Table II-3: Example of Procedural HAZOP Worksheet for Light Ends

10:33:09 AM	Worksheet Report						5/22/01
Node: 1. Cooling water on light ends condenser EX-102.							Drawing:
Type: Procedural Step							Fig A 2.1
1.1. No cooling water to EX-102							
Causes	Consequences	Safeguards	S	L	RR	Recommendations	Responsible
1. Frozen pipeline in winter	1.1. Can't start-up	1.1. 3/4" Bypass line which is electrically traced	1	4	4	1. Ensure that there is a low point drain and a high point vent.	Eng. Dept.
			1	4	4	2. Introduce procedure for cooling water drainage of EX-102 on shut down.	Operations
1.2. Sooner/Later cooling water to EX-102							
Causes	Consequences	Safeguards	S	L	RR	Recommendations	Responsible
1. Failure loss of supply of key utilities	1.1. Can't start-up	1.1. None	1	3	3	3. Ensure that cooling water is available before start-up.	Operations
1.3. More cooling water to EX-102							
Causes	Consequences	Safeguards	S	L	RR	Recommendations	Responsible
1. No con-sequences of concern							
1.4. Less cooling water to EX-102							
Causes	Consequences	Safeguards	S	L	RR	Recommendations	Responsible
1. See No cooling water to EX-102							
2. Operator fails to open up butterfly valves sufficiently	2.1. Imposes a limit on performance of EX-102	2.1. Normal operating practices	1	4	4	4. (Safeguards are adequate)	
1.5. Reverse cooling water to EX-102							
Causes	Consequences	Safeguards	S	L	RR	Recommendations	Responsible
1. No causes							

1.6. Other than cooling water to EX-102							
Causes	Consequences	Safeguards	S	L	RR	Recommendations	Responsible
1. Cooling water valves hard to access	1.1. . Operator may have to engage in unsafe practices to open valve due to over reaching.	1.1. None	1	4	4	5. Locate cooling water valves in readily accessible locations. alternatively provide extended valve operator.	Eng. Dept.
2. Operator runs alternative equipment	2.1. Imposes a limit on performance of EX-102	2.1. Normal operating practices	1	4	4	4. (Safeguards are adequate)	
1.7. As well as cooling water to EX-102							
Causes	Consequences	Safeguards	S	L	RR	Recommendations	Responsible
1. No causes							

Node: 2. Bleed non condensibles to flare via bypass around PV-106							Drawing:
Type: Procedural Step							Fig A 2.1
2.1. No Bleed non condensibles to flare via bypass around PV-106							
Causes	Consequences	Safeguards	S	L	RR	Recommendations	Responsible
1. Operator fails to open by pass valve around PV-106 sufficiently	1.1. Pressure build-up in column	1.1. PSV-105	1	3	3	6. Operating manual to allow 50% opening on by-pass around PV-106.	Operations
2.2. Sooner/Later Bleed non condensibles to flare via bypass around PV-106							
Causes	Consequences	Safeguards	S	L	RR	Recommendations	Responsible
1. Failure to open by pass valve around PV-106 in time	1.1. Pressure build-up in column	1.1. PSV-105	1	3	3	7. Operator to regularly check column pressure on start-up PIC -106.	Operations.
2.3. More Bleed non condensibles to flare via bypass around PV-106							
Causes	Consequences	Safeguards	S	L	RR	Recommendations	Responsible
1. Operator opens by-pass valve fully	1.1. Column may not build-up pressure adequately.	1.1. None	1	3	3	6. Operating manual to allow 50% opening on by-pass around PV-106.	Operations
			1	3	3	7. Regular check on column pressure on start-up PIC -106.	Operations
2.4. Less Bleed non condensibles to flare via bypass around PV-106							
Causes	Consequences	Safeguards	S	L	RR	Recommendations	Responsible
1. Same as 2.2 above							

2.5. Reverse Bleed non condensibles to flare via bypass around PV-106							
Causes	Consequences	Safeguards	S	L	RR	Recommendations	Responsible
1. Flow back from flare system via by pass around PV-106	1.1.Minor contamination: inconsequential	1.1 N/A					
2.6. As well as Bleed non condensibles to flare via bypass around PV-106							
Causes	Consequences	Safeguards	S	L	RR	Recommendations	Responsible
1. Air being pushed to the flare line when feed is being introduced.	1.1. Potential explosions in flare line	1.1. None	4	4	16	8. Recommend purging column with Nitrogen at start-up prior to introduction of feed material.	Eng. Dept. + Operations
2.7. Other than Bleed non condensibles to flare via bypass around PV-106							
Causes	Consequences	Safeguards	S	L	RR	Recommendations	Responsible
1. No cause							

Node: 3. Minimum feed supply from V-101 to the column C-101							Drawing:
Type: Procedural Step							Fig A 2.1
3.1. No Minimum feed supply from V-10No1 to the column C-101							
Causes	Consequences	Safeguards	S	L	RR	Recommendations	Responsible
1. Battery limit valve closed on feed to column.	1.1. Can't start-up the column.	1.1. None	1	4	4	9. Check positions of battery limit valves before start-up.	Operations
3.2. More Minimum feed supply from V-101 to the column C-101							
Causes	Consequences	Safeguards	S	L	RR	Recommendations	Responsible
1. Operator sets setpoint on FRC-101 too high.	1.1. Rapid start-up and possible loss of control.	1.1. None	1	3	3	10. Update operating instructions to manually crack open steam on reboiler at start prior to feed entering.	Operations
	1.2. Without steam there will be no vaporization in the column during start-up early phase.	1.2 None	1	3	3	11. Start-up at 25% feed rate to be included in the operating procedure.	Operations
3.3. Less Minimum feed supply from V-101 to the column C-101							
Causes	Consequences	Safeguards	S	L	RR	Recommendations	Responsible
1. Operator sets feed too low at start-up	1.1. Lengthen start-up period, inconsequential.	1.1 N/A					

3.4. Reverse Minimum feed supply from V-101 to the column C-101							
Causes	Consequences	Safeguards	S	L	RR	Recommendations	Responsible
1. No causes							

3.5. Other than Minimum feed supply from V-101 to the column C-101							
Causes	Consequences	Safeguards	S	L	RR	Recommendations	Responsible
1. Gas breakthrough from upstream feed drum V-101	1.1. Over-pressurizing of the column	1.1. PSV-105	1	3	3	12. Operating procedure to include checking of level of liquid in upstream feed from V-101.	Operations

3.6. As well as Minimum feed supply from V-101 to the column C-101							
Causes	Consequences	Safeguards	S	L	RR	Recommendations	Responsible
1. Vapor/Liquid mixture from upstream feed drum V-101	1.1. Over-pressurizing of the column	1.1. PSV-105	1	3	3	12. Operating procedure to include checking of level of liquid in upstream feed from V-101.	Operating
			1	3	3	13. Ensure that there is a vortex breaker in the upstream vessel V-101 bottoms.	Eng. Dept.

Advantages of Procedural HAZOP

1. Good for HAZOPing batch operations.

2. Good for HAZOPing Operating Manuals, including Start-up, Shutdown etc.

Disadvantages of Procedural HAZOP

1. Limited for HAZOPing continuous operations.

2. Can be time consuming.

D. Knowledge Based HAZOP

This methodology typically is sometimes applied in place of the Guide Word Methodology. Some assumptions are:

- Extensive design standards and procedures are in place.
- HAZOP team has experience with similar designs.
- Process being HAZOPed is well established.

Basis is to use detailed Knowledge Based Checklists and brainstorm process for possible deficiencies.

Table II-4: Example of Knowledge Based Checklist for Centrifugal Compressor

```
TYPE = COMPRESSOR (CENTRIFUGAL)
COMPRESSOR SUCTION

QUESTION = Suction side overpressured from backflow/ leakage of recycle valve
on compressor shutdown?

QUESTION = Interstage equipment overpressured from backflow/ leakage of
recycle valve on compressor shutdown?

QUESTION = Suction side overpressured from backflow or recycle leakage with
parallel compressors?

QUESTION = Interstage equipment overpressured from backflow or recycle leakage
with parallel compressor?

QUESTION = Does suction side have permanent strainer with local pressure
indication downstream?

QUESTION = Does suction side have low-pressure alarm and, possibly, trip at
low pressure?

QUESTION = Do suction/ interstage knockout drums have high liquid level alarms
and trips at high high liquid level?

QUESTION = Will the compressor be shut down at low suction pressure?

QUESTION = Are air compressors intakes protected against contaminants
(flammables, carbon monoxide, etc.)?
```

Table II-5: Applicability of Different Types of HAZOP

	BATCH	CONTINUOUS	NEW PROCESS	EXISTING PROCESS	OPERATING MANUAL STARTUP & SHUTDOWN
PARAMETRIC DEVIATION BASED HAZOP	✓	✓✓✓	✓✓✓ (If Continuous)	✓✓✓ (If Continuous)	✓
GUIDE WORD (ELLIS KNOWLTON METHOD)	✓✓✓	✓	✓✓✓ (If Batch)	✓✓✓ (If Batch)	✓✓✓
PROCEDURAL STEP	✓✓✓	✓	✓ (If Batch)	✓✓✓ (If Batch)	✓✓✓
KNOWLEDGE BASED HAZOP	✓	✓✓✓	✓	✓✓✓	✓

Note: More ticks are better

SUGGESTED READING (Note: URLs current at date of publication)
"Apply the HAZOP Method to Batch Operations" by R.L.Collins, Chemical Engineering Progress, April 1995, pages 48 to 51 www.che.com/
"Guidelines for Hazard Evaluation Procedures" by AIChE, CCPS, 2nd edition, 1992 plus "Guidelines for Hazard Evaluation Procedures" by AIChE, CCPS, 1st edition, 1985 www.aiche.org/pubcat/seadtl.asp?Act=C&Category=Sect4&Min=20
"A Manual of Hazard & Operability Studies – The Creative Identification of Deviations and Disturbances", published by Chemetics International, 1992 www.kvaerner.com/companies/companiesdetail.asp?id=796
"DOE Handbook – Chemical Process Hazards Analysis", (Website) http://tis.eh.doe.gov/techstds/standard/hdbk1100/hdbk1100.pdf

References

(i) Regulations & Recommended Practices:

U.S. Regulations

1. *21 CFR Parts 808, 812, and 820 Medical Devices; Current Good Manufacturing Practices (CGMP); Final Rule,* Food and Drug Administration (7 October 1996)
2. *29 CFR 1910.119, Process Safety Management of Highly Hazardous Chemicals*, Occupational Safety & Health Administration (February 24, 1992)
3. *29 CFR 1910.252, Welding, Cutting, and Brazing*, Occupational Safety & Health Administration (September 29, 1986)
4. *40 CFR Part 68, Risk Management Programs under the Clean Air Act, Section 112(r)(7) Final Rule,* Environmental Protection Agency (20 June 1996)
5. *40 CFR Parts 9 and 68, List of Regulated Substances and Thresholds for Accidental Release Prevention; Requirements for Petitions under Section 112(r) of the Clean Air Act as Amended,* Environmental Protection Agency (January 31 1994)

U.K. Regulations

6. *Control of Major Accident Hazards (COMAH) Regulations,* Safety Policy Directorate, Health & Safety Executive (1999)
7. *Control of Substances Hazardous to Health (COSHH) Regulations,* Health Directorate, Health & Safety Executive (1999)

European Commission Regulations

8. *Council Directive (Seveso I Directive) 82/501/EEC on the major-accident hazards of certain industrial activities,* European Union (5 August, 1982)

9. *Council Directive (Seveso II Directive) 96/82/EEC on the control of major-accident hazards involving dangerous substances,* European Union (9 December, 1996)

U.S. Recommended Practice & Standards

10. *API Recommended Practice 520, Sizing, Selection, and Installation of Pressure – Relieving Devices in Refineries Part 1 – Sizing and Selection*, American Petroleum Institute, American Petroleum Institute, 6th Edition (March 1993)

11. *API Recommended Practice 521, Guide for Pressure-Relieving and Depressuring Systems*, American Petroleum Institute, 4th Edition (March 1997)

12. *API Recommended Practice 650, Welded Steel Tanks for Oil Storage*, American Petroleum Institute, 10th Edition (November 1998)

13. *API Recommended Practice 752, Management of Hazards Associated with Location of Process Plant Buildings*, American Petroleum Institute, CMA Manager's Guide, 1st Edition (May 1995)

14. *ANSI/ISA-SP-84.01, "Application of Safety Instrumented Systems for the Process Industries," Instrument Society of America Standards and Practices,* (1996)

15. *MIL-STD-1629A Procedures for performing a failure mode, effects and criticality analysis,* U.S. Department of Defense (1980)

16. *SAE APR 5580 Recommended Failure Modes and Effects Analysis (FMEA) Practices for Non-Automobile Applications*, The Engineering Society for Advancing Mobility Land Sea Air and Space, SAE (July 2001)

17. *SAE J1739 Potential Failure Mode and Effects Analysis in Design (Design FMEA), Potential Failure Mode and Effects Analysis in Manufacturing and Assembly Processes (Process FMEA) and Potential Mode and Effects Analysis for Machinery (Machinery FMEA),* The Engineering Society for Advancing Mobility Land Sea Air and Space, SAE (June 2000)

HSE Guidelines

18. *A Guide to the Control of Major Accident Hazards Regulations,* HSE Books (1999)

19. *COMAH Safety Report Assessment Manual*, Issue 2.0, HSE (January 2002)

20. *Major Accident Prevention Policies for Lower-Tier COMAH Establishments*, HSE (March 1999)

21. *COSHH Essentials: Easy Steps to Control Chemicals: Control of Substances Hazardous to Health Regulations*, HSE Books (1999)

22. *A Step-By-Step Guide to COSHH Assessment*, HSE (1993)

23. *Technical Basis for COSHH Essentials; Easy Steps to Control Chemicals*, HSE (1999)

International Standards

24. *IEC 61508, Functional Safety of Electrical / Electronic / Programmable Electronic Safety-related Systems, Parts 1-7*, Geneva, Switzerland, International Electrotechnical Commission (1998)

25. *IEC 61511, Functional Safety Instrumented Systems for the Process Industry Sector, Part 1-3* Geneva, Switzerland, International Electrotechnical Commission (2003)

26. *ISO 9001 Quality Management Systems – Requirements,* Geneva, Switzerland, International Organization for Standardization (2000)

27. *ISO/TS 16949 Quality management systems – Particular requirements for the application of ISO 9001:2000 for automotive production and relevant service part organizations,* Geneva, Switzerland, International Organization for Standardization (2002)

(ii) Books & Publications:

Center for Chemical Process Safety (CCPS) Publications

1. Center for Chemical Process Safety (CCPS), *Guidelines for Hazard Evaluation Procedures*, 2nd Edition (1992)
2. Center for Chemical Process Safety (CCPS), *Guidelines for Safe Process Operations and Maintenance* (1995)
3. Center for Chemical Process Safety (CCPS), *Guidelines for Evaluating Process Plant Buildings for External Explosions and Fires* (1995)
4. Center for Chemical Process Safety (CCPS), *Guidelines for Implementing Process Safety Management* (1994)
5. Center for Chemical Process Safety (CCPS), *Guidelines for Engineering Design for Process Safety* (1993)
6. Center for Chemical Process Safety (CCPS), *Guidelines for Hazard Evaluation Procedures*, (1992)
7. Center for Chemical Process Safety (CCPS), *Guidelines for Investigating Chemical Process Incidents*, (1992)
8. Center for Chemical Process Safety (CCPS), *Guidelines for Technical Management of Chemical Process Safety*, (1992)
9. Center for Chemical Process Safety (CCPS), *Guidelines for Process Safety Documentation*, (1995)
10. Center for Chemical Process Safety (CCPS), *Guidelines for Chemical Process Quantitative Risk Analysis* (1989)

11. Center for Chemical Process Safety (CCPS), *Layer of Protection Analysis, Simplified Process Risk Assessment* (2001)

Other Books & Publications

12. Bhimavarapu, K, *Use Layer of Protection Analysis (LOPA) to Comply with Performance-based Standards*, Control Engineering, February 1, 2000

13. Cox, A.W., Lees, F.P. & Ang, M.L., *Classification of Hazardous Locations*, Inst. of Chemical Engineers (1992)

14. Crumpler, D. K. and Whittle, D. K., *Effective Revalidation of Process Hazard Analyses (PHAs)*, ABS consulting publication

15. Dow Chemical Company, *Fire and Explosion Index Hazard Classification Guide*, 7th Edition., AIChE, New York (1994)

16. Dow Chemical Company, *Chemical Exposure Index Guide*, AIChE, New York (1994)

17. Dowell, A.M. and Green, D.L., *Formulate Emergency Shutdown Systems by Cookbook*, Chemical Engineering Progress (April 1998)

18. Dowell, A.M. and Hendershot D.C., *Simplified Risk Analysis – Layer of Protection Analysis (LOPA)*, AIChE 2002 National Meeting (2002)

19. Eisenberg, N.A., Lynch, C.J. and Breeding, R.J., *Vulnerability Model: A Simulation System for Assessing Damage Resulting from Marine Spills Rep. CG-D-136-75*. Enviro Control Inc., Rockville, MD (1975)

20. Federal Emergency Management Agency, *Handbook of Chemical Hazard Analysis Procedures* (ARCHIE Manual), Washington, DC (1989)

21. Finney, D.J., *Probit Analysis*, Cambridge University Press, London (1971)

22. Greenberg, H.R. and J.J. Cramer, *Risk Assessment and Risk Management for the Chemical Process Industry*, Van Nostrand Reinhold, New York (1991)

23. Hyatt, R.N. & N., Mulvey, N.P., *Using Computer Software for Process Hazards Analysis (PHA)*, Chemputers IV (1996)

24. Kletz, T.K., *HAZOP & HAZAN*, 3rd Edition, Inst. of Chemical Engineers (1992)

25. Kletz, T.K., *What Went Wrong?*, Gulf Publishing (1985)

26. Knowlton, R. E. *A Manual of Hazard & Operability Studies – The Creative Identification of Deviations and Disturbances*, Chemetics International Ltd. (1992)

27. Knowlton, R.E., *Hazard and Operability Studies, The Guide Word Approach*, Chemetics International (1981)

28. Lees, F.P., *Loss Prevention in the Process Industries*, Vols. 1, 2 & 3 Butterworth - Heinemann, London, 2nd Edition (1996)

29. Louvar, J. F. and Louvar, B.D., *Health and Environmental Risk Analysis: Fundamentals with Applications*, Prentice Hall PTR (1998)

30. Mecklenburgh, J.C., *Process Plant Layout*, John Wiley (1985)

31. Parry, S.T., *A Review of Hazard Identification Techniques and their Application to Major Accident Hazards*, Safety & Reliability Directorate, UKAEA (1980)

32. Philley, J. and Moosemiller, M., *PHA Revalidation: A Six-step Approach*, Chemical Process Safety Report (February 1997)

33. Smith D.J. and Simpson K.G.L., *Functional Safety A Straightforward Guide to IEC61508 and Related Standards*, Butterworth-Heinemann (2001)

34. Stephen, M. M., *Minimizing Damage to Refineries*, U.S. Dept. of the Interior, Office of Oil & Gas (February 1970)

35. Summers A.E., *What Every Manager Should Know About The New SIS Standards*, Copyright ISA.

36. Sutton, I., *Management of Change,* Morris Publishing, 2nd edition (1997)

37. Technica Ltd., *Manual of Industrial Hazard Assessment Techniques*, World Bank (1985)

38. Wells, G.L., *Safety in Process Plant Design*, Halsted Press (1980)

Note: For specific material and references see the lists of suggested reading material (with Website URLs applicable at the time of publication) included at the end of each chapter.

Index

DYADEM

DYADEM

DYADEM

DYADEM

DYADEM

DYADEM

DYADEM

DYADEM

DYADEM

DYADEM